BIRDS of AUSTRALIA

BIRDS OF AUSTRALIA

A PHOTOGRAPHIC GUIDE

IAIN CAMPBELL

SAM WOODS AND NICK LESEBERG

PHOTOGRAPHY SUPPLIED BY GEOFF JONES

PRINCETON UNIVERSITY PRESS

PRINCETON AND OXFORD

Published by Princeton University Press, 41 William Street, Princeton, New Jersey 08540

In the United Kingdom: Princeton University Press, 6 Oxford Street, Woodstock, Oxfordshire OX20 1TW

press.princeton.edu

Cover image: Spotted Pardalote (*Pardalotus striatus*). Photograph copyright © and courtesy of Iain Campbell.

ISBN (pbk) 978-0-691-15727-6

Library of Congress Control Number: 2014933938

British Library Cataloguing-in-Publication Data is available

This book has been composed in Perpetua (main text) and Avenir (headings and captions)

Printed on acid-free paper. ∞

Images previous page: Rainbow Bee-eater (*Merops ornatus*), top left; Squatter Pigeon (*Geophaps scripta*), top right; Emu (*Dromaius novaehollandiae*), bottom left; New Holland Honeyeater (*Phylidonyris novaehollandiae*), bottom right. All photographs copyright © and courtesy of Iain Campbell.

Designed by D & N Publishing, Baydon, Wiltshire, UK

Printed in China

10 9 8 7 6 5 4 3 2 1

CONTENTS

ACKNOWLEDGMENTS

Many people have contributed to this photographic guide, none more so than Geoff Jones, who made his complete collection of Australian bird photos available to the authors. For this we are very grateful. Descriptions of seabirds from Keith Barnes formed the base for that section of the book, which the authors freely admit is not their realm of expertise. Alan Davies helped with many of the shorebird descriptions. Rebecca Hinkle looked at the manuscript many times to help make it more accessible to non-Australian birders, for whom words like 'billabong' and 'Outback' are not second nature.

We would also like to thank Tropical Birding Tours for allowing us the time to get some of the required shots and to write this book. We give many thanks to editor Amy K. Hughes for turning our untidy submission into a proper manuscript, and to Robert Kirk and Ellen Foos at Princeton University Press, and to David and Namrita Price-Goodfellow and their design team at D & N Publishing. Finally, we want to thank our families for dealing with some late nights and temperamental moods when we were sorting through thousands of photos. We hope this book does everyone proud.

A Red-collared Lorikeet coming in to a dripping tap in the dry tropics

INTRODUCTION

Australia is a vast country with more than 700 regular bird species, most of which are found nowhere else. Cassowaries, chowchillas, sittellas, and other names are unfamiliar to many visitors from North America or Europe, and this can make bird identification seem daunting. This guide, with its easy-to-understand writing style, detailed photos, and clear distribution maps, is designed to help you identify the birds you see—and make your experience of the Australian bush much more fulfilling.

The goal of this book is to make birding and bird identification accessible to the vast majority of people, while still providing a resource to more experienced birders. To achieve this, the text is written in a casual style, in the way one birder would likely describe a bird to another birder; the most important feature first, be it the plumage, a restricted range, a habitat requirement, or a bizarre habit. Then a general description is given, along with the bird's distribution, and how and where to find it. The key to birding in Australia is in understanding the habitats, so to really get the most out of this book, it is very important to read 'The Habitats of Australia' section.

There are a few different checklists and taxonomic treatments of Australian birds. This guide almost always follows the International Ornithological Congress taxonomy (*IOC World Bird List*, version 3.3), as it is the most advanced in dealing with the Australian bird families. There are a few notable deviations from the IOC taxonomy, essentially predictive in nature, where it is expected that certain birds will be described as separate species in the near future. These extra splits include Naretha Parrot (from Bluebonnet), Golden-backed Honeyeater (from Black-chinned Honeyeater), Mallee Whipbird (from Western Whipbird), Western and Northern Shriketits (formerly lumped with Eastern Shriketit as Crested Shriketit), and Silver-backed Butcherbird (from Grey Butcherbird). The sequence is mostly taxonomic, but we made exceptions when putting similar-looking birds together (e.g., stork, pelican, and cranes on the same plate) makes for greater ease of identification.

Birds are rarely encountered at a standard distance or in perfect lighting, so although we are using photos for comparison of birds, the species do not appear in similar positions. The photos show the diagnostic features, but the birds are in a variety of positions, because you will be watching birds that are constantly moving. The presentation in most field guides looks very static, so we have put the chaos of nature into the book. The sizes of the images and the relative sizes of the birds within them have no relation and should not be used as an indication of species' sizes. Photos that appear on left-hand pages are for artistic as much as identification purposes, and sometimes are used to show a bird in its habitat.

Bird calls and songs are very important for identification of a range of species and are well worth learning. We think that onomatopoeic descriptions do not do these vocalisations justice, and a birder is far better served learning the calls though recordings. We have elected not to devote much text to describing bird vocalisations, which do little to aid practical identification of the species.

The geographical range covered includes all of mainland Australia and Tasmania (TAS). The Cocos (Keeling) Islands, which are much closer to Java than to the Australian mainland; Lord Howe Island, 600km (375mi) east of the mainland; and Norfolk Island, which is closer to New Zealand than to Australia, are excluded, as they are very different from mainland Australia and better covered in individual volumes. The book deliberately excludes rare vagrants to the country; for the vast majority of users these would add confusion to the birding experience. The general rule is that unless recorded a few times a year, such a visitor is skipped. Most of these vagrants are hard to identify shorebirds or seabirds, and there are some very good resources for identifying them.

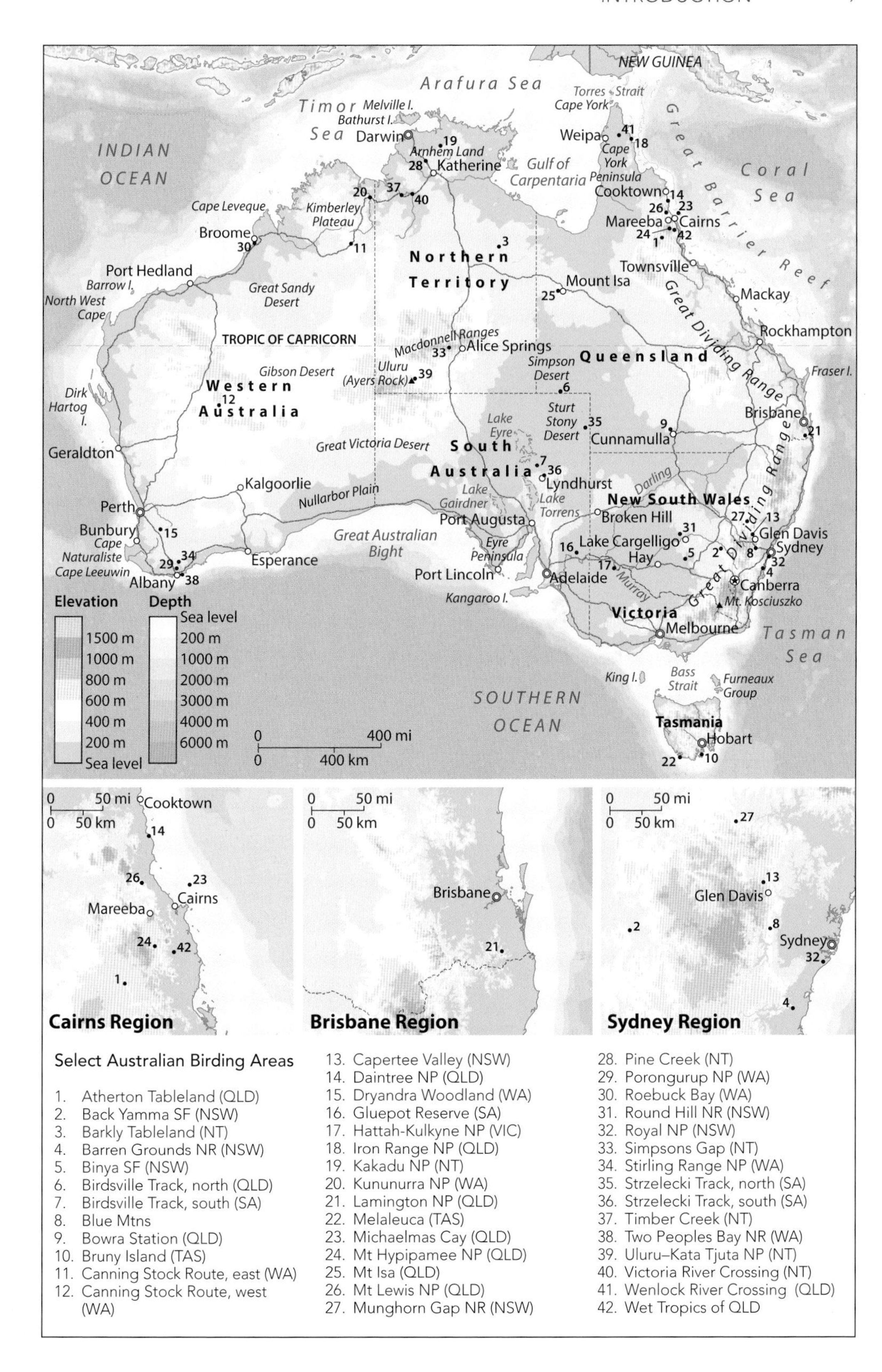
NEW GUINEA
Arafura Sea
Timor Sea
Torres Strait
Cape York
Melville I.
Bathurst I.
Darwin
INDIAN OCEAN
Arnhem Land
Katherine
Gulf of Carpentaria
Weipa
Cape York Peninsula
Coral Sea
Great Barrier Reef
Cooktown
Cape Leveque
Kimberley Plateau
Mareeba
Cairns
Broome
Northern Territory
Townsville
Port Hedland
Barrow I.
North West Cape
Great Sandy Desert
Mount Isa
Mackay
Great Dividing Range
TROPIC OF CAPRICORN
Macdonnell Ranges
Alice Springs
Rockhampton
Simpson Desert
Queensland
Fraser I.
Gibson Desert
Uluru (Ayers Rock)
Western Australia
Dirk Hartog I.
Lake Eyre
Sturt Stony Desert
Brisbane
Cunnamulla
Geraldton
Great Victoria Desert
South Australia
Lyndhurst
Kalgoorlie
Nullarbor Plain
Lake Gairdner
Lake Torrens
Darling
New South Wales
Perth
Port Augusta
Broken Hill
Glen Davis
Bunbury
Cape Naturaliste
Cape Leeuwin
Great Australian Bight
Eyre Peninsula
Lake Cargelligo
Hay
Sydney
Esperance
Port Lincoln
Adelaide
Murray
Canberra
Albany
Kangaroo I.
Mt. Kosciuszko
Victoria
Melbourne
Tasman Sea
Elevation
1500 m
1000 m
800 m
600 m
400 m
200 m
Sea level
Depth
Sea level
200 m
1000 m
2000 m
3000 m
4000 m
6000 m
0 400 mi
0 400 km
SOUTHERN OCEAN
King I.
Bass Strait
Furneaux Group
Tasmania
Hobart
Cairns Region
Brisbane Region
Sydney Region
0 50 mi
0 50 km
Select Australian Birding Areas
1. Atherton Tableland (QLD)
2. Back Yamma SF (NSW)
3. Barkly Tableland (NT)
4. Barren Grounds NR (NSW)
5. Binya SF (NSW)
6. Birdsville Track, north (QLD)
7. Birdsville Track, south (SA)
8. Blue Mtns
9. Bowra Station (QLD)
10. Bruny Island (TAS)
11. Canning Stock Route, east (WA)
12. Canning Stock Route, west (WA)
13. Capertee Valley (NSW)
14. Daintree NP (QLD)
15. Dryandra Woodland (WA)
16. Gluepot Reserve (SA)
17. Hattah-Kulkyne NP (VIC)
18. Iron Range NP (QLD)
19. Kakadu NP (NT)
20. Kununurra NP (WA)
21. Lamington NP (QLD)
22. Melaleuca (TAS)
23. Michaelmas Cay (QLD)
24. Mt Hypipamee NP (QLD)
25. Mt Isa (QLD)
26. Mt Lewis NP (QLD)
27. Munghorn Gap NR (NSW)
28. Pine Creek (NT)
29. Porongurup NP (WA)
30. Roebuck Bay (WA)
31. Round Hill NR (NSW)
32. Royal NP (NSW)
33. Simpsons Gap (NT)
34. Stirling Range NP (WA)
35. Strzelecki Track, north (SA)
36. Strzelecki Track, south (SA)
37. Timber Creek (NT)
38. Two Peoples Bay NR (WA)
39. Uluru–Kata Tjuta NP (NT)
40. Victoria River Crossing (NT)
41. Wenlock River Crossing (QLD)
42. Wet Tropics of QLD

We generated the maps by using trip reports from birding tours, commercial private surveys, and records from *The Atlas of Australian Birds* from BirdLife Australia. Any survey or sighting is limited to accessible areas, which can be very erratic in remote regions. Large tracts of aboriginal, private, and government lands are off-limits to most people and have not been represented in any survey. There is also a very strong bias of records to more populated centres of the country and places that are birded frequently. We have tried to overcome this bias and lack of records from remote locations by extrapolating from known areas across the blank areas where the suitable habitat occurs. We use a mapping system in which the map does not show absolute population abundance of the bird, but rather shows relative abundance throughout its Australian range; faint shading shows where the bird is less common and darker shading shows where the bird is relatively more common. Some species, such as Australian Masked Owl, have such a sporadic distribution over a vast region that a standard type of map becomes almost useless, and distribution would be better represented by spots showing recent sightings. For a few species, such as the Australian Ringneck, we have mapped ranges of the individual subspecies. This is because the subspecies of this species look very different from one another, have overlapping ranges, and may be split out as separate species in the future.

Bird descriptions have been made as simple as possible; for easy groups such as honeyeaters the descriptions can be easily understood by even the novice birder. Some groups, such as shorebirds and especially seabirds, are notoriously difficult to identify even to the most experienced birder; here the terminology required to describe the birds becomes far more complex, and many words used will be new and unintelligible to casual birders. We have avoided overly complicated or convoluted descriptions where possible, but for those birds that require them, the glossary covers all terms used. A casual birder is highly unlikely to find him- or herself 100km (60mi) offshore in the Southern Ocean pouring through prion flocks, or in similar extreme conditions, so the simplified descriptions will suffice in most situations.

Grassland fires are common in inland Australia

AUSTRALIAN CLIMATE AND RAINFALL

A basic understanding of climate and rainfall patterns within Australia helps us comprehend why habitats and hence Australian birds are distributed the way they are. Australia's climate is primarily influenced by its position under the belt of high pressure that encircles the globe at latitude 30° south, a result of the earth's orientation relative to the sun. Furthermore, the earth's rotation around the sun throughout the year causes this belt to move and creates the regular seasons and rainfall patterns we are familiar with.

The primary impact of the high-pressure belt is stable, dry conditions that do not promote rainfall. As this belt moves throughout the year, these calm, dry conditions influence different parts of the continent, hence the differences in climate between the tropical north and temperate south. From December to March the belt of high pressure sits over the s. part of Australia, promoting hot, dry conditions (summer) and opening the n. part of the continent to the effects of the equatorial monsoons (the wet season). From May to August it moves northward, meaning stable, dry conditions over n. Australia (the dry season), and allowing the southerly parts of the continent to experience the cold and wet conditions created by cold fronts coming off the Southern Ocean (winter). Around 70 percent of Australia is desert, and most of c. and w. Australia is considered arid or semi-arid and receives very little rain; this area is generally referred to as the Outback. Most important, the rainfall that does occur is very unpredictable, and parts of the desert may go for years without significant rainfall. There are several reasons for this. Even though the high-pressure belt moves throughout the year, the majority of Australia's landmass sits under it year-round, so very little moist air makes it to the inland. Additionally, the interior and w. parts of Australia are very flat, so no lifting occurs to cause what moisture there might be to fall as rain. Rain falls only when large amounts of very moist air reach the interior, usually as a result of large monsoonal systems from the north during the summer or the tips of very strong southerly cold fronts during the winter. Only the most severe weather systems penetrate far enough into the inland and still hold enough moisture when they arrive to create rainfall, explaining the infrequency and unpredictability of rainfall across most of c. and w. Australia.

In far n. Australia the primary cause of rainfall is the annual monsoons. During this period n. Australia receives most of its rain, and this annual rain supports the swathe of grassy tropical savannahs across n. Australia. Eastern Queensland (QLD) experiences additional rain due to the Great Dividing Range. The prevailing winds during the middle of the year at these latitudes are easterly, and coming off the warm Coral Sea this moisture-laden air strikes the dividing range and releases plenty of rain. Combined with rain from the monsoon season, this makes ne. QLD the wettest part of Australia, and creates the conditions that support wet habitats such as tropical rainforests.

In s. Australia, conditions during the summer are generally warm and dry as the high-pressure belt sits over that part of the continent. When the belt moves northward during the winter, s. Australia comes under the influence of frontal weather systems that come up from the Southern Ocean. This regular cycle of warm, dry summers and cold, wet winters is called a Mediterranean climate, and is experienced in the areas of sw. Western Australia (WA), se. South Australia (SA), Victoria (VIC), and s. New South Wales (NSW). Again the mountains of the Great Dividing Range play an important role, as it is around these mountains that most of the rain falls. As you move inland away from the mountains there is progressively less rain and the vegetation changes from tall wet forests on the mountains to dry open forests and eventually the arid habitats that predominate further inland.

HABITATS OF AUSTRALIA

Becoming familiar with the variety of habitats within Australia and how and why they occur where they do can help us understand how birds are distributed across the landscape. It will help you learn what birds to expect in a new area, how to find certain birds with specific habitat preferences, and how to plan a trip to a new area so you can maximise the number of species you will see.

Many factors determine how various habitats are distributed throughout the landscape. The physical environment is very important, with availability of water, soil type, and the corresponding availability of nutrients, plus prevailing weather conditions, all playing important roles. Disturbance is another very important factor, either through natural causes such as fire or human-induced changes, such as clearing land for agriculture. The most influential factor in determining the distribution and make up of habitats within Australia is its extreme aridity, the reasons for which were discussed in the preceding section on climate. This aridity has had a noticeable effect on the vegetation types that predominate in Australia. 'Sclerophyll' is a term given to plants that develop hard, often small or thin leaves, an adaptation to harsh conditions with little moisture. Unsurprisingly, this condition is prevalent in much of Australia's native vegetation, which has evolved under such conditions. All eucalypts, for example, fall under this classification, even though many have subsequently adapted to higher rainfall environments. Characteristic flora of Australia, eucalypts are woody plants with capsule fruits known as 'gum nuts'. They form the basis of many of the habitat types in Australia, but are highly varied and look very different from the introduced monotypic eucalyptus plantations that people may be familiar with in California, the Mediterranean, s. Africa, or the dry Andean regions.

These factors combine to create a variety of habitats, which are usually classified based on the vegetation they support. The level of classification depends on the level of detail required. An experienced botanist may recognise 30 different types of rainforest in ne. QLD, but as birdwatchers, we don't need that level of detail. Instead we may need to recognise only a difference between lowland and montane rainforests. Sometimes the type of vegetation present is not important. Along the coast we may need to recognise only a basic physical difference; for example, we learn to look for Sooty Oystercatchers along rocky coastlines and Pied Oystercatchers along open sandy beaches.

With that in mind, this section is not intended to provide an exhaustive, botanically accurate analysis of every habitat found in Australia, but rather an introduction to some of the general habitats across the continent that are important for us as birders. The more experienced you become, the more differences you will recognise, even within the broad categories presented here, and the more you will see how different birds exploit and are distributed within these microhabitats.

MARINE AND COASTAL HABITATS

The following habitats are those associated with Australia's extensive coastline and offshore waters. As an island continent Australia's mainland has around 35,000km (22,000mi) of beaches, bays, inlets, estuaries, and rocky coastlines, plus many offshore islands. The coastline of Australia's temperate south is quite different from that of the tropical north. Australia's n. coastline generally experiences calmer waters due to features such as the Great Barrier Reef in the east and the sheltered Torres Strait across the north. This results in larger areas of open mudflats and sand flats and in some areas extensive mangrove forests. In the temperate south the coastline is generally more exposed, with sandy surf beaches and rugged rocky headlands. Here mudflats tend to occur only in sheltered bays and inlets. Shorebirds and marine species predominate around the coast, some with specific habitat preferences and others with a more general coastal distribution.

SOUTHERN ROCKY COASTLINES AND SANDY BEACHES

Most of w., s., and se. Australia (including TAS) is dominated by erosional terrains, which form often dramatic, rocky coastlines interspersed with nutrient-deficient, steep-sloping sandy beaches that are usually open to the surf. These habitats tend to be the domain of shorebirds and near-shore seabirds. Rocky headlands are the preferred habitat of resident shorebirds such as Sooty Oystercatchers and, in the summer, migratory species including Whimbrel and Ruddy Turnstone. Sandy beaches provide habitat for Pied Oystercatcher and Hooded Dotterel. They are not the homes only of shorebirds; in WA and SA they are also important environments for Rock Parrot, while some near-shore seabirds, such as cormorants and gulls, can be found almost anywhere along the coast. Coastlines are often prone to human disturbance by recreation and holiday home developments, which have caused significant problems for beach-nesting birds, most notably Hooded Dotterel and Fairy Tern. Rocky headlands can also be good places to look for pelagic species (birds that mainly come ashore only to breed), such as albatrosses, gannets, and shearwaters. Popular headlands for land-based sea-watching include Magic Point in Sydney (NSW), Bass Point near Shellharbour (NSW), Cape Nelson near Portland (VIC), Eaglehawk Neck (TAS), Bruny Island (TAS), and Cape Naturaliste (WA).

Rocky coastline, Albany, WA

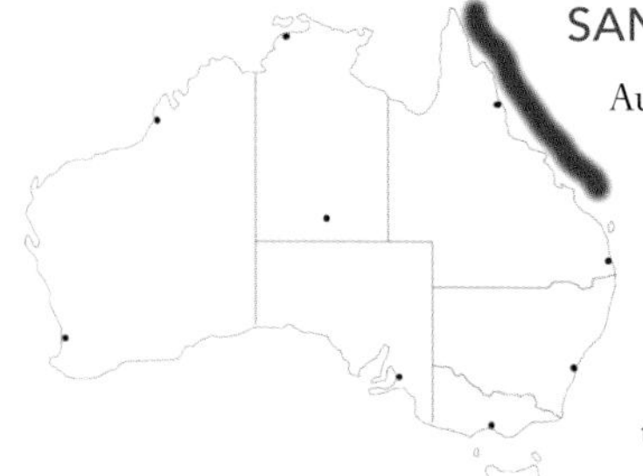

SANDY CAYS AND BARRIER REEFS

Australia has two large barrier reef systems. One is located along the nw. Australian coastline, and is inaccessible and rarely visited. The other, the Great Barrier Reef, extends from n. Cape York Peninsula in ne. QLD southward to about Bundaberg in s. QLD. Parts of this massive reef system are very accessible and include some of the most popular tourist attractions in the country. In some localities within this large system the coral sands become concentrated to form small, sparsely vegetated sandy cays or islands. These remote cays are very important for seabirds, as they provide safe breeding areas free from land predators. Some species, such as Sooty Terns and Brown Noddies, which require little more than a patch of open ground, breed on nearly all the sandy cays. Red-footed Boobies and such burrow-nesting species as shearwaters and petrels nest on the larger islands, including Raine Island in the n. Great Barrier Reef (extreme ne. QLD), which support the vegetation required to provide cover and nesting sites. Although tourist resorts are scattered throughout the Great Barrier Reef, access to the important, densely packed breeding seabird colonies is generally restricted to researchers and park rangers. The most accessible is Michaelmas Cay, a 1.6-ha (4-acre) island about 30km (20mi) off the coast from Cairns, which is widely considered the premier cay for visiting birders. It can hold up to 20,000 breeding seabirds. More than 30 seabird species have been recorded there; Sooty Terns and Brown Noddies provide the greatest number of nesting birds in the colony. Other species seen on the cay include Greater Crested, Lesser Crested, Little, Bridled, Roseate, and Black-naped Terns; Lesser and Great Frigatebirds; Black Noddies (in small numbers); and Brown Boobies.

Seabird colony, Michaelmas Cay, QLD

PELAGIC WATERS

About 70 percent of the globe is covered by ocean, and these oceans support a surprising diversity of birds that spend much of their lives on the open seas. As the only continent completely surrounded by ocean, Australia has a diverse pelagic bird fauna and is widely considered one of the best seabird-viewing countries in the world. To see these birds requires a pelagic boat trip, usually a day-long activity. These trips focus on reaching the edge of the continental shelf, where the upwelling of nutrient-rich, colder waters produces substantial concentrations of sea life, from microscopic organisms right up the food chain to large pelagic fish. This concentration of sea life also attracts the greatest diversity of pelagic birds, including shearwaters, petrels, storm petrels, prions, albatrosses, and gannets. The seasonal patterns of pelagics are important when planning a seabird trip, as different species are present at different times of year. Generally, the austral winter (ca. Jun–Aug) is best for pelagic groups such as prions, albatrosses, and some of the more southerly breeding petrel species (e.g. Cape Petrel, giant petrels, and storm petrels). The Australian spring and summer (Oct–Feb) are better for such birds as shearwaters, northerly breeding petrels (e.g. Providence, Great-winged, and most other *Pterodroma* petrels), and tropical-breeding species such as tropicbirds and boobies. The change in diversity and numbers of birds on pelagic trips can be dramatic, and is also dependent on weather conditions; generally, the calmer the sea, the duller the birding, as many of these species require updrafts produced from waves to fly and feed. A pelagic excursion on a windy day may record a high diversity of species, while a week later in calm conditions the same birds may be nowhere to be seen. Most pelagic tours depart from s. ports, and there are regular trips from Southport and Mooloolaba (QLD), Sydney and Wollongong (NSW), Eaglehawk Neck (TAS), Portland (VIC), and Albany and Perth (WA). There has been a recent increase in the number of extended pelagic trips being run, and multi-day trips embark from several ports, including those of TAS and of Broome (WA) in nw. Australia. These trips can go far offshore and have been turning up some unexpected and unusual species.

TIDAL MUDFLATS AND BEACHES

Tidal mudflats and beaches are found all around Australia. These are areas of calm deposition most of the year and tend to have many more organisms and biomatter in the silts and sands than the sandy beaches described above. They are most prevalent around estuaries and sheltered bays and include beaches in the north of the country where the seas are calm. They are important roosting and feeding areas for many shorebirds. The vast majority of species are migratory waders, which breed in Siberia or Alaska and are therefore present only during the austral spring–summer (Oct–Mar), but some species are resident, such as Beach Stone-Curlews, Pied Oystercatchers, and Red-capped Plovers. These habitats also provide important resting and feeding areas for terns and gulls. Tidal flats vary in structure, and some species prefer flats with a sandy substrate (Sanderling, Greater and Lesser Sand Plovers), while others prefer the muddy substrates of estuaries. Different shorebirds also feed in different areas of the flats; species segregate on the shoreline depending on their leg and bill structure, which determines the depth of water in which they can feed. It can be difficult to see shorebirds at low tide, as they are often quite a distance away. The best viewing opportunities are generally as the tide comes in and birds are forced up the shore by the incoming water. Once the tide has covered the feeding areas, the birds will travel to high-tide roosts, where thousands of waders may gather waiting for the next low tide. This can be the best opportunity to view them, with many birds in the one place allowing direct comparison of different species. There are many good places to view waders around Australia, with the most accessible areas including Roebuck Bay near Broome (WA), Lee

Extensive mudflats, Broome, WA

Point and Buffalo Creek near Darwin (NT), the Cairns Esplanade (QLD), the Hunter River Estuary in Newcastle (NSW), and Port Phillip Bay near Melbourne (VIC).

MANGROVES

Mangroves are trees that are adapted to a saline environment. They grow in intertidal situations, including many large inlets and tidal river systems. Species diversity is highest in the north, and up to 30 species of mangroves occur across the coastlines of n. Australia, while at these trees' s. limits in VIC and sw. WA there are only one or two species. There is considerable structural variation, both within and between mangrove species, and different species tend to grow in different conditions. Indeed, it is this variety in both species and structure that appears important in determining the bird communities found, rather than just the species composition of the mangrove forests themselves. Thus in mangrove communities across n. Australia, which have a variety of species and complex structures, there is a more diverse community of mangrove birds than in, say, NSW, where the mangrove habitat has fewer species and less structural diversity. Within northern mangrove stands the distribution of some bird species is highly stratified; for example, in Broome the White-breasted Whistler prefers short stands of mangroves closer to the water's edge, while the sympatric Mangrove Golden Whistler prefers taller stands farther inland. Some bird species, including those mentioned above, are highly specialised to this habitat, while other mangrove birds, although still closely linked to mangroves, will venture into other habitat if it is in close proximity. Australia's mangrove specialists include Chestnut Rail, Collared Kingfisher, Dusky Gerygone, Mangrove Gerygone, Mangrove Honeyeater, Red-headed Myzomela, White-breasted Whistler, Mangrove Golden Whistler, Canary White-eye, and Mangrove Fantail. Good mangrove systems can be found around Darwin (NT), Brisbane (QLD), Broome (WA), and Cairns (QLD); some of these areas have boardwalks that allow access to both the mangroves and their birds.

Mangroves, Cairns, QLD

TROPICAL HABITATS

This section covers those habitats found across n. Australia in an area generally experiencing a monsoonal climate. The habitats in this region are generally dependent on the regular and often large amounts of rain associated with the annual wet season. The region affected by this monsoonal climate extends across n. Australia and down the e. coast of QLD to around Brisbane. At this southern limit the summer rainfall is not as regular or extensive, but it does locally support subtropical rainforest and tropical wetland habitats.

TROPICAL RAINFORESTS

Tropical rainforest is confined to the lowlands of coastal ne. QLD, where warm weather and high rainfall support this mega-diverse plant community. It is characterised by a large variety of evergreen trees, including Copper Laurel (*Eupomatia laurina*), Queensland Nutmeg (*Myristica insipida*), and pepper tree (*Tasmannia*), that form a thick canopy and, owing to poor light penetration, a relatively open understorey. This forest appears much like the picture-postcard images taken in rainforests in the tropics of the world, such as in New Guinea, Borneo, and the Amazon. The birdlife is generally highly specialised and restricted to rainforests. In Australia these habitats are remnants of rainforests that were once more widespread and that have more in common with the rainforests of New Guinea than forests farther south on the continent. This is reflected in some of the bird species shared by n. Australia and s. New Guinea, such as Red-bellied Pitta, Eclectus Parrot, Green-backed Honeyeater, White-faced Robin and Magnificent Riflebird. A handful of these rainforest species, such as Buff-breasted Paradise Kingfisher and Red-bellied Pitta, even migrate across the Torres Strait annually to winter in New Guinea, though they breed in Australia. While this rainforest was always a restricted habitat, it has been further reduced in modern times by human activity, such as

Lowland rainforest, Cairns, QLD

sugar-cane farming and clearance for urban development. Good examples of this habitat can still be found in the Daintree area near Cairns, and here you will find many of the species typical of the rainforests of ne. QLD, such as Wompoo Fruit Dove, Double-eyed Fig Parrot, Victoria's Riflebird, and Spotted Catbird. Only farther north (e.g., at Iron Range NP) are there birds endemic to lowland tropical rainforests in Australia, such as Magnificent Riflebird, White-faced Robin, Northern Scrub Robin and Green-backed Honeyeater.

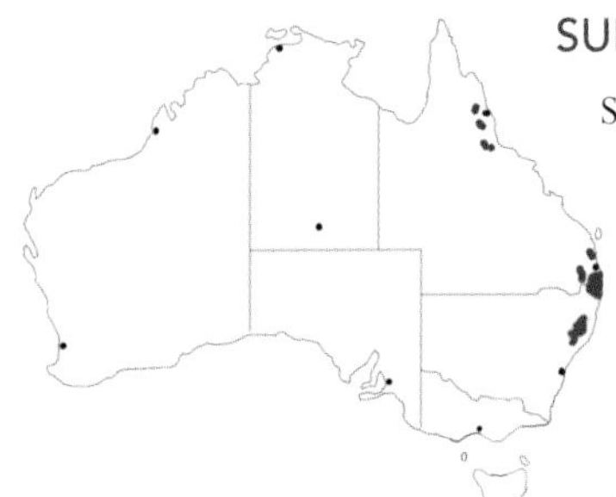

SUBTROPICAL RAINFORESTS

Subtropical rainforest is very similar to tropical rainforest. It also requires high rainfall, but can occur in cooler conditions. In appearance it is generally similar to lowland rainforest, but it has slightly lower diversity, and on closer inspection a more extensive understorey. The dominant trees include species such as White Booyong (*Argyrodendron trifoliolatum*), White Beech (*Gmelina leichhardtii*), Red Cedar (*Toona ciliata*), and Moreton Bay Fig (*Ficus macrophylla*). It is found in two main areas, each with a unique suite of birds. In ne. QLD it occurs in the cooler upland areas from the Atherton Tableland south to the Paluma Range near Townsville. In this area there is a gradual transition between lowland tropical rainforests and the montane subtropical rainforests, with the transition occurring around 600m (2,000ft), although this is not clearly defined and varies from location to location. There are several species that are limited to this habitat and therefore highly desired by visiting birders: Tooth-billed and Golden Bowerbirds, Mountain Thornbill, Atherton Scrubwren, Fernwren, Bridled Honeyeater, Grey-headed Robin, Bower's Shrikethrush, and Chowchilla. Accessible examples of this habitat can be found at the Curtain Fig near

Subtropical rainforest, Mt Hypipamee NP, QLD

Yungaburra, Mt Lewis NP near Julatten, Mt Hypipamee NP, and Paluma Range NP north of Townsville. There is a significant gap south of Townsville before this habitat occurs again, from the mountains west of Mackay, and then in patches south as far as the c. coast of NSW. It is generally found at lower elevations, from the coast to the lower reaches of the mountains, with higher elevations supporting temperate rainforests (see later in this section). Birds found in subtropical rainforests at these more southerly latitudes include Marbled Frogmouth, Regent Bowerbird, Australian Logrunner, Green Catbird, and Paradise Riflebird. Most subtropical rainforests in this area have been cleared, but there are still accessible areas near Eungella NP west of Mackay (QLD) and the ranges west of Brisbane including Lamington NP (QLD) and the Dorrigo NP area (NSW).

DRY RAINFORESTS AND MONSOON VINE FORESTS

Dry rainforests and the similar-looking monsoon vine forests (both of which support the same bird communities) occur only in small pockets but are worth mentioning because there are several species of birds that show a definite preference for these habitats where they occur. They are found only in areas with pronounced wet and dry seasons, so occur across far n. Australia and down the e. coast. Throughout this range, they can be found in close proximity to the coast, often on sandy soils. Across n. Australia they may also be found along dry creek beds or sheltered gullies, and particularly in the NT and n. WA they may also be found in sheltered areas at the bases of sandstone escarpments. A thick, low- to medium-height canopy, with trees such as Native Apple (*Syzygium eucalyptoides*) mixed with she-oaks (*Casuarina*), paperbarks (*Melaleuca*), and

Monsoon vine forest, Timber Creek, NT

OPPOSITE PAGE TOP: Tropical savannah, Katherine, NT

OPPOSITE PAGE BOTTOM: Tropical savannah, Pine Creek, NT

palms, often with some emergent trees, characterises monsoon vine forests; they possess a shady, vine-dominated, and open understorey with a deep and extensive layer of leaf litter. Birds found in this localised habitat differ across its range. In the NT they include Rainbow Pitta, Large-tailed Nightjar, Green-backed Gerygone, and the striking but localised Black-banded Fruit Dove (found only around sandstone escarpments). On Cape York Peninsula (QLD), where this habitat occurs in coastal depressions, the species composition reflects that found in the surrounding rainforests and includes Magnificent Riflebird, Fawn-breasted Bowerbird, Yellow-billed Kingfisher, Palm Cockatoo, and Green-backed Honeyeater. In e. QLD this is a preferred habitat of Fairy Gerygone. Several accessible areas of this habitat can be found close to Darwin (NT), at East Point, Buffalo Creek, and Lee Point, for example. It can also be found throughout Kakadu NP (NT), including around the base of Nourlangie Rock. There are scattered small patches in e. Australia, including around Iron Range NP on Cape York Peninsula and Inskip Point (QLD).

TROPICAL SAVANNAHS OR TROPICAL EUCALYPT WOODLANDS

These open woodlands are restricted to the humid tropical north of Australia, occurring most extensively in n. WA, the Top End of the NT, and the Cape York Peninsula (QLD) south to c. QLD. These areas have a monsoonal climate, with very dry winters but wet summers with rainfall concentrated in short, though very heavy bursts, often in the form of thunderstorms. As the influence of the monsoons lessens to the s. and w. limits of the tropical savannahs, there is a gradation from this habitat into more arid habitats in a patchwork

OPPOSITE PAGE TOP: Bottom of sandstone escarpment, Kakadu NP, NT

OPPOSITE PAGE BOTTOM: Top of sandstone escarpment, Kakadu NP, NT

fashion (see map). This eucalypt-dominated habitat is characterised by sparse tree cover and an extensive, dense ground layer of grasses, which are often seasonally burnt. A new flush of grasses sprouts after each burn, providing feeding areas for many seed-eating species, such as finches and pigeons. Tropical savannahs are also affected by human-caused fires, and if such fires are considerably more frequent than the natural burn cycle, they can adversely affect the bird species present, as they tend to homogenise habitat, lowering plant species diversity. The key to birding this extensive habitat is finding the patches of bird activity, which is concentrated around areas of flowering and water. The monsoonal climate leads to a concentration of bird activity especially towards the end of the dry season when only limited water remains in a few waterholes. These watering points attract flocks of finches, parrots, and pigeons, and make their locations and movements easier to predict during this period, which is a great time to track down such birds as Gouldian Finches. The avifauna is remarkably similar across the tropical savannahs, and many birds found around Derby (WA) extend through the NT into the drier sections of the Cape York Peninsula (QLD). Tropical savannah specialists include Partridge Pigeon, Varied Lorikeet, Hooded and Golden-shouldered Parrots, Northern Rosella, Banded Honeyeater, and Gouldian, Long-tailed and Masked Finches. This habitat is widespread and easily accessible across n. Australia. It is the dominant habitat throughout the Top End (n. NT), including Kakadu NP, and from Darwin south to Mataranka and beyond. In n. QLD it reaches its easterly limit at the w. edge of the Atherton Tableland, where it can be easily accessed for birding around Mareeba and north to Mt Carbine.

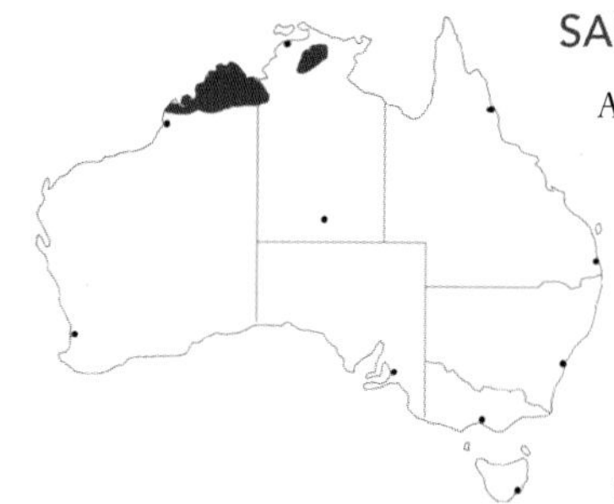

SANDSTONE ESCARPMENTS

Although their reach as habitat is not extensive, sandstone escarpments are included here because they are the preferred environment for a small suite of birds. These rich red-stained rocks form dramatic escarpments, which in some places rise up out of the surrounding tropical savannahs and in others are the result of millions of years of erosion of the surrounding less-resistant rock types. They are restricted to tropical n. Australia, in a band from the Kimberley in n. WA eastward to the sw. Top End (n. NT), then north-eastward to the Arnhem Land escarpment (nc. NT). (Other, similar rocky terrain in Australia—e.g., Flinders Ranges, SA; MacDonnell Ranges, NT—could be placed within this habitat category, but we have excluded them because the bird assemblages are less distinct from the surrounding habitats they occur within.) In most places with sandstone escarpments, the surrounding habitat is tropical savannah. The bases of the rocky slopes at the bottom of the escarpments may support monsoon vine forests or even spinifex grasslands (see further below in this section), while the tops of the escarpments may support a cover of spinifex with scattered stunted eucalypts. While relatively depauperate of birds, this spinifex and eucalypt assemblage is home to some of Australia's most restricted-range bird species. Some species, including Black Grasswren and Kimberley Honeyeater, are restricted to the w. escarpments of the Kimberley, while Chestnut-quilled Rock Pigeon, White-lined Honeyeater, and White-throated Grasswren are restricted to the e. escarpments of Arnhem Land. Others, such as Sandstone Shrikethrush and White-quilled Rock Pigeon, are more widespread. The most accessible e. escarpments occur in Kakadu NP (NT) at such sites as Nourlangie Rock and Gunlom. The w. escarpments in the n. Kimberley (WA) are quite remote, but some places (e.g., Mitchell Falls), are accessible by four-wheel-drive vehicle.

TROPICAL WETLANDS

Tropical wetlands are found across far n. and e. Australia in areas with a monsoonal climate. These are often large expanses of water and exist as part of river floodplains, billabongs (oxbow lakes), and even man-made wetlands. Features of these tropical wetlands, particularly in shallow areas, are mats of floating vegetation, which are sometimes bordered by reeds in shallow areas or by palm-like trees of the genus *Pandanus* where there are steeper banks. Water lilies are often a major component, and this habitat is important for such birds as Comb-crested Jacana and Green Pygmy Goose, and may also support massive congregations of Magpie Geese and whistling ducks. The vegetation surrounding these wetlands is also important, and some birds, such as Crimson Finch rely on it. Some of these shallow wetlands are ephemeral, filling up during the wet season (Dec–Apr) before drying out over the rest of the year, perhaps completely before the rain arrives again. Others are deep enough or large enough to retain water year-round, and provide an important refuge for birds of many species. At the end of the dry season, these wetlands may hold thousands of herons, egrets, ducks, and geese, and particularly in far n. Australia they are important for all birds, and may provide a focal point for not just waterbirds but also pigeons, parrots, finches, and other birds, which come to drink. There are several such wetlands close to Darwin (NT), such as Fogg Dam and Knuckey Lagoon, while perhaps the most famous attraction in Kakadu NP is Yellow Water billabong, an enormous tropical wetland. In ne. QLD, Lake Mitchell and Mareeba Wetlands on the Atherton Tableland are good examples.

Tropical billabong, Kakadu NP, NT

TEMPERATE HABITATS

These habitats predominate in the cooler, s. parts of the continent, including sw. WA, se. SA, VIC, TAS, and se. NSW. The climate here is Mediterranean, characterised by warm, dry summers and cool, wet winters. This regular cycle can sustain tall, dense forests and even rainforests in the wettest areas around the mountains, but away from the mountains this changes to drier, open woodlands.

Temperate billabong, inland NSW

TEMPERATE WETLANDS

Wetlands are widespread in temperate s. Australia and take a variety of forms, including permanent depressions, river floodplains, oxbow lakes (billabongs), man-made lakes, and the ephemeral lakes of inland SA and WA, such as Lake Eyre. Temperate wetlands generally lack the mats of floating vegetation that characterise the tropical wetlands of the north, and attract a very different suite of birds. All temperate wetlands are subject to the vagaries of the climate, and while most are permanent, they will still experience fluctuations in water levels, filling up after rain and slowly evaporating during times of drought. Bird populations fluctuate similarly, and while some species will stay, others are nomadic in nature, roaming great distances depending on water levels in the region. Therefore, while some species may be present at certain sites year-round for years at a time, they may also be absent for similar periods in response to changing water availability on a regional scale. The lakes of far inland Australia, including Lake Eyre, spend most of their time empty and may fill up only every few years, during which time they become important nesting areas for birds such as Australian Pelican and Banded Stilt. The permanent lakes sometimes have a thickly vegetated edge, which acts as cover for secretive species, such as Australian,

Spotless, and Baillon's Crakes, Black-tailed Nativehen, and Australasian Bittern. The open waters are often permanent and attract species such as Hardhead, and Pink-eared, Blue-billed, and Musk Ducks. The reedy edges may also be home to such passerines as Australian Reed Warbler and Little Grassbird. Most cities and small towns have a sewage pond or town weir that will hold some wetland birds, and some examples of accessible permanent wetlands with a good diversity of species are Fivebough Swamp near Leeton (NSW), Gum Swamp near Forbes (NSW), the Werribee Sewage Plant (Western Treatment Plant) near Melbourne (VIC), and Herdsman Lake in Perth (WA).

TEMPERATE RAINFORESTS

A common misconception is that rainforests require warm temperatures. In fact, the primary requirement is large amounts of rain, and so rainforests can occur in very cool climates. Temperate rainforests are found primarily in s. Australia, although the highest mountain peaks of the Atherton Tableland in ne. QLD do support some warm-temperate rainforests. Generally, though, they are found from se. QLD south to VIC, as well as in TAS. They are closely associated with the Great Dividing Range and northward occur only on the cool peaks of the ranges west of Brisbane (QLD) but become more widespread farther south and in TAS. These rainforests are characterised by lower plant diversity than tropical and subtropical rainforests but are still dominated by non-eucalypt trees such as Leatherwood (*Eucryphia lucida*), Sassafras (*Atherosperma moschatum*), and Huon Pine (*Lagarostrobos franklinii*); they have a simpler structure, with fewer large vines, more ferns, and in cooler regions hanging mosses. Although bird diversity in temperate rainforests is generally low, certainly much lower than in tropical and subtropical rainforests, there are several birds that prefer this habitat. Rufous Scrubbird prefers temperate rainforests throughout its range, and Olive Whistler is found in only this habitat at the n. extremity of its range. Other species often found in temperate rainforests throughout their ranges include Pink Robin and Bassian Thrush. There are good examples of temperate rainforests in Lamington NP (QLD) and Barrington Tops NP (NSW).

WET SCLEROPHYLL OR TALL OPEN FORESTS

The term 'wet sclerophyll forest' applies to those tall, open, eucalypt-canopy-dominated forests that form in areas of fertile soil and high rainfall. Wet sclerophyll forests are primarily associated with the Great Dividing Range and occur along this range from c. QLD south to VIC and also TAS. The Karri (*Eucalyptus diversicolor*) forests of sw. WA, and the tall forests along the w. edge of the Atherton Tableland in QLD also fall under this classification. They are the tallest forests in Australia, and include some of the tallest trees in the world. Although these wet sclerophyll forests may be quite dense, the trees' sclerophyllous leaves still allow much light through the canopy, resulting in a dense and diverse understorey comprising many plant families. This understorey often has a similar composition to that of temperate rainforest, with many of the same understorey and ground vegetation species represented. Because of this shared lower vegetation, wet sclerophyll forest can look similar to temperate rainforest at ground level, but the two appear very different at the canopy level. Rainforest and wet sclerophyll forest often occur beside and blend into each other, with rainforest occurring in wet gullies or on cool ridges and wet sclerophyll dominating on slopes. In many areas

TOP: Karri forest, Pemberton, WA

ABOVE: Wet sclerophyll forest, Bruny Island, TAS

wet sclerophyll forest is a buffer between higher-elevation rainforests and drier, more open woodlands. In areas such as the Blue Mtns (NSW) and Snowy Mtns (NSW) wet sclerophyll forests are more extensive, forming large stands. Few bird species are completely restricted to this habitat; the similar nature, both physically and in terms of plant species composition, of the understorey to that of the rainforests leads many of the understorey birds to occur in both habitat types, while a similar canopy structure ensures that the canopy species found in each habitat are also quite similar. Some birds typical of wet sclerophyll forests include Crimson Rosella, Sooty Owl, Superb Lyrebird, Red-browed Treecreeper, Crescent Honeyeater, Bell Miner, Eastern Spinebill, Satin Flycatcher, Rose Robin and Eastern Whipbird. Wet sclerophyll forest is well represented in parks and reserves, and many of the national parks along the Great Dividing Range in NSW and VIC are good locations to access this habitat.

OPPOSITE PAGE TOP: Open woodland, Dryandra, WA

OPPOSITE PAGE BOTTOM: Open woodland, Toowoomba, QLD

DRY SCLEROPHYLL OR OPEN EUCALYPT FORESTS

These forests cover large areas of coastal and sub-coastal se. and sw. Australia. They occur on extremely low-nutrient soils, resulting in a forest, usually dominated by just a few eucalypt species, with regularly spaced trees that do not form an interlinking closed canopy. The open nature of the canopy enables an extensive understorey of shrubs and sometimes also a mixture of shrubs and hardy grasses. Like the canopy, this understorey is also composed of sclerophyllous species, and thus is very different from the lush understorey associated with wet sclerophyll forests. Dry sclerophyll forests often occur in close proximity to wet sclerophyll forests, and along the Great Dividing Range often occur on both the lower inland slopes of the mountains and the coastal plains. Fire plays an important role in the ecology of dry sclerophyll forests, and many of the understorey shrubs are specially adapted to regenerate after fire, while most eucalypts are able to sprout after fire. There are no birds restricted to the dry sclerophyll forests of se. Australia, but typical species include Painted Buttonquail, White-throated Treecreeper, Buff-rumped Thornbill, Spotted Quail-Thrush, and a range of honeyeaters including Regent Honeyeater. In sw. WA, several endemic species such as Rufous Treecreeper, Blue-breasted Fairywren, and Western Yellow Robin are most common in this habitat. Dry sclerophyll forest is the most extensive habitat throughout much of the area where it occurs, so is easily accessible. Many of the coastal national parks in se. Australia contain areas of this forest, as do the well-known birding sites Capertee Valley (NSW) and Chiltern–Mt Pilot NP (VIC).

BELOW: Dry sclerophyll forest, Lamington NP, QLD

OPEN WOODLANDS

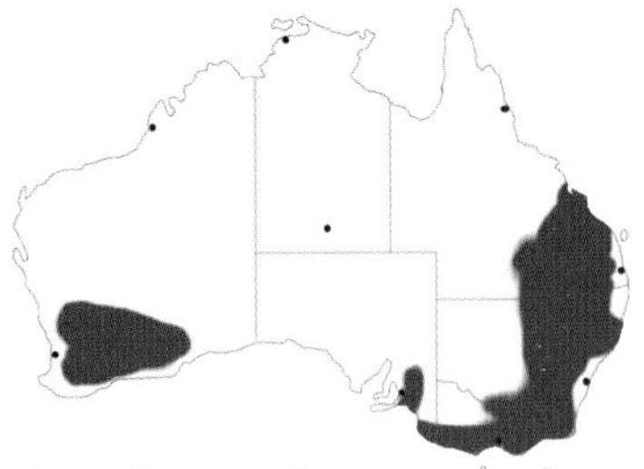

This is probably the least specific of the habitat classifications presented in this section, and is really a catch-all term for those open woodlands that characterise the regions in e. and sw. Australia between the dry sclerophyll forests of the Great Dividing Range and sub-coastal areas, and the truly arid habitats of the interior. These woodlands occur in regions of decreasing rainfall and so are typically quite open. The understorey often depends on the surrounding landscape. Woodlands on floodplains generally have a grassy understorey, while those on elevated areas often support a shrubbier understorey. Unlike in wet and dry sclerophyll forests, the dominant trees are not necessarily only eucalypts but often include a mix of eucalypts, acacias, and

Callitris (conifers of the cypress family). This habitat includes associations such as brigalow and gidgee woodlands (in both of which the dominant trees are acacias) of inland QLD, and the open ironbark and box woodlands (where the dominant trees are eucalypts) of inland se. Australia. Many species of birds that occur in open woodlands also occur in either more arid environments or wetter environments but rarely both. Species with a preference for open woodland habitat include Bluebonnet, Mallee Ringneck, Mulga Parrot, Speckled Warbler, Painted Honeyeater, Rufous Whistler, Hooded Robin and Jacky Winter. Well-known birding locations with examples of this habitat include Bowra Station (sw. QLD), Weddin Mtns NP (NSW), Binya SF (NSW), and Terrick Terrick NP (VIC).

COASTAL AND ALPINE HEATHLANDS

Heathlands are widespread throughout Australia, tending to occur on nutrient-poor soils that are either very shallow or heavily leached, and are often clay-deficient and sand-dominated. They are most prevalent in coastal s. Australia, particularly areas with a Mediterranean climate, such as s. coastal NSW and s. coastal WA, but they also occur in some mountainous areas on poor soils. In general appearance and structure, heathlands are similar across Australia, but despite the nutrient-poor conditions in which they grow, they are actually very floristically diverse. The plants are typically hardy, with adaptations to survive in tough conditions. Low woody shrubs often dominate, and genera such as *Banksia* are quite common. There are sometimes emergent trees throughout the heathlands, but these are often short and stunted. Fire plays a very important role in

Coastal heathland, Fitzgerald River NP, WA

succession in heathlands, and some plant species rely on fire to promote germination. Over time, heathlands will thicken, and fire is important for maintaining a mosaic of heath plants at different stages of succession. Some birds, such as ground parrots, rely on this succession to maintain suitable habitat. The most obvious residents of heathlands are nectarivorous birds such as Little and Western Wattlebirds and New Holland, White-cheeked, and Tawny-crowned Honeyeaters, particularly in spring when the heathlands burst into spectacular flower. Other species that occur in heathlands include Southern Emu-Wren, Beautiful and Red-eared Firetails, and Western and Eastern Bristlebirds. Some of the most accessible areas of heathland occur in Cooloola Recreation Area (Great Sandy NP) near Gympie (QLD), Royal NP and Barren Grounds NR near Sydney (NSW), and the Stirling Range NP and Fitzgerald River NP (WA).

The only areas of alpine vegetation within Australia occur in the south-east from the highest mountains of ne. NSW south along the Great Dividing Range to ne. VIC, and also TAS. This habitat is generally subject to very cold and windy conditions in the winter, including regular snowfall, and comprises low, often stunted eucalypt woodlands, which above the snowline give way to open grasslands, often with areas of dense low shrubbery. Alpine areas are generally low in species diversity, particularly during winter, but some birds such as Flame Robin are present in summer. It is the primary habitat found at high elevations in Kosciuszko NP (NSW), Alpine NP (VIC), and even on Mt Wellington behind the city of Hobart (TAS).

ARID AND SEMI-ARID HABITATS

The majority of the Australian continent, around 70 percent, is considered arid or semi-arid, and so the habitats in this section cover most of Australia. These habitats have all evolved to make the most of the nutrient-poor soils and low rainfall that typify the interior of Australia, and while many do not have what would be considered a rich bird fauna, they shelter a number of species that are found nowhere else on earth. Typically, the vegetation in the arid inland is sparse and hardy. Many areas are too poor in moisture and nutrients to support trees, and hardy shrubs such as mulga (various acacia species) or salt-tolerant chenopods and saltbush predominate. Most vegetation is sclerophyllous and has adapted to the unpredictable nature of the rainfall. When these rains do arrive, the flora and fauna burst into life to make the most of the good conditions before the next long dry spell settles in.

ARID OPEN WOODLANDS

This habitat is not dissimilar to the open woodlands of temperate e. Australia. It often occurs with an understorey of grasses or small shrubs, but the most obvious vegetation is widely spaced trees, often she-oaks (*Casuarina*). It is not a widespread habitat, found from arid n. WA south and east across the arid interior in isolated patches to w. NSW. There are no birds restricted to it, although it is perhaps best known as the preferred habitat of the rare and elusive Princess Parrot. Much of this habitat is inaccessible, existing in remote, rarely visited deserts such as the Great Sandy Desert (WA) and the Great Victoria Desert (WA–SA). It is possible to access this habitat from Alice Springs (s. NT), where it can be found in the area to the south-west around Kings Canyon (Watarrka NP) and Uluru–Kata Tjuta NP, and to the north-west in Newhaven Sanctuary.

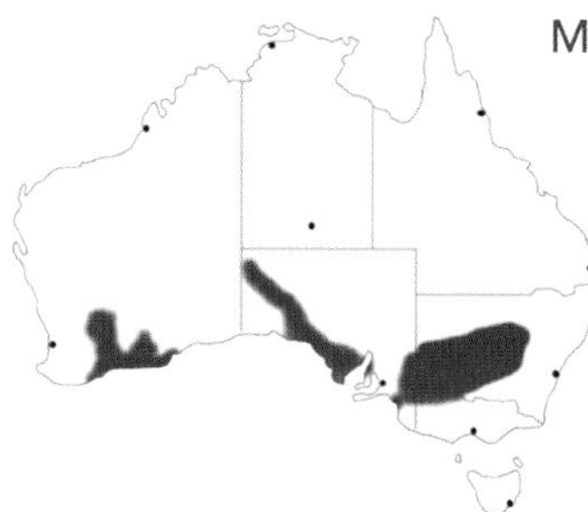

MALLEE

Mallee is a type of scrub-like woodland dominated by several unique species of eucalypts that form a distinctive structure. Mallee occurs mostly in areas of Mediterranean climate and was once widespread throughout s. Australia in a broad swathe arcing southward from the rangelands of inland NSW down to nw. VIC and se. SA and across to the WA border and the WA wheat belt. Mallee generally occurs on nutrient-poor soils on flat, open ground or gentle dunes. The lack of nutrients and low rainfall restricts the growth of the eucalypts, which develop in a curious low formation rarely exceeding 3–6m (10–20ft) high and often in a multi-stemmed coppice-like fashion. It is a dangerously uniform habitat in which it is remarkably easy to get lost. Although appearing quite dense from the edges, mallee may be quite open inside and can be quite easy to walk around in, depending on the understorey, which is often limited because of the nutrient-poor soils. There is usually a scattering of leaf litter that is neither deep nor extensive. Some mallee associations may have little ground cover, while others may have a covering of spinifex grasses or sparse shrubs. Mallee birding experiences can be extremely varied, as many of the species are blossom nomads (e.g., Black, White-fronted, and Yellow-plumed Honeyeaters, Masked and White-browed Woodswallows), present in an area of mallee only when certain plants are in bloom, such as *Eremophila* species or mistletoes. Therefore, the birding in the same area varies greatly; a site can be alive with birds feasting on the abundant available nectar crop on one visit, and appear dead and largely devoid of birds on another visit when few flowers are blooming. There are many bird species that rely heavily on mallee, including Malleefowl, Chestnut-backed Quail-Thrush, Red-lored Whistler, Black-eared Miner, and Shy Heathwren. The mallee habitat has been significantly cleared, however, to support agriculture, and generally only small tracts remain; some of the best-known areas include Nombinnie NR (NSW), Little Desert NP (VIC), Hattah-Kulkyne NP (VIC), Gluepot Reserve (SA), and the Stirling Range NP (WA).

Mallee, Round Hill, NSW

MULGA

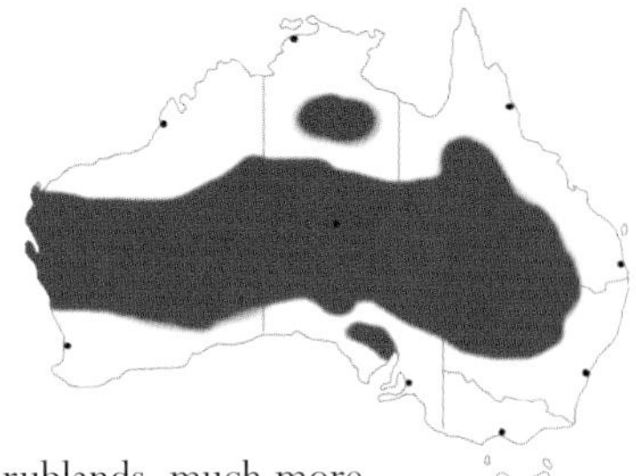

Mulga is an acacia-dominated habitat found across inland Australia that gets its name from the dominant species in these associations, *Acacia aneura*, commonly known as Mulga. It is found on a variety of substrates, from rocky ranges to sandy plains, and occurs in a variety of forms, which differ in both structural composition and plant diversity. In some areas, particularly at the edges of the arid inland, it forms quite dense stands formed almost entirely of *Acacia aneura* that are sometimes called mulga woodlands. More common are mulga shrublands, much more open associations with open stands of *Acacia aneura*, often interspersed with other shrubs such as *Eremophila* or even eucalypts, and sometimes a spinifex or other grass cover. While there is much overlap in terms of the avifauna between scrublands and woodlands, there are a small number of species restricted to either one or the other of these similar habitats. Hall's Babbler is one species that is more common in dense mulga woodlands, while Bourke's Parrot, Grey Honeyeater, and Slaty-backed Thornbill prefer the more open mulga habitat. Other species that can be found in mulga include Mulga Parrot, Chestnut-breasted Quail-Thrush, and Splendid Fairywren. Mulga is widespread and often encountered across the arid interior, including around Alice Springs and Uluru–Kata Tjuta NP (NT). It is also a common habitat around Bowra Station in sw. QLD.

Mulga, Cunnamulla, QLD

SPINIFEX GRASSLANDS

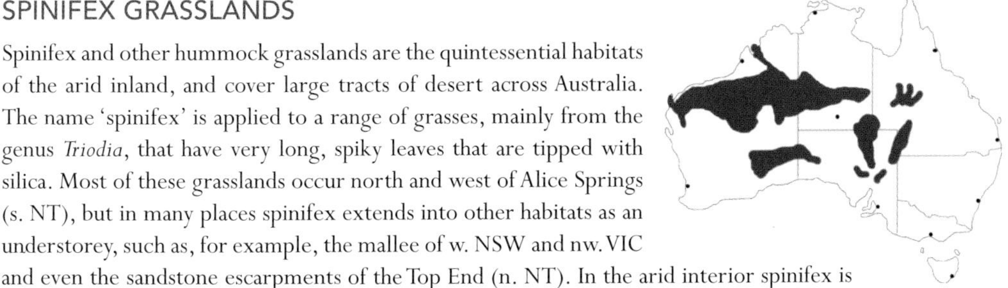

Spinifex and other hummock grasslands are the quintessential habitats of the arid inland, and cover large tracts of desert across Australia. The name 'spinifex' is applied to a range of grasses, mainly from the genus *Triodia*, that have very long, spiky leaves that are tipped with silica. Most of these grasslands occur north and west of Alice Springs (s. NT), but in many places spinifex extends into other habitats as an understorey, such as, for example, the mallee of w. NSW and nw. VIC and even the sandstone escarpments of the Top End (n. NT). In the arid interior spinifex is

often the dominant vegetation type across large areas, and it is a characteristic ground cover on sandy soils and rocky desert ranges. It has a curious and unique growth form, occurring in large, low hummocks that are usually not very high but may be up to 1.5m (5ft) wide. It is quite spiny and thus unattractive to grazing animals, but is important as a sheltering site for many small animals and birds. Spinifex grasslands are the preferred habitat of several bird species including the Rufous-crowned Emu-Wren, Striated and Dusky Grasswrens, Painted Finch, Spinifexbird, and Night Parrot. Any visitor to the arid interior is likely to come across spinifex grasslands, particularly around Alice Springs, where they are widespread in the West MacDonnell NP, Watarrka NP (Kings Canyon), and Uluru–Kata Tjuta NP (NT). In nw. QLD, spinifex grasslands can be found in the Mt Isa area and in parts of Bladensburg NP, while in the n. NT they are widespread around the escarpments of the sw. Top End, such as those of Judbarra/Gregory NP.

OPPOSITE PAGE TOP: *Spinifex on rocky terrain, Lyndhurst, SA*

OPPOSITE PAGE BOTTOM: *Spinifex covered sand dunes, Strzelecki Track, SA*

TUSSOCK GRASSLANDS

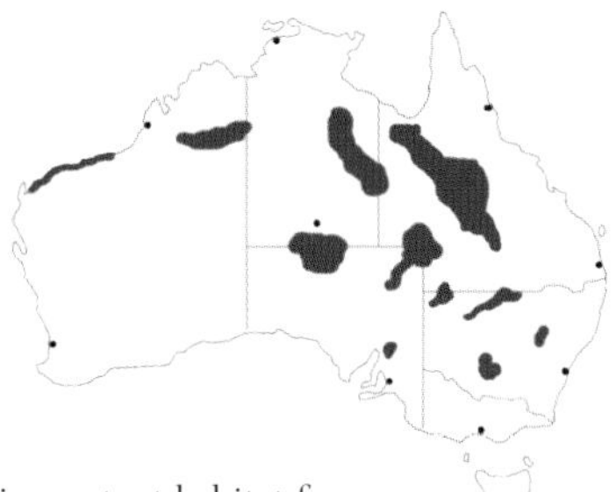

Tussock grasslands, dominated by species such as Mitchell grasses of the genus *Astrebla*, occur on more fertile soils than spinifex grasslands, mainly in ne. inland Australia, particularly nw. QLD and the s. NT. Tussock grasslands often grow in large areas of cracking clay soils high in certain silicates (such as montmorillonite) and are capable of withstanding significant grazing, so are important for the pastoral industry. The tussocks themselves are well spaced, usually about 1m (3ft) apart, and are up to 50cm (20in) high. Tussock grasslands are an important habitat for several species including Letter-winged Kite and Flock Bronzewing. Although the habitat is widespread, most of the areas where it occurs are relatively remote, including the Barkly Tableland of s. NT and Bladensburg NP (nw. QLD).

BELOW: *Gibber*
BOTTOM: *Bluebush and saltbush shrublands*

GIBBER PLAINS

This desert environment is defined less by the plant community present than by the scarcity of plant life. It is an open habitat, its vegetation dominated by low open shrubs and sometimes by perennial tussock grasses, particularly after a rare wet period. The low shrubs are generally chenopods or saltbush (*Chenopodium*, *Atriplex*, and other genera), types of vegetation adapted to the saline soils that predominate. These plains get their name from a local term for the layer of small pebbles and cobbles that cover the ground. Over time, the exposed stones have been varnished with silica, giving them a shiny, smooth texture, and they form an almost impenetrable layer to the fine red sands below. These plains appear to sustain remarkably little wildlife, but they support some very special species, including Gibberbird, Orange Chat, Chestnut-breasted Whiteface, Thick-billed Grasswren, and Inland Dotterel, all of which are very sparsely distributed, highly nomadic, and very hard to find, even within this open habitat. Good examples of gibber plains can be found along the Strzelecki (SA) and Birdsville (SA–QLD) Tracks, around Coober Pedy (SA), and also in several of the national parks in far w. NSW, including Mutawintji NP and Sturt NP.

Gibber plains, Strzelecki Track, SA

MAN-MADE HABITATS

URBAN PARKS AND GARDENS

Although their underlying natural habitat is severely altered, urban parks and gardens in cities across Australia still provide habitat for many birds. The warm weather and summer rainfall of many of the n. cities, including Brisbane (QLD), Cairns (QLD), and Darwin (NT), mean they can support quite a variety of birdlife. Centenary Lakes in downtown Cairns is a fantastic birding destination, and Australasian Figbirds, Australian Brushturkey and Rainbow Lorikeets are common in a number of Brisbane suburbs. The planted parks and gardens of many of the s. cities, such as Sydney (NSW), Melbourne (VIC), Hobart (TAS), Adelaide (SA), and Perth (WA), provide plenty of food for different species of lorikeet, and the botanic gardens in the centre of Sydney often have a resident Powerful Owl. Such species as Tawny Frogmouth, Laughing Kookaburra, Australian Magpie, and Willie Wagtail can still be found in many cities and towns, while Little Corellas, Eastern Rosellas, and Yellow-rumped Thornbills are just a few of the species common around some of the towns in the south-east.

LARGE-SCALE CROP FARMING

Large-scale crop farming, and especially wheat farming, has drastically altered massive tracts of habitat, primarily throughout inland se. and sw. Australia. This has resulted in severe fragmentation of habitats, such as mallee in NSW, VIC, and WA. While some species have lost within this equation, the new croplands and permanent water sources have benefited some species. Country roads within inland Australia in these wheat belts have now become established strongholds for birds such as Rufous and Brown Songlarks, Horsfield's Bush Larks, Nankeen Kestrels, Australian (a.k.a. Black-shouldered) Kites, and a host of parrot species, including Australian Ringnecks, Bluebonnets, and Red-rumped Parrots. Conversely, Malleefowl, Gilbert's Whistler, and Turquoise and Superb Parrots are among the species that have greatly deteriorated from the expansion of these agricultural operations.

PHOTOGRAPHS FOLLOWING PAGES: PAGE 38: Red-browed Pardalote (top left), male Scarlet Robin (top right), and male Splendid Fairywren (bottom) PAGE 39: male Red-capped Parrot

GRAZING LANDS

Grazing has occurred in some form across nearly the whole of the Australian continent, even the arid inland. The intensity of this grazing depends on the suitability of local conditions, and while its effect has not been significant on some inland areas where grazing is light, the impact has been much greater in other regions, particularly in e. Australia. Intense sheep farming is concentrated in the s. rangelands (i.e., the Great Dividing Range and its inland w. slopes), while the significantly less intense cattle farming is concentrated within the inland of the north. Dairy farming is limited to humid coastal regions throughout the south-east, south-west, and the Wet Tropics of QLD. These modified habitats are important to the birder, as almost anywhere outside of urban or protected areas of sub-coastal and inland n. Australia are cattle concessions. Much of these areas are unfenced, open-range, low-density operations, somewhat reminiscent of the vast ranches of the American South-west. Northern Australia, even with the presence of widespread cattle farming, still appears much wilder and less obviously human-modified than the south of the country, which has undergone dramatic and conspicuous modification through commercial wheat farming and high-intensity grazing from sheep. An unseen effect of grazing lands, whether in the south or north, has been the redistribution of water, which has had significant consequences for the birdlife. In the north the bores and dams associated with cattle farms can be highly productive for birds, which gather to drink there in significant numbers, especially in the late afternoon to dusk. Indeed, this can be one of the best ways to find many northern species, such as Flock Bronzewing, Hooded Parrot, Gouldian Finch as well as other finches, and a variety of honeyeaters. Southern grazing lands also provide habitat for Emus, songlarks, Banded Lapwings, buttonquail, and even Plains-wanderer in some specific areas of NSW and VIC.

Species Accounts

CASSOWARY; EMU

1 SOUTHERN CASSOWARY *Casuarius casuarius* 170–175cm/67–69in

A massive, heavyset, flightless rainforest bird weighing up to 86kg (190lb), Southern Cassowary (family Casuariidae) has a coarsely feathered black body and a striking blue and red naked neck and head, which bears a distinctive, large triangular knob or helmet on top. Cassowaries have powerful legs with three-toed feet that possess a 13cm- (5in-) long claw on the innermost toe. Juveniles are buff-coloured with chocolate stripes. The adult's black colouration and the habitat separate this bird from the only other large flightless bird in Australia, Emu, which is never found in rainforests. Southern Cassowary is a shy and very local bird in tropical and subtropical rainforests and in palm groves north of Townsville (QLD); it has also become habituated around Mission Beach and Kuranda (QLD), where it can be seen readily.

2 EMU *Dromaius novaehollandiae* 150–190cm/59–75in

Emu, the only member of the family Dromaiidae, is the national bird, Australia's answer to the Ostrich (*Struthio camelus*) of Africa. It is one of only two huge, flightless land birds in the country, along with Southern Cassowary. It stands nearly 2m (6.5ft) tall, has a stride over 2.7m (9ft) long, and weighs up to 45kg (99lb). Emu is a massive brown bird of open country, differing in colour and habitat from the black, rainforest-dwelling cassowary. Emu is found in most open habitats but is especially common in non-irrigated croplands and pastures. Typically seen crossing red dirt roads or feeding in small groups (two to six birds), it is seemingly quite unafraid when approached slowly by a vehicle, though typically quickly runs away when pursued on foot.

left 1

ad. & imm. 2

MEGAPODES (MEGAPODIIDAE)

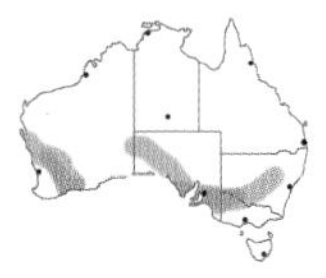

1 MALLEEFOWL *Leipoa ocellata* 55–60cm/21.5–23.5in

Malleefowl is a megapode, a member of the Megapodiidae, a family of birds also known as mound-builders. It builds a large earth mound, in which it lays its eggs, and lets rotting vegetation provide the heat to incubate them. It moderates the temperature by regularly adding and removing vegetation over the eggs. If seen well, Malleefowl is easy to identify: cryptically marked above with intricate chestnut, white, and black markings across the upperparts, though much plainer and pale below, with a distinctive black stripe running down the throat to breast. The best way to see these shy birds is to visit an active mound and observe from a distance. Although Malleefowl will occupy traditional territories for long periods, the territory can contain multiple mounds, of which only one is usually active. This species requires undisturbed areas of mallee woodlands and scrub in its range in s. Australia, and is best seen within protected areas in nw. VIC and e. SA.

2 AUSTRALIAN BRUSHTURKEY *Alectura lathami* 60–70cm/23.5–27.5in

A large, turkey-like megapode, or mound-nest-building bird. Brushturkeys are mainly black, with bright red bare skin on the head and neck and a vivid lemon-yellow collar. They are poor flyers, spending most of their time on the ground, and preferring to run from danger or flap clumsily away. They are commonly found in the rainforests and thick woodlands of e. Australia, from extreme n. QLD down to ne. NSW. Unlike many other rainforest creatures, this bird is not shy, and can even be bullish and sometimes a pest around picnic areas. Any picnic ground within its range is likely to have at least a few resident birds.

3 ORANGE-FOOTED SCRUBFOWL *Megapodius reinwardt* 42–47cm/16.5–18.5in

A small, sooty-grey megapode with rusty-brown upperparts, a crested head, and conspicuous orange legs and feet. Its powerful feet are used to build truly massive mounds, larger in size than those of all other Australian mound-builders, measuring some 12m (39ft) across and up to 5m (16.5ft) high. Scrubfowl are common birds within botanical gardens, town parks, forests, and tropical woodlands in humid parts of tropical n. Australia, and have become more abundant over the last decade. Easily found in parks within the cities of Cairns (QLD) and Darwin (NT).

2

1
2
3

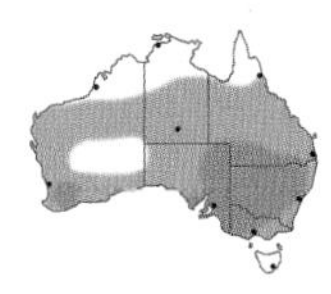

1 STUBBLE QUAIL *Coturnix pectoralis* 18–20cm/7–8in

This quail is usually found in drier, more open, and shorter grasslands than the other two species. It is generally brown, with usually prominent, heavy pale streaks on the back and breast. If the bird is seen well on the ground the obvious white eyebrow and perhaps the male's buff-coloured throat may be seen. In flight it has relatively long wings and usually flies away fast and low over the grass, rocking gently from side to side. The streaking on the back is usually quite obvious. It is found across Australia except the humid tropics, occasionally irrupting in large numbers, and is quite common in rural areas of se. and sw. Australia.

1

2 BROWN QUAIL *Coturnix ypsilophora* 18–22cm/7–8.5in

This quail is usually found in denser and wetter grasslands than Stubble Quail, often around the edges of swamps and wetlands or in coastal heaths. It is quite variable in colour, but is generally brown and covered in fine streaks, though these are much less obvious than on Stubble Quail and often not seen without good views. When flushed it jumps quite high in the air before whirring away on short, rounded wings, usually in a straight line. It often glides for the final section of its flight before plunging back to cover. In flight it looks back-heavy or pear-shaped. Common around most coastal regions, ranging into the main croplands and pastoral lands of the continent, it occurs in a wide variety of dry and wet grasslands, swamps, and heaths. The easiest of any quail or buttonquail in the country to see, it often feeds in coveys on the sides of country roads.

3 KING QUAIL *Excalfactoria chinensis* 11.5–14cm/4.5–5.5in

This quail is quite a bit smaller than the other two species, and is usually found in dense, wet, swampy grasslands and heaths, a habitat it often shares with Brown Quail. The female is quite similar to Brown Quail but is generally darker above. The male is blue-grey with brownish wings and a chestnut belly, and has an intricate black-and-white pattern around the throat. King Quail are difficult to flush, preferring to run away through the dense undergrowth. When they do flush they are feeble flyers and look small and dark in flight. They fly with a strange vertical posture and drop suddenly back into cover after a short flight. The species occurs in the n. Kimberley and along the e. seaboard. In the east it prefers wet heaths, swamp margins, and dense grasslands. In the north it also extends into tropical grasslands. Usually found fairly close to the coast, this quail is quite rare and difficult to see, although there are several locations around Brisbane where it can be found.

1
1
2
2
♂ 3
♀ 3

1 HOARY-HEADED GREBE *Poliocephalus poliocephalus* 28–30cm/11–12in

A small, compact grebe with a relatively short bill and with a large blackish-grey head streaked or frosted with white in breeding plumage. That head pattern is distinctive, but this species can be easily confused with the smaller Australasian Grebe when both are in non-breeding plumage and appear as greyish grebes with a two-tone neck and head; in Hoary-headed the frontal section of the neck is all whitish while the nape is dark, and the head is clean whitish in the lower third but has a dark cap. Both species display a pale bill at this time, while when breeding both have a dark bill with a small pale tip. Non-breeding Hoary-headed Grebe can be differentiated by its eye, which is largely dark compared to the beady, pale yellow-orange eye of non-breeding Australasian. Both species show a dark wing with a broad white wing bar in flight. Immatures of both Hoary-headed and Australasian Grebes have multiple white stripes on the head, which in Hoary-headed are broader and bolder. Hoary-headed Grebe is found over most of Australia, though is generally much less common in n. and inland Australia. It is found most often on bodies of open fresh water but can also occur in calm bays and inlets. It tends to avoid smaller water bodies, such as farm dams, unlike Australasian Grebe.

2 AUSTRALASIAN GREBE *Tachybaptus novaehollandiae* 23–25cm/9–10in

The smallest Australian grebe, it has a distinctive breeding plumage: A striking yellow eye and a diagonal golden-yellow patch below it are set in a dark head, which also shows a bold chestnut stripe that runs down each side of the neck. In non-breeding plumage it is dull and indistinctive, and readily mistaken for the larger Hoary-headed Grebe (see more details of how to distinguish them in that species' account), although Australasian displays a strikingly bold yellow-orange eye in all plumages. Immatures of both Australasian and Hoary-headed are similar, with white-striped heads, although the striping in Australasian is narrower and less striking. Australasian Grebe is found across much of the continent, although it is more abundant in the south-east, on a variety of freshwater wetlands, including small farm dams. Rarely, it also occurs in tidal areas.

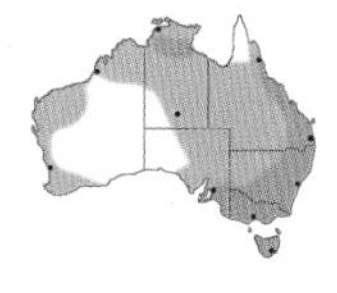

3 GREAT CRESTED GREBE *Podiceps cristatus* 48–61cm/19–24in

A large, striking grebe that shows prominent chestnut neck frills and a bold, dark crest. Great Crested Grebe retains the adult plumage in both breeding and non-breeding seasons in Australia, acquiring no dull non-breeding plumage as birds occurring in the Northern Hemisphere do. The dramatic, synchronised courtship dance involves the male and female of a pair facing each other, bobbing their heads regularly and touching bills at times, while flaring the crest and neck frills; the dance climaxes with the pair rising high up in the water, with vegetation in their bills, and treading water. The species is patchily distributed in e. Australia (where most abundant in the south-east), rarely in the north, on large freshwater wetlands and occasionally also in tidal bays. It is restricted to large bodies of water.

1 br.

br. 1
nbr. 1
br. 2
nbr. 2
3

DUCKS, GEESE, AND SWANS (ANATIDAE)

1 MAGPIE GOOSE *Anseranas semipalmata* 70–99cm/27.5–39in

A huge pied goose with some odd physical features, such as the hooked bill and a strange, helmet-like knob on its crown, which leads to its classification within its own, one-species family, Anseranatidae. It is a large waterbird with a wingspan of over 1.5m (5ft), especially conspicuous in the tropical north of Australia, where it often congregates in truly massive numbers, sometimes flocking in groups of thousands. Although currently still most abundant in the tropical north (n. WA, n. NT, and n. QLD), it is slowly spreading southward and can be found in smaller numbers in s. QLD and n. NSW. Kakadu NP and Fogg Dam (near Darwin) in NT are both excellent sites to find Magpie Geese in good numbers.

1

2 CAPE BARREN GOOSE *Cereopsis novaehollandiae* 75–90cm/29.5–35.5in

A large, pale grey goose with few striking markings save a few rows of dark spots on the wings and a dark bill possessing a conspicuous lime-green cere. It also displays a diffuse white crown, black undertail coverts and tail, and a broad black trailing edge to the wings, most visible in flight. Due to its adaptability to feeding on agricultural lands, after recent declines the species has recovered and now appears to have a stable, if local, population in s. Australia. Large grazing flocks occur on the VIC coast during the non-breeding season, although when breeding the birds split off into smaller groups or pairs to breed on offshore islands. It is most readily seen on TAS.

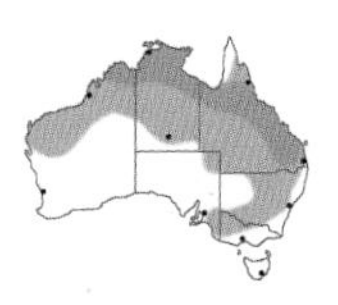

3 PLUMED WHISTLING DUCK *Dendrocygna eytoni* 41–60cm/16–23.5in

A tall, pale brown duck with prominent creamy-white plumes that protrude from its flanks. Like all whistling ducks, it has a long neck and long legs relative to many other duck species. At a distance it can be confused with Wandering Whistling Duck, with which it sometimes occurs. However, Wandering lacks the conspicuous flank plumes, and Plumed has a uniform pale brown neck and head, lacking the bold dark cap and nape that Wandering shows. Behaviourally, these species differ further in that Plumed tends to forage on aquatic vegetation by grazing around the edges of wetlands, whereas Wandering generally spends more time in the water. Plumed is a widespread species found in a variety of wetland areas and swamps through the tropical north of Australia and down the e. side of the continent. It is often encountered in huge single-species flocks numbering hundreds of birds.

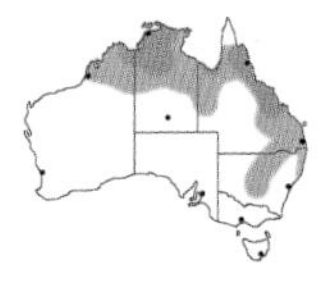

4 WANDERING WHISTLING DUCK *Dendrocygna arcuata* 56–60cm/22–23.5in

A tall duck with the long legs and elongated neck that characterise all whistling ducks. A darker bird than Plumed, it has a deep rufous body, with only short, inconspicuous plumes along the flanks, and a pale foreneck that contrasts strongly with a dark cap extending as a stripe down the back of the neck, quite unlike Plumed. It further differs from Plumed in showing a dark eye and dark legs. Furthermore, Wandering also feeds predominantly at night on open water, where it dabbles and dives for aquatic vegetation. It is a common bird of tropical n. Australia and is also found along the e. side of the continent. However, it travels in large flocks that are sporadic, varying from being numerous at times to completely absent at other times within the same sites.

1
1
2
2
3
4

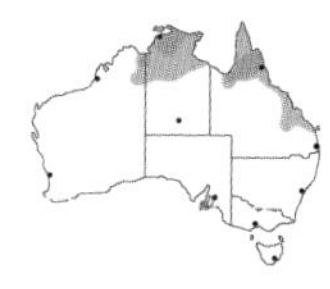

1 RAJA SHELDUCK *Tadorna radjah* 48–61cm/19–24in

A small white shelduck, closer to most *Anas* ducks in size. Boldly marked and patterned quite unlike any other duck in Australia, it is white-headed and has dark chestnut upperparts and clean white underparts except for a conspicuous chestnut breast band. At a distance, it essentially looks black and white. The male has a broader dark chestnut breast band than the female, although otherwise the sexes look very similar. These shelducks feed around the margins of tropical wetlands, including pools, lagoons, tidal areas, and swamps, and are often seen perched on banks and dykes around the fringes of these areas. They are found only in n. Australia, and are common in parts of the NT but quite local and scarce in e. QLD. Good sites are Fogg Dam (NT) and the Yellow Water, accessible by river cruise, in Kakadu NP (NT). The species is nomadic and less reliably seen in QLD.

1

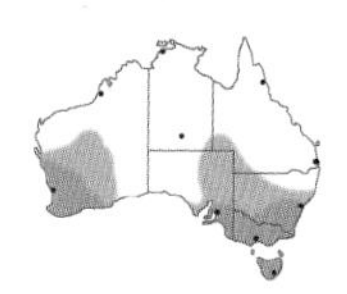

2 AUSTRALIAN SHELDUCK *Tadorna tadornoides* 56–71cm/22–28in

This plump duck is blackish-bodied, with a bright buff area on the lower neck in the male, and bright, rich chestnut on the lower neck and chest in the female. Both sexes have a dark blackish hood; however, the male has an all-dark head with a subtle green sheen and no pale markings around the eye, whereas the female has a variable amount of white around the eye. In flight birds show white flashes at the front of the wing. Australian Shelduck occurs around swamps, dams, and streams over much of the south, although is much scarcer inland. Good places to find it include the shorelines of TAS, sw. WA, and sw. VIC. Like many wetland birds in Australia, it responds to changes in local water levels by migrating over large areas in order to seek out suitable wetlands.

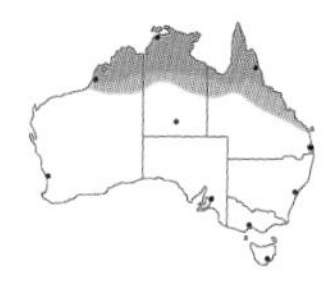

3 GREEN PYGMY GOOSE *Nettapus pulchellus* 32–38cm/12.5–15in

A tiny duck (the smallest in Australia) with green-glossed upperparts and conspicuous white cheek patches. It is sexually dimorphic; males have a dark neck and a reduced white area on the face, while females have a paler neck and larger facial patch. It is usually observed swimming among mats of floating vegetation, is rarely seen on land, and prefers to perch on submerged logs and branches well away from the edges of wetlands. It is locally common on large wetlands with abundant floating vegetation, such as water lilies, in the tropical north of Australia from Broome (n. WA) eastward to around Rockhampton (QLD). It is especially common at Mareeba Wetlands (n. QLD) and the Yellow Water in Kakadu NP (NT).

4 COTTON PYGMY GOOSE *Nettapus coromandelianus* 34–38cm/13.5–15in

A tiny, mostly white duck (one of the smallest in the world) with dark, green-glossed upperparts in males (brownish in females), a dark cap, and a solid but indistinct green breast band (very indistinct or lacking in females and immatures). Males have a clean white face and neck; females display a dark line through the eye. Both sexes have a small, dark bill. In flight they show dark greenish wings with a conspicuous white wing bar. Possible confusion species include Raja Shelduck, which is also largely white but lacks a dark cap, shows no green above, and has much more white in the wing in flight, showing two bold white wing flashes. Green Pygmy Goose is similarly sized but has a largely dark head and neck. Although widespread in Southeast Asia, Cotton Pygmy Goose is limited to n. coastal areas of QLD, and most regular from Townsville to the Atherton Tableland. Much less common than Green Pygmy Goose, it seems to prefer deeper lagoons than that species.

1
1
♀
♂ 2
♂ 3
♀ 3
♂ centre; 2 ♀s 4
♀ 4

1 BLACK SWAN *Cygnus atratus* 110–140cm/43.5–55in

This is the only swan species in the world that is predominantly black; the other eight species are completely or nearly all white in colour. Otherwise it is typically swan-like in appearance, with a long graceful neck and rotund body. It has a bold reddish bill and broad white fringes on the wings, which are normally only noticeable when the bird takes flight. Black Swan is a truly massive waterbird, with a wingspan of up to 2m (6.5ft). It is common and conspicuous on larger wetland areas throughout most of Australia (including TAS), except for the interior of the west. In n. Australia it tends to favour freshwater environments, although in the south, where more numerous, it can also occur in saltwater habitats.

2 FRECKLED DUCK *Stictonetta naevosa* 51–58cm/20–23in

A sooty-grey duck covered with mid-grey vermiculations. The high peak at the back of the crown and prominent ski-jump bill give the head a very distinctive shape. In flight, the armpit and leading part of the underwing show an off-white colour that contrasts with the otherwise dark bird. This endangered duck is usually found in se. and sw. continental Australia on well-vegetated swamps with standing dead wood, such as Gum Swamp near Forbes.

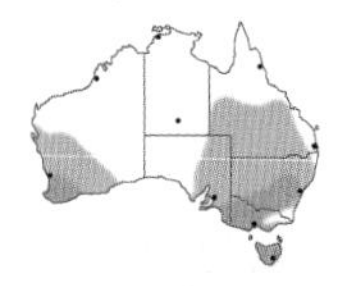

3 MUSK DUCK *Biziura lobata* 51–71cm/20–28in

This is a bizarre-looking duck, somewhat reminiscent of an eider. It is a large, low-sitting, sooty-black bird with a wedge-shaped bill and a spiky, fan-shaped tail that sits flat on the water or raised over the back in display. The males have a large dewlap flap on the bottom of the bill. Females are smaller and lack the flap. In se. and sw. Australia, Musk Duck occurs in many types of water habitats, from well-vegetated streams to open lakes. In TAS it also lives along the coast in bays and in the Derwent River estuary.

3 ♀

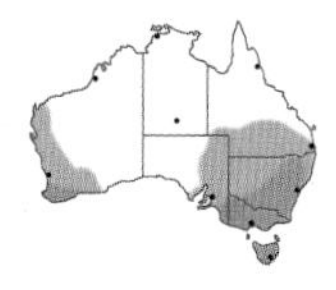

4 BLUE-BILLED DUCK *Oxyura australis* 36–44cm/14–17.5in

Blue-billed is a distinctive diving duck, similar in appearance to its American cousin, Ruddy Duck (*O. jamaicensis*). Both are deep red-wine-coloured ducks with a black head and a conspicuous powder-blue bill. Females are less distinctive: all dull brown with indistinct pale barring. Blue-billed Ducks sit low in the water, usually with the tail submerged, but they can cock the tail, particularly in display, when it may also be fanned. When they realise an observer is nearby, these ducks will sneak into the reeds, but generally will come out if the observer remains silent. Uncommon, they inhabit large, well-vegetated wetlands in s. Australia (regularly in s. WA, locally in SA, VIC, NSW, and TAS), where they are typically secretive, usually sticking close to the vegetated edges, and can be hard to track as they dive frequently.

1
2
♂ 3
♀ 3
♂ 4
♀ 4

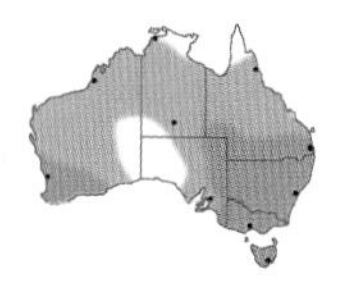

1 GREY TEAL *Anas gracilis* 42–44cm/16.5–17.5in

A uniformly small teal, this species is also the most uniform and nondescript, with a uniformly greyish-brown body and a two-tone head—dark on the cap, whitish on the throat and neck—and a blood-red eye. Males and females are identical. The contrasting throat can be used to differentiate this species from the female Chestnut Teal. A common species, Grey Teal can be found on any type of wetland throughout Australia and TAS; it is most often seen in flocks. Grey Teal is usually shyer than other teal species.

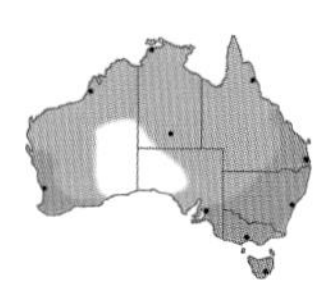

2 PACIFIC BLACK DUCK *Anas superciliosa* 47–60cm/18.5–23.5in

Australia's most common dabbling duck, similar to the familiar female Mallard (*A. platyrhynchos*) of America and Eurasia. A brown duck with a striking face pattern, it possesses a black cap and two horizontal black lines across the side of the buff face. It has an iridescent, glossy deep purplish-green wing patch (known as a 'speculum'), best seen in flight although sometimes revealed in swimming birds as a small, square patch towards the rear of the body sides. Males and females are alike. This is one of Australia's most familiar waterbirds, found around a variety of wetlands across the entire continent, though generally much more common in the south-east and south-west.

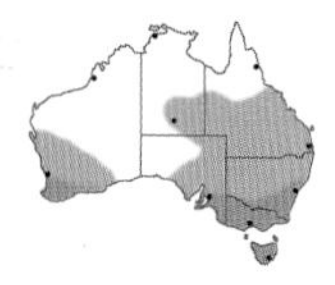

3 CHESTNUT TEAL *Anas castanea* 36–50cm/14–19.5in

A strikingly dimorphic duck: Males are distinctive, with a deep green head, chestnut underparts, and a striking white flank patch. Females are wholly greyish brown, like many other female ducks, particularly Grey Teal, although Chestnut is warmer-toned and has a uniformly coloured head and neck. Chestnut Teal is a southern species, found in the sw. and se. mainland and TAS; it is most abundant in saline and coastal areas.

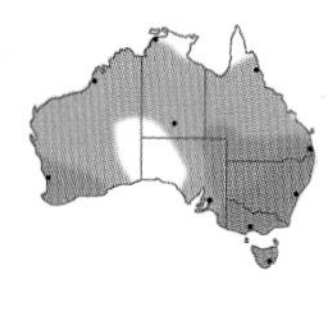

4 HARDHEAD *Aythya australis* 43–50cm/17–19.5in

Very easy to identify, Hardhead is rich brown all over and has a white undertail. If the bird is seen out of the water or flying, the white patch on the underside is obvious. Also in flight, the combination of white underwing and an upperwing with the trailing half white fringed in black makes identification easy. Although Hardhead is sometimes called White-eyed Duck, only the male has a white eye; the female's eye is dark. This common duck is found on a wide variety of deep still waters throughout most coastal and semi-arid regions of Australia. It can also be found in coastal swamps and suburban wetlands.

1
2
♂ 3
♀ 3
♂ 4
♀ 4

DUCKS, GEESE, AND SWANS (ANATIDAE)

1 PINK-EARED DUCK *Malacorhynchus membranaceus* 37–44cm/14.5–17.5in

A peculiar Australian duck with both a bold plumage pattern and a distinctive shape, due to its large, spatulate (shovel-shaped) bill, which it uses for filter feeding. The body is boldly barred, and the head is pale, with a large dark eyepatch that extends as a lateral crown stripe (giving the top of the head a dark appearance in side view) and an indistinct pink ear patch, visible only at very close range. The bold plumage pattern differentiates this species from our only other duck with a similar bill shape, Australasian Shoveler. Pink-eared Duck is widespread but highly nomadic and generally confined to inland Australia (except the driest desert areas), though it wanders coast-ward during exceptionally dry periods inland. It occurs on a variety of wetland habitats, from large permanent lakes to temporary flood waters and sewage ponds, where it is most often encountered in large flocks, usually dabbling on the water's surface or perched on emergent logs and branches.

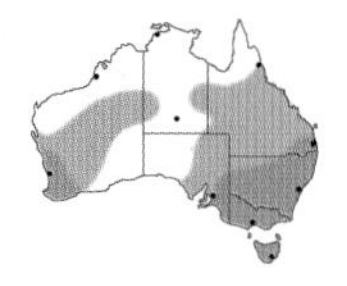

2 AUSTRALASIAN SHOVELER *Anas rhynchotis* 44–55cm/17.5–21.5in

A distinctively shaped duck with a long, spatulate (shovel-shaped) bill; the male in breeding plumage is boldly patterned, with a dark green head, beady yellow eye, and striking white half crescent on the lores. It also displays deep chestnut flanks, on the lower portion of which is a conspicuous white patch, easily visible in a bird swimming on the water, as it is normally seen. Females, immatures, and non-breeding males are dull greyish-brown ducks with scattered dark markings, best differentiated from other similarly patterned female ducks by the shape of the bill; only one other Australian duck species possesses a long, spatulate bill, Pink-eared Duck, which shows a strikingly patterned body with bold flank stripes, quite unlike Australasian Shoveler. This species is moderately common across most of s. Australia, excluding the Nullarbor region, preferring large, well-vegetated wetlands though also found in many non-coastal sewage ponds. It is common at Fivebough Swamp near Leeton (NSW). *(Illustrated below with Silver Gull and two Red-necked Avocets)*

2

3 MANED DUCK *Chenonetta jubata* 44–50cm/17.5–19.5in

Also known as Australian Wood Duck. This is a small-billed, grey-bodied grazing duck with bold black stripes on its back and a distinctive rusty-coloured, maned head in males (pale brown in females). Bold scaling on the lower neck and upper breast extends all the way down onto the belly in females. Immatures are like a washed-out version of the female. The common suburban duck of se. Australia, it is a familiar species commonly found over much of the country, except for humid n. areas and interior desert areas. It is easily found in a variety of habitats, especially around small farm dams and wooded areas near water. It is the most land-based of the Australian ducks, often seen grazing out of water.

1
1
2
2
♂
♀ 3

GANNET; TROPICBIRDS

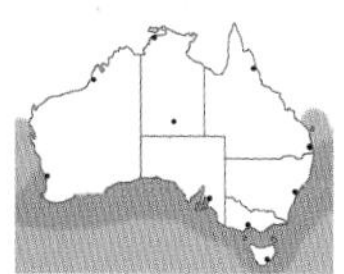

1 AUSTRALASIAN GANNET *Morus serrator* 84–91cm/33–36in

Australasian Gannet is the only regularly occurring gannet species in Australia; like the boobies, it is in the family Sulidae. It is almost all white, except for a broad black border to the trailing edge of the wing and a mainly black, pointed tail, both of which are best viewed in flight. It also has a black outline to the front of the face and a yellow head. These birds are superficially similar to some of the graceful albatrosses when seen at distance, most notably Wandering Albatross. They are best separated by shape: Albatrosses have longer, narrower wings that are more uniform in width along most of their length, squarish tails, and thick, uniform-width bills that are blunt-ended. Gannets, by comparison, have shorter, broader wings that gradually come to a pointed tip, long pointed tails, and sharp, triangular bills. These features combine to give albatrosses and gannets different profiles in flight, which is how they are most often seen at sea. On close inspection, the gannet also has a yellow head that albatrosses do not possess. Furthermore, although gannets do glide for short periods, they tend to flap regularly in flight, while the albatrosses can glide great distances without flapping. Australasian Gannet is a s. coastal species occurring from the wc. coast of WA south- and eastward and around to the s. QLD coastline. These pelagic birds are most likely to be seen in flight either when blown in close to shore during windy conditions or from a boat trip into deeper waters, where they are more regular than along the coastlines. It is seen away from coasts only exceptionally.

1 ad.

2 RED-TAILED TROPICBIRD *Phaethon rubricauda* 70–99cm/27.5–39in

These large seabirds appear white at a distance and have a strange, stubby-tailed silhouette; the red tail streamers are obvious only at close range. The adults have a red bill, while young birds have a dark bill and a complex pattern of black bars and speckles on the upperparts. They are rarely seen close to the mainland; spotting them usually requires a summer pelagic birding trip or a visit to one of their offshore breeding islands. There is a small breeding population at Sugarloaf Rock near Cape Naturaliste in WA that can be seen from the mainland; the birds can also be seen on both Lord Howe and Norfolk Islands. Tropicbirds are in family Phaethontidae.

3 WHITE-TAILED TROPICBIRD *Phaethon lepturus* 70–99cm/27.5–39in

Similar in shape to Red-tailed Tropicbird, this species is smaller and has more black on the upperparts and white tail streamers. The adult has a yellow bill, while young birds have a dusky-yellow bill with a darker tip. White-tailed Tropicbird can be seen on some islands off nw. Australia, where it breeds; otherwise it is usually seen far offshore around n. Australia.

ad. 1
imm. 1
2
ad. 3
imm. 3

PIGEONS AND DOVES (COLUMBIDAE)

1 WHITE-HEADED PIGEON *Columba leucomela* 41cm/16in

A large forest pigeon with a prominent dirty-white head and underparts, contrasting with a sooty-black back and tail. It is a fairly common species in a narrow strip along the e. coast. It prefers rainforests and wet sclerophyll forests but may venture into country towns to feed if decent patches of forest occur nearby. Far ne. NSW appears to be a stronghold and a good place to find these birds.

2 BROWN CUCKOO-DOVE *Macropygia phasianella* 38–44cm/15–17.5in

A large, very long-tailed, all rusty-brown dove that feeds both in trees and on the ground. In its range there are no other all-brown pigeon species. This is the only cuckoo-dove in Australia. This common dove is confined to the e. coastal forest belt (from n. QLD south to e. NSW, rarely to VIC), where it occurs around the margins of temperate and tropical rainforests and wet sclerophyll eucalypt forests.

3 PACIFIC EMERALD DOVE *Chalcophaps longirostris* 23–28cm/9–11in

An attractive forest pigeon with glistening green upperparts and a distinctive double white band on its black rump that is visible in flight as it flashes low through the understorey. It is otherwise rich salmon-coloured below. This beautiful dove is highly terrestrial, foraging mostly on the ground, often in the shade of the forest. It is found in the tropical north: n. WA, n. NT, and n. QLD, and south down the coastal zone to s. NSW, rarely into VIC. It is mainly a bird of forest edges, including both tropical rainforests and wet sclerophyll forests, and also occurs in scrub, coastal heaths, and mangroves. It is most often encountered when single birds are flushed suddenly out of the leaf litter at close range, which often startles the observer. It is best found feeding around the verges of rainforest-lined parking lots or feeding along forested country roads at dawn.

4 WHITE-QUILLED ROCK PIGEON *Petrophassa albipennis* 28cm/11in

A very dark red-brown long-tailed ground pigeon with a conspicuous large white wing patch, especially visible in flight. It differs from the very similar Chestnut-quilled in having a white, not chestnut wing patch and fine mottling concentrated on the face and throat. It is found on spinifex-dominated sandstone plateaus and escarpments through the Kimberley (WA) east to Victoria River Crossing (NT). This bird hides under rocks and in crevices during the heat of the day, coming out to sun itself in the morning and feeding intensively in the evening.

5 CHESTNUT-QUILLED ROCK PIGEON *Petrophassa rufipennis* 28–31cm/11–12in

This long-tailed pigeon is very dark brown over the whole of its body, with indistinct white flecking over some of the head and upperparts. The only bold marking is a conspicuous chestnut marking on the outer wing, usually visible only in flight. It is restricted to rocky areas around sandstone escarpments in the tropical north of the NT, and is best found by walking across rocky areas, where singles or pairs may be found scurrying across the rocks or may burst into flight when disturbed with a loud whirr of wings. During the distinctive low flight interspersed with long glides, it displays its most prominent feature: the chestnut wing patch. It is very local and uncommon, and best looked for at Gunlom Falls in Kakadu NP.

ad. 1
imm. 1
2
3
4
5

PIGEONS AND DOVES (COLUMBIDAE)

1 COMMON BRONZEWING *Phaps chalcoptera* 30–36cm/12–14in

A largish pigeon that at distance appears as an unattractive, largely greyish-brown bird. However, up close a subtle beauty is exposed in its bold white facial markings: a prominent white forehead, a white stripe below the eye, and delicate scaling on the upperparts. There is also intricate patterning on the wings, and parts of the wing show a bronze metallic sheen. It is most often first seen when disturbed, as it bursts into the air with a clatter of wing noise, exposing rich cinnamon underwings as it flies off at high speed. Found in open forests and woodlands all over the continent except the most arid desert areas, it is often seen feeding along the edges of dirt roads or around farm dams during late afternoons.

2 BRUSH BRONZEWING *Phaps elegans* 28–32cm/11–12.5in

A ground pigeon with a dark brown back and metallic green-purple wing bars. It has grey underparts with a chocolate crescent on the sides of the breast. Like Common Bronzewing it has a buff crown, but it has a chocolate eye stripe extending back to the nape. Often seen in flight, it is dark above with prominent chestnut shoulders and neck and a slate-grey body, but differing from the paler back of Flock Bronzewing. Brush Bronzewing prefers denser habitats than Common Bronzewing, such as heaths and coastal woodlands with a scrubby understorey, and can be difficult to see well, scurrying away on trails through the forest or flushing suddenly off the ground. It is generally uncommon, but can be locally abundant. It is fairly easy to find in TAS and coastal areas of sw. WA.

3 FLOCK BRONZEWING *Phaps histrionica* 28–30cm/11–12in

This rare bird is the quintessential inland nomad, turning up unexpectedly in sometimes huge numbers when conditions are good, before disappearing just as quickly to parts unknown. It is easy to identify, being a plump pigeon with sandy-brown upperparts, grey underparts, and a black-and-white-patterned head. Females are duller than males and young birds are duller still. Flock Bronzewings are usually seen in flight, often in tight flocks flying fast on long wings, when the rufous underwings are obvious. The Barkly Tableland in the NT is a stronghold for these birds, and they are occasionally recorded from sw. QLD, but they may turn up anywhere in nc. Australia and along the Strzelecki (SA) and Birdsville (SA–QLD) Tracks.

4 SQUATTER PIGEON *Geophaps scripta* 25–30cm/10–12in

This is an uncommon ground-feeding brown pigeon with the brown upper breast and grey-brown lower breast framed in a white V running from sides to white belly. The contrasting black-and-white markings on face, chin, and sides of neck form an unusual pattern: a large white cheek patch dissected by a black vertical line below the eye and a horizontal one behind the eye. There are two subspecies, which differ in the colour of the eye ring: Birds in n. QLD have a red eye ring, while southern birds possess a pale bluish-grey eye ring. These pigeons are most often encountered in small parties feeding on the ground, where they may allow close approach. The species occurs in e. Australia from n. QLD southward into the far north of NSW; it is usually found near water sources, in open woodlands, tropical savannahs, around rural homes, and in open grassy areas. A good method of finding Squatters is to drive along dirt roads in the late afternoon, when small groups of them can often be found quietly feeding along the verges; another is checking around small dams, where they may gather to drink late in the day.

5 PARTRIDGE PIGEON *Geophaps smithii* 25–28cm/10–11in

Very similar in appearance and behaviour to Squatter Pigeon, but Partridge Pigeon possesses a large patch of bare skin around the eye, which is red in most of its range but yellow in the north-western subspecies of the Kimberley region (WA). Partridge Pigeons are sandy-brown in colour, with white flanks and belly. They occur in open woodlands with open rocky ground or short grass, and are usually seen in small groups walking around on the ground. They are generally uncommon, but the more easterly subspecies of the NT may be seen in open woodlands throughout Kakadu NP. Partridge Pigeon, which occurs in the tropical north of the NT and WA, replaces Squatter Pigeon in nw. Australia.

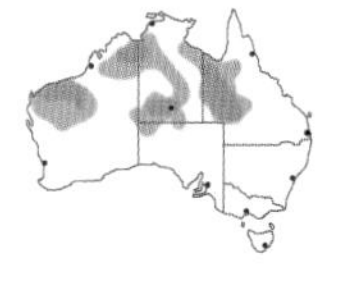

6 SPINIFEX PIGEON *Geophaps plumifera* 19–23cm/7.5–9in

A handsome, reddish inland species, Spinifex Pigeon is mostly ruddy-coloured and has a pointed rufous topknot and a striking face pattern: Bare red skin surrounds a pale eye, and the face is striped black and white, with some subtle blue markings too. Bold black bars are spread across the wings and sides of the mantle, and some subspecies also show a bold white bar across the chest. It is inconspicuous when it forages on the ground, and is well camouflaged, as its reddish colouration mirrors the red dirt and the rocky outcrops within the arid landscapes it inhabits: rocky and hilly areas and spinifex grasslands within the north of the Outback. It occurs patchily within the NT, c. and n. WA, far w. QLD, and far n. SA. Spinifex Pigeon is never far from water in its arid environment, and is therefore best located around shrinking water sources late in the dry season, when these nomadic birds become more concentrated.

♂ 1
♀ 1
♂ 2
♀ 2
3
n. QLD 4
Kimberley 5
NT 5
nw. 6
c. 6

PIGEONS AND DOVES (COLUMBIDAE)

1 PEACEFUL DOVE *Geopelia placida* 19–23cm/7.5–9in

A heavily barred dove, with blackish bars on the nape, down the back, and extending around onto the throat, Peaceful Dove also has a ring of pale blue skin around the eye. This tame dove is very terrestrial, often observed feeding on the ground, and once disturbed flies up with a whir of rapid wing beats, showing a long black tail with white sides as it flies away at high speed. The bird's distinctive, oft-repeated *doodley-doo* call is the common and iconic sound of the Australian countryside. Peaceful Dove is a common species over much of n. and e. Australia, found in a wide variety of habitats, including open woodlands, tropical savannahs, open eucalypt forests, edges of rainforests, agricultural lands, and parks and gardens.

2 BAR-SHOULDERED DOVE *Geopelia humeralis* 25–29cm/10–11.5in

A significantly larger dove than both Diamond and Peaceful Doves, with heavily barred brown upperparts, a contrastingly pale blue-grey head, a distinct chestnut nape, and a lack of any barring on the throat. The size, chestnut nape patch, and plain unbarred throat help to identify this species from its smaller cousin, Peaceful Dove. The widespread Bar-shouldered Dove is found through much of n. and e. Australia in open eucalypt forests, tropical savannahs, scrublands, mangroves, and parks and gardens. It is most likely to be found feeding on the ground in small groups.

3 DIAMOND DOVE *Geopelia cuneata* 19–23cm/7.5–9in

A small grey inland dove subtly dotted with white on its back and dull grey wings, forming a distinctive pattern. Diamond Dove also shows a white belly and a distinct red ring around the eye. When disturbed it flies up with rapid wing beats, making a noticeable whirring sound, and displays a large chestnut flash in the outer wing and a long, white-sided tail. It is found in drier areas through much of the Australian mainland, normally is not found along the wetter coastal regions, and is absent from extreme s. Australia. It is a common and widespread dove, though notably shyer than Bar-shouldered Dove, and generally found in more open habitats. Diamond Dove is irruptive, and its numbers fluctuate dramatically in response to local weather conditions.

4 CRESTED PIGEON *Ocyphaps lophotes* 30–36cm/12–14in

Crested is a largely grey-coloured pigeon with pink hues on the underparts, red skin around the eye, and a thin black topknot, quite unlike the rich ochre-coloured Spinifex Pigeon. It makes a loud and distinctive whirring sound, reminiscent of a washboard being played quickly, when it flies. Crested Pigeons generally forage on the ground, usually in groups, bobbing their heads as they walk along. When disturbed, they fly with a rapid, low flight characterised by long glides, tilting regularly. Crested Pigeon is a widespread, common, and conspicuous species found over much of the continent, except n. Cape York Peninsula (QLD), TAS, and the tip of the NT. It occurs in a range of open habitats from farmlands to gardens and golf courses, avoiding few habitats except closed forests.

5 SPOTTED DOVE *Spilopelia chinensis* 29–33cm/11.5–13in

A large, brownish introduced Asian species, Spotted Dove has a distinct black patch boldly spotted with white on the sides of the neck and pale grey underparts washed with soft pink. The neck patch distinguishes this dove from all other pigeon species in Australia. When it takes flight it shows a long blackish tail with a striking, broad white band at the tip. It is a common species in the coastal regions of e. Australia, from ne. QLD south to VIC and west into se. SA. There is also an isolated population in extreme sw. WA. Spotted Doves are found largely in urban environments, and are most commonly encountered around towns and cities, including Brisbane (QLD), Sydney (NSW), and Perth (WA).

6 LAUGHING DOVE *Spilopelia senegalensis* 27cm/10.5in

Laughing Dove is a small pigeon with a rusty-coloured head and back, pale grey wings, and a pale belly. Adult birds have a patch of black speckling across the upper breast. This introduced species from Africa and India is unlikely to be seen away from human habitation, and is usually found in parks or gardens or along roads. It is found in sw. WA, where it is quite common around towns and cities.

1
2
3
4
5
6

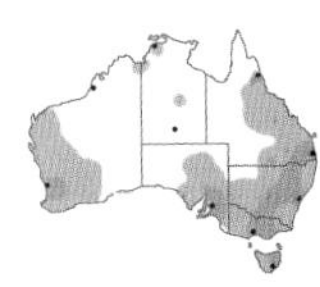

1 ROCK PIGEON *Columba livia* 30–36cm/12–14in

Also known as Feral Pigeon. The common town pigeon throughout the world, Rock Pigeon, introduced into Australia by early European settlers, is a feral descendant of the Rock Dove of Eurasia. Among its many colour variations, the most common is dull grey with a paler grey back, solid black bars in the wings, and subtle metallic green and purple on the neck. In flight it shows a bright white rump. Less common bright rufous variants occur, as do other colour variations, but all are very different from colour patterns seen in any other pigeons. The classic urban bird that has adapted extremely well to living alongside humans, Rock Pigeon is familiar to most people, as it inhabits many cities and towns across the globe. In Australia it occupies the south and east from c. WA to s. WA and eastward from c. SA to VIC and NSW; it also occurs in s. QLD and TAS, and is largely absent from the NT and n. WA.

2 TOPKNOT PIGEON *Lopholaimus antarcticus* 41–61cm/16–24in

This large pale grey pigeon is most often seen in flight, usually in flocks high over the forest. It has a distinctive silhouette, with a heavy body, small head, and broad wings. The long, broad tail is dark with a pale band just before the tip. When the bird is perched, a viewer might be able to pick up the unusual rusty-coloured topknot or crest, but it is not prominent. Although most often seen flying over rainforests, Topknot Pigeon at times will wander into other coastal woodlands searching for fruiting trees. It is quite nomadic but fairly common in coastal areas; the rainforests of the Atherton Tableland and Lamington Plateau in QLD are good places to look for it.

3 TORRESIAN IMPERIAL PIGEON *Ducula spilorrhoa* 38–42cm/15–16.5in

Also known as Pied Imperial Pigeon. Imperial pigeons are a group of large Australasian pigeons. Torresian is the only regularly occurring imperial pigeon in Australia, and is one of our largest pigeons. It is mainly ivory white, except for contrasting black on the outer wings, tail, and undertail. It occurs in rainforests, mangroves, offshore islands, eucalypt woodlands, and parks and gardens, specifically in coastal areas in the far north of the mainland. These birds are often observed in small flocks flying around coastal towns (e.g., Cairns, QLD; and Darwin, NT) or perched on overhead cables.

4 WONGA PIGEON *Leucosarcia melanoleuca* 38–44cm/15–17.5in

A large, boldly patterned, plump pigeon of the e. forests, Wonga Pigeon is all powdery slate blue above, and has a bright white underside scaled with dark markings that form a pattern quite unlike that on any other bird. The patterning of the white, when viewed front-on, is somewhat suggestive of the bird being cradled in a human hand, with two white 'fingers' positioned on either side of the neck. During the day these vocal pigeons often give an incessant *woop* call. Wonga Pigeons occur in e. Australia from the se. corner of QLD south to e. VIC; they favour wet sclerophyll forests and temperate rainforests. This is a terrestrial species, most often observed foraging along forest edges at dawn. At O'Reilly's Rainforest Retreat in Lamington NP (QLD) a number of birds have become remarkably habituated.

4

1
1
ad. 2
imm. 2
3
3
4
4

1 WOMPOO FRUIT DOVE *Ptilinopus magnificus* 36–50cm/14–19.5in

A strange, massively oversized fruit dove, often referred to as Wompoo Pigeon. Decorated with many striking colours (as are all fruit doves), it has green upperparts; a pale powder-grey head; deep maroon throat, chest, and belly; a contrasting bright lemon-yellow vent; and a scattering of yellow markings that form an incomplete bar on the wing. It is named for its distinctive, guttural *woolompoo* call, which is given frequently and is most often heard when clusters of these birds gather in fruiting trees. Wompoo Fruit Dove is an eastern species, found in closed forests from n. QLD south along the coast to n. NSW. It is fairly common and most often encountered in groups in the canopies of large fruiting trees, when the distinctive calls and regular loud fluttering of wings often draw attention to the birds.

2 ROSE-CROWNED FRUIT DOVE *Ptilinopus regina* 22–25cm/8.5–10in

A small, beautiful green dove with a powder-blue head, conspicuous rose-pink crown, and a bright orange belly fading into yellow on the vent. It could be confused with another small green fruit dove, Superb (which occurs in some of the same areas), although that species has a white belly, and its males have a prominent black band on the chest. Immature Rose-crowned is generally all green, lacking the crown colour of the adults, and may be confused with female Superb, which is also largely green and lacks the crown colouration. However, Rose-crowned always shows an orange to yellow belly, whereas all plumages of Superb display a white belly and vent. Rose-crowned Fruit Dove is a tropical species found within monsoon vine forests in n. WA and n. NT and rainforests of the e. seaboard south to n. NSW. Like other fruit doves it is most likely to be found in the canopy of fruiting trees, where the birds draw attention to themselves with their soft cooing calls but can be frustratingly difficult to locate as they are small, blend in well with the large leaves around them, and move infrequently.

2

3 SUPERB FRUIT DOVE *Ptilinopus superbus* 23–24cm/9–9.5in

The male is a gorgeous, strikingly patterned fruit dove with a deep pink crown patch, a bold orange-red half-collar on the nape and sides of the neck, and a deep purple band across its chest (which generally appears blackish). This combination of colours, and the prominent breast band in particular, should aid in distinguishing it from Rose-crowned Fruit Dove. Females and immatures are more confusing, lacking a prominent crown patch and appearing essentially green over most of the body, and are therefore easily confused with immature Rose-crowned. However, Superb at all ages shows a white belly, while Rose-crowned is always yellow to orange in this area. Superb Fruit Dove is a fairly common, though shy canopy species in most of the coastal rainforests and wet sclerophyll forests of e. QLD and n. NSW. Its presence can be easy to verify from its loud *whoop* calls, although it can still be difficult to locate, as it often sits motionless, high in the forest canopy, surrounded by leaves that are larger than it is.

4 BLACK-BANDED FRUIT DOVE *Ptilinopus alligator* 33–42cm/13–16.5in

Also known as Banded or Black-backed Fruit Dove. A striking, pied fruit dove: Both perched and in flight it essentially appears as a black bird with a creamy white hood. The tail is dark and concolourous with the upperparts, except for a broad, contrastingly pale tip, easily visible both when perched and in flight. Black-banded Fruit Dove is a shy bird that is often hard to locate except when moving between fruiting trees, such as figs. It is uncommon and highly localised, confined to monsoon vine forests and eucalypt woodlands around the w. edge of the Arnhem Land escarpment, at sites such as Nourlangie Rock within Kakadu NP in the NT. It is most likely to be found by following its low hooting calls, or listening for the flapping of wings, which may betray its presence within fruiting trees.

1
1
2
2
♂ 3
imm. 3
4
4

FROGMOUTHS (PODARGIDAE)

1 TAWNY FROGMOUTH *Podargus strigoides* 36–51cm/14–20in

Frogmouths are a nocturnal group of birds in cryptic camouflage that sit dead still to hide their presence during the day, much in the same way that potoos (*Nyctibius*) do in the Neotropics. They hunt in similar ways to large kingfishers, taking prey from the ground. Tawny is the only frogmouth species in Australia with big yellow eyes. The most widespread of the three Australian species, it is found throughout the mainland and also in TAS, occurring in all habitats except within dense rainforests. The other two frogmouth species are confined to e. rainforests. Tawny Frogmouths are best found when roosting during the day or when nesting, as they often return to traditional sites to sleep and nest year after year.

2 PAPUAN FROGMOUTH *Podargus papuensis* 51–60cm/20–23.5in

A huge, highly camouflaged nocturnal bird that is the largest Australian frogmouth. Like all frogmouths, it has rufous and grey forms and hunts by seizing lizards and small mammals from low branches and trunks or by pouncing on them on the ground. This species has distinctive orange-red eyes. It is a forest frogmouth confined in Australia to ne. QLD, and although it is most often found in rainforests, it also occurs in mangroves and palm forests. These birds are best located in traditional nesting areas; local people can often take visitors to their favoured haunts (e.g., on Daintree River cruises, QLD).

3 MARBLED FROGMOUTH *Podargus ocellatus* 41–48cm/16–19in

The first indication of this bird's presence is usually its incredible gobbling call, often given in a duet between male and female and surely one of the most distinctive calls of any Australian bird. It inhabits rainforests and occurs in two distinct populations, one on n. Cape York Peninsula and one in coastal se. QLD and ne. NSW. The southern subspecies was once split, as Plumed Frogmouth, because of the large bristly plumes above the bill. Inexperienced observers may find Marbled difficult to separate from Tawny and Papuan Frogmouths by sight, but eye colour is a good guide: Tawny tends to have yellow eyes, Papuan has orange-red eyes, and Marbled has orange eyes. In the areas where Papuan and Marbled occur together on Cape York, the Papuan Frogmouth is much larger. The southern population can be confused only with Tawny Frogmouth, which tends to avoid the subtropical rainforests preferred by Marbled Frogmouth. Tawny also tends to be a plainer more uniform grey, while Marbled appears a mixture of greys and browns.

1
cryptic pose 1
grey 2
rufous 2
south 3
north 3

OWLET-NIGHTJAR; NIGHTJARS

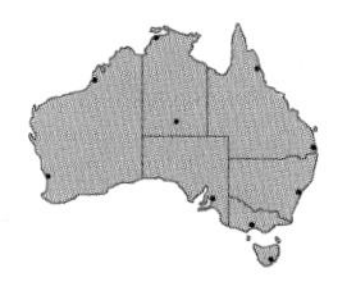

1 AUSTRALIAN OWLET-NIGHTJAR *Aegotheles cristatus* 22–24cm/8.5–9.5in

A tiny, cat-faced, cryptically patterned night bird, this is the smallest of all the nocturnal species in Australia. Unlike the eyes of other nocturnal birds (frogmouths, nightjars, and owls), the eyes of Australian Owlet-Nightjar do not glow in a spotlight, which makes them especially tricky to find. The sole representative of its family (Aegothelidae) in Australia, Australian Owlet-Nightjar is fairly widespread across the continent, found throughout the mainland and TAS wherever there are wooded areas or even small clusters of trees in open country. These birds are not readily found however, and dedicated searches are usually required to locate them. Australian Owlet-Nightjars nest and sleep in tree hollows, where they are perhaps most likely seen sunning themselves at the entrance during early mornings, before going to sleep for the day.

1

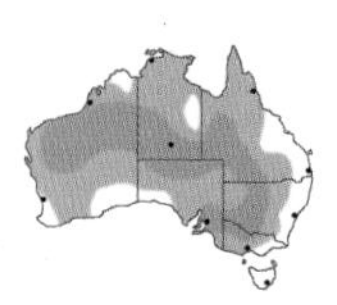

2 SPOTTED NIGHTJAR *Eurostopodus argus* 30cm/12in

Eurostopodus and *Caprimulgus* nightjars are in the family Caprimulgidae. This widespread species is found throughout most of Australia west of the Great Dividing Range (QLD). It prefers drier habitats than other nightjars, such as mallee, mulga, and savannah woodlands. Spotted Nightjar is smaller than White-throated, and in flight is fairly easy to identify by the white patch in the outer wing. It is common across its range and is often seen at night sitting on roads. *(Flying image computer-generated)*

3 LARGE-TAILED NIGHTJAR *Caprimulgus macrurus* 28cm/11in

This is the smallest nightjar likely to be seen in Australia, and is found across far n. Australia and down the e. coast of QLD. It has a distinctive flight pattern that is more erratic and buoyant than that of other nightjars and, more important, it has white patches on both the outer wings and the corners of the tail. The call is unusual, a recurring *chonk-chonk-chonk*, often repeated continuously. Its preferred habitat is monsoon rainforests, but during the night it will forage in nearby woodlands, parks, and gardens. Buffalo Creek, Lee Point, and East Point near Darwin (NT) are all good places to see it. *(Flying image computer-generated)*

4 WHITE-THROATED NIGHTJAR *Eurostopodus mystacalis* 33cm/13in

This bird is often first detected by its call, which sounds like a weird, maniacal laugh. In flight it is large and dark all over, with no real distinguishing field marks. This nightjar is a migrant, wintering in n. QLD and Papua New Guinea and spending the summer in se. Australia. It is the species most likely to be seen along the e. coast, and can be spotted in reliable locations around Brisbane (QLD), Sydney (NSW), and Melbourne (VIC). *(Flying image computer-generated)*

SWIFTS (APODIDAE)

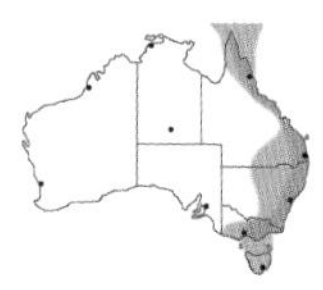

1 WHITE-THROATED NEEDLETAIL *Hirundapus caudacutus* 20cm/8in

White-throated Needletail is a huge migrant swift from Asia, occurring in Australia between October and April. Swifts have narrow, scythe-like, pointed wings, and tiny feet that prevent them from perching on anything but vertical surfaces; they mate, drink, and feed in flight. This boldly patterned species is unlikely to be confused with any other aerial bird: Larger than other swifts, it is dark brown with a bright white horseshoe shape on the undertail and a clean white throat. The 'needles' on the tail are fine and unlikely to be seen. The needletail is found in open skies above a variety of habitats over e. Australia.

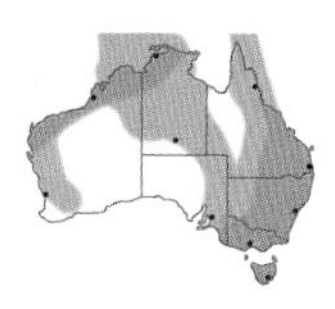

2 PACIFIC SWIFT *Apus pacificus* 18cm/7in

This swift is a summer migrant, arriving in nw. Australia around October and spreading throughout the country as summer progresses. Although Pacific Swifts tend to concentrate in w. and inland areas, they can be seen almost anywhere in small numbers, particularly in late summer. Shape is the best way to separate them from White-throated Needletails. Pacific Swifts have a more slender, cigar-shaped body and a longish tail that is usually held closed and looks long and pointed. With good views you may be able to see a shallow fork in the end when the tail is spread. Pacific also has narrower wings that look longer in proportion to the body, compared to White-throated Needletail.

3 AUSTRALIAN SWIFTLET *Aerodramus terraereginae* 11.5cm/4.5in

These swifts have a distinctive, erratic and flickering flight, are constantly turning and changing direction, and often glide on stiff, down-pointed wings. Flight pattern and smaller size are a good way to separate them from other similar birds, such as swallows and martins. They have no real distinctive plumage characteristics, being plain dark grey with a pale rump. They can be told from Pacific Swift by tail shape; the larger Pacific has a long, forked tail, while the swiftlet has a short, squarish tail. This small swift is found in ne. QLD, where it breeds in caves. The birds are common close to these caves, but can also be found across a range of other habitats. They are quite common and easily seen around suburban Cairns.

1
1
2
2
3
3

STORM PETRELS (HYDROBATIDAE)

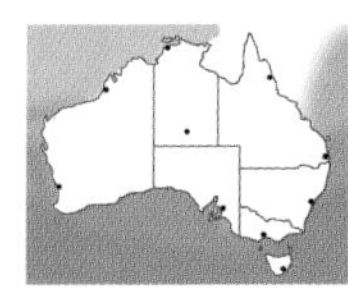

1 WILSON'S STORM PETREL *Oceanites oceanicus* 15–20cm/6–8in

A small and compact, all blackish-brown storm petrel with a short, square-cut black tail. A bold white V shape on the rump extends to vent sides on the otherwise all-dark body. The inner webs of the feet, although seldom seen, are distinctively yellowish. Flight is direct and undeviating, undertaken with shallow, swallow-like wing beats. The bird stops frequently to forage, dangling its feet in the water and holding its wings aloft in a shallow V shape. One of the most common and widespread of all birds, this species can be seen on many autumn, winter, and spring pelagic boat tours and is the most likely storm petrel to be seen from land.

2 GREY-BACKED STORM PETREL *Garrodia nereis* 16–19cm/6.25–7.5in

A small, slate-hooded storm petrel, with the back, mantle, and rump variably slate grey to ashy grey. The square tail is always greyish with a black tip. The belly and central underwings are white, while a distinctive, broad black leading edge and dark secondaries give the underwings a dark-outlined appearance. Direct flight is mostly fast and undeviating, but flight may occasionally appear unsteady on account of the short wings. The bird stops frequently to forage, dangling its feet in the water. The closest breeding sites to mainland Australia are Macquarie Island, in the sw. Pacific (about 1500km/950mi south-east of TAS), and s. New Zealand. The species is most often seen from pelagic boat tours off se. Australia in winter.

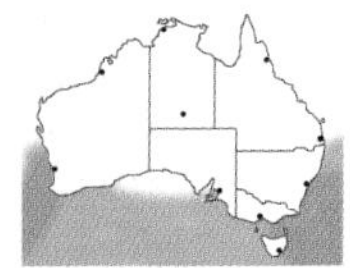

3 WHITE-FACED STORM PETREL *Pelagodroma marina* 18–22cm/7–8.5in

A large, long-legged, white-faced storm petrel. The distinctive head has a white face cut by a dark grey-brown eye stripe that creates a characteristic white supercilium and frons. Upperparts, including crown and neck, are grey-brown. The flight and tail feathers are darker, and the rump is paler. The throat, belly, and central underwings are white. An indistinct narrow black leading edge and broad dark secondaries give the underwings a dark-outlined appearance, narrow on the leading edge and broad on the trailing edge. Compared to other storm petrels, this species' long legs and wings give it a less gainly appearance when foraging. Its wheeling flight is direct and prion-like, with its legs protruding. This storm petrel breeds on offshore islands south of the continent, and can be seen on most s. Australian pelagic tours in the summer months. The taxon *albiclunis*, which breeds on the Kermadec Islands (New Zealand), has a white rump and is occasionally considered a separate species by some authorities.

3

4 BLACK-BELLIED STORM PETREL *Fregetta tropica* 19–22cm/7.5–8.5in

A medium-sized and stout storm petrel with shortish, rounded wings and a square-ended tail. The head, mantle, upperwings, and tail are sooty black, and the bird is hooded in appearance. A lateral dark belly stripe cuts distinctly through the white underparts and joins the black vent and tail; occasionally the dark belly stripe is absent. In fresh plumage, the wing coverts are pale-edged, giving the upperwings a scaly appearance. The white rump cuts across and on to the flanks and extends through the armpits onto the underwings, which are bordered by broad black leading and trailing edges. Flight is erratic and meandering with a strong zigzag pattern. This storm petrel is a rare visitor, found far offshore on pelagic tours. It is most common off s. Australia in spring and summer, and is seen off the e. coast on pelagic tours in winter and early spring.

1
1
2
2
3
3
4
4

ALBATROSSES (DIOMEDEIDAE)

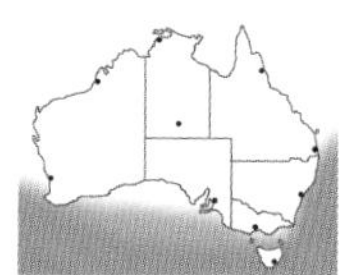

1 WANDERING ALBATROSS *Diomedea exulans* 109–130cm/43–51in

Enormous size and pinkish bill distinguish *Diomedea* albatrosses from mollymawks (*Thalassarche*). This is the largest and whitest species of the 'wandering albatross complex', a series of very closely related birds regarded as a variable 'superspecies' by some and as a range of different species by others (Antipodean is the other Australian representative). Possessing the most massive bill of the group, this long-winged seabird has an effortless, dynamic soaring flight. The mature adult always has an almost entirely white back; its upper- and underwings are white with a black trailing edge on the two-thirds nearest the body and are strongly, contrastingly black on the outer third. Some breeding adults have a pink-stained rear ear covert patch. Wandering differs from Antipodean Albatross in having a more massive bill and fewer vermiculations on the underside. Juveniles are dark chocolate brown with a contrasting white face. As they mature, immature birds become whiter and progressively mottled in stages, until eventually they attain full adult plumage. Juveniles/immatures are difficult to separate from others in the wandering albatross complex. The large pink bill does not have a strong black cutting edge to upper mandible as in Northern and Southern Royal Albatrosses. White tail has black edges and tips in all stages except oldest males (all white). The species is regular in open ocean and occurs year-round but is more common in the austral winter.

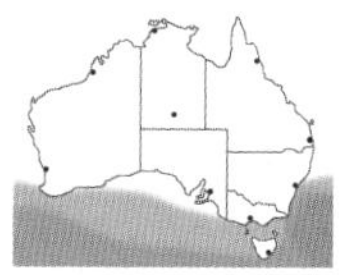

2 ANTIPODEAN ALBATROSS *Diomedea antipodensis* 109–130cm/43–51in

This is part of the wandering albatross complex. The mature adult always has an almost entirely white back, and its upper- and underwings are white with a dark trailing edge on the two-thirds nearest the body, and are strongly, contrastingly black towards the tips. Some breeding adults have a pink-stained rear ear covert patch. Antipodean differs from Wandering Albatross in having a smaller bill and more vermiculations on the breast. Juveniles are dark chocolate-brown with a contrasting white face. As they mature, immature birds become whiter and progressively mottled in stages, until eventually they attain full adult plumage. Juveniles/immatures are difficult to separate from others in the wandering albatross complex.

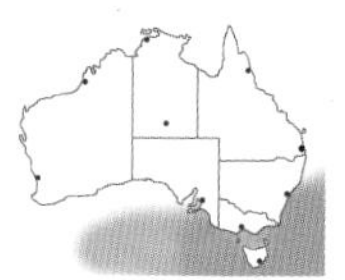

3 SOUTHERN ROYAL ALBATROSS *Diomedea epomophora* 99–119cm/39–47in

Royal albatrosses are told from birds of the wandering albatross complex at close range by the distinctive black cutting edge to the upper mandible. Southern adult has all-white head, mantle, and tail, and the under- and upperwings are white on the three-quarters nearest the body, and the wing has a dark trailing edge. Immature bird has all-dark upperwings, with vermiculations spreading over the shoulders and onto the back, and a black-tipped white tail. The immature has a white leading edge between the shoulder and the wrist that separates it from Northern Royal Albatross. Southern Royal Albatross is regular in open ocean and occurs year-round but is more common from July to October.

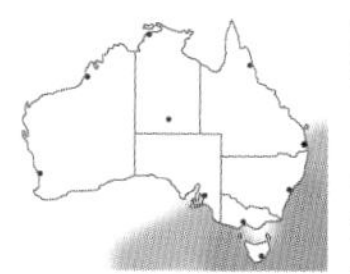

4 NORTHERN ROYAL ALBATROSS *Diomedea sanfordi* 99–119cm/39–47in

Royal albatrosses are told from birds of the wandering albatross complex at close range by their smaller size and the distinctive black cutting edge to the upper mandible. Northern adult has all-white head, mantle, body, and tail, and strongly sooty-black upperwings from the shoulder across the length of the wing. The upperwings lack any mottling or chequering typical of other species. The immature has all-dark upperwings with extensive smudgy vermiculations spreading over the shoulders and onto the back, and a black-tipped white tail. The immature lacks the white leading edge between the shoulder and wrist that distinguishes the Southern Royal Albatross. Northern Royal Albatross is a regular visitor in open ocean, mostly from May to September.

1
1
2
2
3
3
4
4

ALBATROSSES (DIOMEDEIDAE)

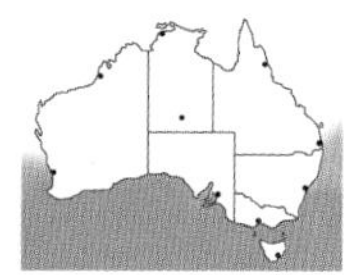

1 SHY ALBATROSS *Thalassarche cauta* 89–99cm/35–39in

A large, bulky mollymawk (as birds of the genus *Thalassarche* are known) with an ochre-coloured bill lacking a dark spot at the tip of the lower mandible. Shy Albatross adult has all-white head, rump, and underside of body; the pale grey cheeks and dark eyebrow give it a 'white-capped' appearance. The upperwings are blackish, the mantle is grey, and the tail is black. Underwings are the whitest of all mollymawks, with a very narrow black edging and a 'thumbprint' on the underwing at the shoulder. The black of the wing tip is restricted compared to that of all other mollymawks, and the large amount of white on the primary bases is a distinctive ID feature for all age classes. At fledging, the head, neck and throat are often off-grey, giving the juvenile a hooded appearance. A common breeding resident, this species occurs closer to the mainland than any other large albatross, and is the most likely albatross to be seen from land along Australia's s. coast.

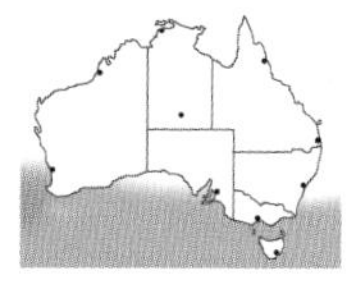

2 SALVIN'S ALBATROSS *Thalassarche salvini* 89–99cm/35–39in

A large, bulky mollymawk with an olive-bone bill showing a dark tip on the lower mandible. The adult has a distinctive all-grey hood, cheeks, and throat but a 'white-capped' forehead. The upperwings are blackish, the mantle is grey, the tail black. Underwings are white with a very narrow black edging, a 'thumbprint' on the underwing at the shoulder, and a black wing tip. At fledging, the head, neck, and throat are often silvery grey, giving the juvenile a hooded appearance, with little contrast between head, mantle, and shoulder. The area of black at the wing tip is large compared to that shown by Shy Albatross, with most of the visible primaries appearing black; in combination with the heavily grey-hooded appearance, it is a distinctive ID feature for all age classes. Salvin's Albatross is an uncommon visitor to Australian waters from NSW to WA.

3 BLACK-BROWED ALBATROSS *Thalassarche melanophris* 84–94cm/33–37in

A medium-sized mollymawk with a very dark back, tail, and upperwings, and underwings with broad dark edges and wing tips framing a narrow white centre. The adult has a white head and rump, with a dark eyepatch that starts in front of the dark eye and swoops rearward. The distinctive bill, easily seen at close range, is yellow with a reddish tip. The immature has a smudgy grey collar and dark yellow-grey bill with a dark tip. The underwings remain almost entirely dark, making this a distinctive immature mollymawk. Similar adult Campbell Albatross has a yellowish eye and a more obvious eyebrow. Black-browed is one of the common albatrosses seen from land in s. Australia from autumn to spring; it will be encountered on most pelagic boat tours from s. Australia during this period, and can even be seen from the Sydney to Manly ferry in winter.

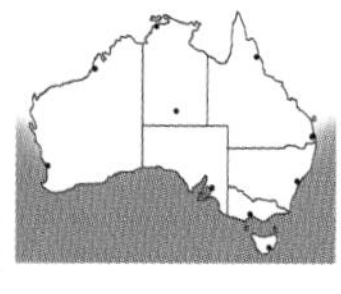

4 CAMPBELL ALBATROSS *Thalassarche impavida* 86–90cm/34–35.5in

A recent split from the very similar Black-browed Albatross, Campbell Albatross is identified by having much more black on the underwing, including the armpits. When seen closely, it has a yellowish eye and more prominent eyebrow than Black-browed. Campbell Albatross breeds on Campbell Island (New Zealand), moving to the s. Australian ocean in winter.

1
1
2
2
3
3
4
4

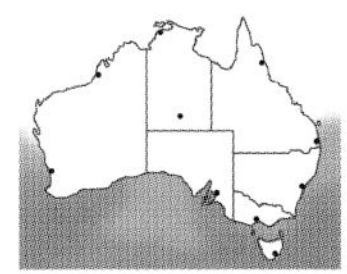

1 INDIAN YELLOW-NOSED ALBATROSS *Thalassarche carteri* 71–81cm/28–32in

The smallest and daintiest of the Australian mollymawks. In the adult, the upper mandible of the long, slim black bill has a bright yellow stripe down the middle from tip to bill base, where it meets the head in a sharp arrow-shaped point. The adult has a white head and neck, black upperwings, a white rump, and a black tail. This species differs from the scarce Atlantic Yellow-nosed Albatross (*T. chlororhynchos*) in that the latter is more grey-hooded, with a more conspicuous black triangular loral patch in front of the eye, and its yellow upper-mandible stripe meets the head as a curve (not an arrow). Typical juvenile and immature Indian Yellow-nosed Albatrosses have whiter heads and reduced loral patches; however, in differential states of wear, juveniles and immatures can be impossible to accurately identify. A common bird off s. Australia in winter, seen from both pelagic boats and from land, this albatross is especially numerous near land around sw. Australia.

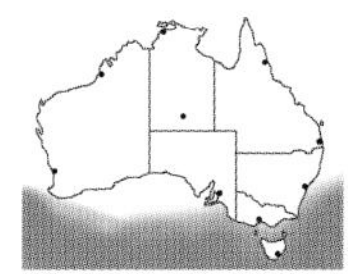

2 GREY-HEADED ALBATROSS *Thalassarche chrysostoma* 71–84cm/28–33in

A small mollymawk with the typical sooty-black upperwings and tail and white rump and body but a distinctive dark hood in both adult and immature forms. Adult has a dark grey hood that merges with the slightly paler nape but has a distinct edge along the white foreneck. The bill is glossy black with a bright yellow ridge along the centre of both upper and lower mandibles, becoming pinkish at the tip. The underwing is white with a broad black leading edge and narrow black trailing edge. Immature is grey-hooded, with a small, indistinct dusky eyepatch, a grey-brown, dark-tipped bill, and almost entirely uniform black underwings, similar to underwings of immature Black-browed Albatross. This species is an uncommon winter visitor to much of s. Australia's offshore waters. It is more common in the south-west than in the e. states.

3 BULLER'S ALBATROSS *Thalassarche bulleri* 76–81cm/30–32in

A small mollymawk with a grey hood and whitish crown giving it a white-capped appearance, and the typical sooty-black upperwings and tail and white rump and lower body. The bill is glossy black with a bright yellow ridge along the centre of both upper and lower mandibles, which broadens towards the base of the upper mandible. The underwing is white, with a broad black leading edge (although narrower than that of Grey-headed) and a narrow black trailing edge. Immature is grey-hooded, with a pale cap and dirty brown collar; its underwings are much like the adult's, very white in the centre with a clean-cut narrow trailing edge and broader leading edge, distinguishing it from the all dark-winged Grey-headed immature. Buller's is an uncommon winter visitor to offshore waters of e. SA, VIC, and s. NSW; it appears to be more common and more likely to be seen from shore in TAS than the mainland.

3

1
1
2
2
3
3

PETRELS AND SHEARWATERS (PROCELLARIIDAE)

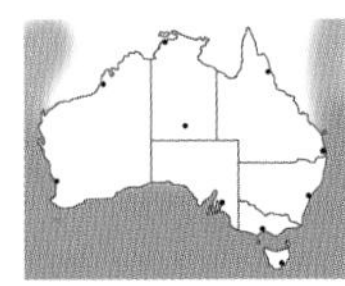

1 SOUTHERN GIANT PETREL *Macronectes giganteus* 81–102cm/32–40in

This large, albatross-sized petrel is told from dark albatrosses by appearing bulkier and less elegant in flight, with more frequent flapping.The adult typically has a paler head and breast than similar Northern Giant Petrel. Southern also has a unique, rare white colour morph. Immatures of the two giant petrels, which are more commonly seen than adults, have almost identical plumage. However, at close range the Southern's pale bill tip is diagnostic in all age groups, although it can be difficult to detect in younger birds. Southern Giant Petrel is primarily an offshore pelagic species that likes to follow boats and squabble for scraps. It has been known to attend seal and penguin colonies, where it scavenges on carcasses in a vulture-like fashion. It frequents Australian waters from May to October; adults are scare, but immatures are more common off s. Australia, becoming rare in the tropics.

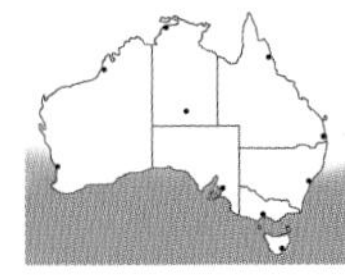

2 NORTHERN GIANT PETREL *Macronectes halli* 89–102cm/35–40in

This large, albatross-sized petrel is told from dark albatrosses by appearing bulkier and less elegant in flight, with more frequent flapping. The adult typically has a darker head and breast than the similar Southern Giant Petrel. Immatures of the two giant petrels, more commonly seen than adults, have nearly identical plumage. However, at close range the Northern's reddish bill tip is diagnostic in all age groups, although it can be difficult to detect in younger birds. This is primarily an offshore pelagic species that likes to follow boats and squabble for scraps and has been known to attend penguin colonies, where it glides overhead searching for dead chicks or abandoned eggs. It frequents Australian waters from May to October; adults are scare, but immatures are more common off s. Australia, becoming rare in the tropics.

2

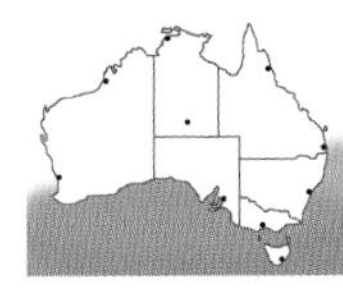

3 CAPE PETREL *Daption capense* 36–43cm/14–17in

A distinctive and compact petrel, this is the only seabird with a strong black-and-white-chequered pattern on the back, rump, and upperwing; the large white windows in the primaries are always very conspicuous. Underparts are mostly white, except the dark hood, dark edges along the underwing, and a dark terminal tail band. New Zealand breeding subspecies *australe* is darker-backed, with reduced chequering. Highly manoeuvrable, Cape Petrel flies low, interspersing rapid wing beats on stiff wings with gliding. It often occurs in noisy flocks, swimming actively around fishing boats while foraging. Primarily an offshore pelagic species, it is a fairly common visitor to Australian waters from April to November, more frequent off s. Australia, becoming scarcer in the tropics.

4 SOUTHERN FULMAR *Fulmarus glacialoides* 46–51cm/18–20in

A large, gull-like petrel with grey upperwings, back, and tail contrasting with a whitish head. The upperwings have a strong white window in the primaries, with the outer webs sooty-edged; the inner upperwing has a dark trailing edge. Underparts are all whitish except dark wing tips. The bill is dull pink with a dark tip. Southern Fulmar uses shallow wing beats interspersed with wheeling glides. Primarily an offshore pelagic species, it occasionally follows boats and squabbles for scraps, but it can equally easily be encountered moving across open ocean. It is an uncommon visitor to Australian waters from May to October, more frequent off s. Australia, and is prone to irruptions in certain years.

1
white morph 1
2
2
3
3
4
4

PETRELS AND SHEARWATERS (PROCELLARIIDAE)

1 BLUE PETREL *Halobaena caerulea* 28cm/11in

A small petrel with blue-grey upperparts and white underparts, Blue Petrel differs from prions in having a dark hood that extends onto the neck in a half-collar. It also has a diagnostic square-cut, white-tipped tail. Upperwings are blue-grey and crossed with a dark M, similar to that seen in prions. It flies low to the waves, with an elegant and buoyant flight. Though recorded most years on pelagic trips off s. Australia, it is still regarded as a 'mega bird' when found.

2 ANTARCTIC PRION *Pachyptila desolata* 25–28cm/10–11in

This bird has the typical prion colour pattern of blue-grey upperparts with an open M connected on the lower back and a whitish supercilium and underparts. Antarctic differs from Slender-billed and Fairy Prions in its larger size and stouter bill. It has a narrower tail band (one-quarter of tail length) than Fairy. Almost identical to Salvin's Prion, it lacks that species' dark undertail and the bill combs are not exposed on closed mandibles; its grey breast patches average larger than Salvin's but are not diagnostic. Antarctic Prion is a common seabird during winter and spring (May–Oct) along the entire s. Australian coastline.

3 SALVIN'S PRION *Pachyptila salvini* 28cm/11in

Salvin's has the typical prion colour pattern of blue-grey upperparts with an open M connected on the lower back and a whitish supercilium and underparts. It differs from Slender-billed and Fairy Prions by being a larger bird with a stouter bill. Compared to Fairy, it has a narrower tail band (one-quarter of tail length). Almost identical to Antarctic Prion, Salvin's has a darker undertail and the bill combs are exposed when its mandibles are closed. Salvin's grey breast patches average smaller than Antarctic's, but this feature is not diagnostic. An uncommon winter visitor to s. Australian waters, this prion can be found on most s. pelagic tours during this period.

4 SLENDER-BILLED PRION *Pachyptila belcheri* 25cm/10in

This species has the typical prion colour pattern of blue-grey upperparts with an open but fainter M connected on the lower back and a whitish supercilium and underparts. It differs from Antarctic and Salvin's Prions in being smaller and having a much weaker M on the upperparts of the wings, and of all prions it has the most distinctive facial pattern, the weakest bill, and the least extensive black tail tip. Slender-billed Prion's flight is daintier and more acrobatic than that of larger prions. An uncommon winter visitor to s. Australian waters, it can be found on most s. pelagic tours during this period. It is more common in the west and south than the east.

5 FAIRY PRION *Pachyptila turtur* 23–28cm/9–11in

The smallest and most petite prion of the region, it has the typical prion colour pattern of blue-grey upperparts with an open M connected on the lower back, a whitish supercilium above a grey eye stripe, and white underparts. The crown and eye stripe are paler than those in all other likely prions, giving its face an indistinct look, and the characteristic dark tip extends about one-third of the way up the tail. Greyish patches on breast sides are poorly developed, less prominent than in other prions. These are the commonest prions off se. Australia and can be seen from land in the winter months.

PETRELS AND SHEARWATERS (PROCELLARIIDAE)

1 WHITE-CHINNED PETREL *Procellaria aequinoctialis* 56cm/22in

A large, uniformly blackish-brown, bulbous-headed petrel with a long and chunky ivory-yellow bill and pale chin patch (not always present). It appears intermediate in size and bulk between other small dark petrels and giant petrels. The bill has a distinctly whitish upper-mandible tip and chin patch, and black skin between the bill plates and nostrils. It flies in a manner simultaneously effortless and lethargic, particularly in strong winds, as the bird sweeps up and down in graceful, dynamic, soaring flight. It is common, especially in late summer, throughout s. Australia, and frequently seen from land.

2 BLACK PETREL *Procellaria parkinsoni* 46cm/18in

A medium-sized, blackish-brown, bulbous-headed petrel with a whitish base to the bill, a distinctly black-tipped upper-mandible and chin patch, and black skin between the bill plates and nostrils, giving the bill a pied appearance. Compared to White-chinned Petrel, Black Petrel is smaller, with a slimmer neck and bill; has a more graceful flight; and is more solitary. Similar Westland Petrel (*P. westlandica*) is a vagrant to Australian waters. Though rare, Black Petrel is recorded in most summers from se. Australian pelagic tours.

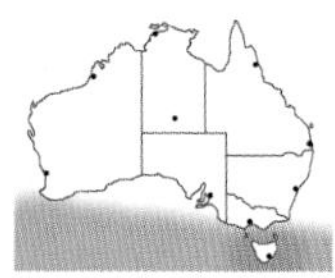

3 GREY PETREL *Procellaria cinerea* 46–51cm/18–20in

A large and chunky petrel with long narrow wings and a slender, greenish-yellow shearwater-like bill and wedge-shaped tail. Ashy grey-brown upperparts are darker on the head, wings, and tail, paler on the mantle and leading edge of wings. Underparts of the body, including throat, breast, and belly, are white. The underwings and tail are blackish with a silvery sheen. Grey Petrel is a strong flyer, with an albatross-like glide and occasional shallow wing beats. It is rare in s. Australian waters; most records are off TAS in winter.

1
1
2
2
3
3

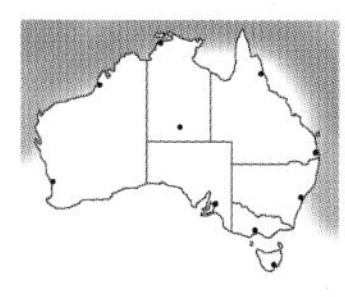

1 STREAKED SHEARWATER *Calonectris leucomelas* 48cm/19in

A large, grey-brown and white shearwater with a streaked and grizzled facial pattern. The upperparts are all grey-brown, with streaks on the neck, crown, and cheeks. Lores and frons are white, giving the bird an open-faced, helmeted appearance. Underparts are white, as are the underwing coverts, while the flight feathers are dark, giving the underwing a broad, dark trailing edge and extensive dark wing tips. The tail is wedge-shaped, and dark-edged below. The bill is yellow-grey with a dark tip. Streaked Shearwater is common off the n. coast of Australia between November and May.

2 WEDGE-TAILED SHEARWATER *Puffinus pacificus* 41cm/16in

The long, pointed, tapering tail and broad wings are characteristic of this shearwater. The bill is dusky, and the feet are pinkish. Both pale and dark morphs occur: Pale morph birds have dark upperparts but whitish underparts, while dark morph birds are sooty brown all over. This shearwater, which glides with its wings held forward in a frigatebird-like pose, is very common off both the e. and w. coasts of Australia in tropical and subtropical waters; it is especially commonly seen on NSW pelagic tours in summer.

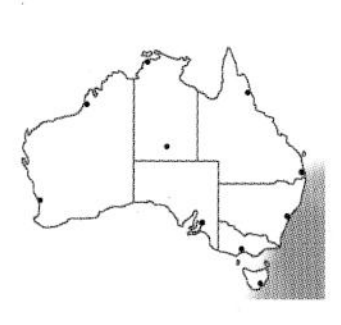

3 BULLER'S SHEARWATER *Puffinus bulleri* 43cm/17in

A large, long-winged, slender-bodied, elegant shearwater. Its head is accentuated by a long neck and long, narrow bill. It has mostly brown-grey upperparts, a blackish cap, and a broad, diffuse M pattern on the back and rump, similar to that of some *Pterodroma* petrels. The grey rump and blackish tail give it a contrasting rear end. Underparts are all whitish. This is a very uncommon bird in e. coast offshore waters in summer. Singletons or small groups sometimes attend fishing boats.

PETRELS AND SHEARWATERS (PROCELLARIIDAE)

1 FLESH-FOOTED SHEARWATER *Puffinus carneipes* 43cm/17in

A large-headed, robust, all chocolate-brown shearwater, characteristically broad-winged and blunt-tailed. It is most readily told by the pink bill with a black terminal tip. It also has pink feet, but these are not always easily seen. It has easy, shallow wing beats, and its flight is slow, often skimming the water surface. Flesh-footed Shearwater is found from sw. WA eastward around s. Australia to subtropical e. waters. Singletons or small groups sometimes attend fishing boats. It is regularly seen on e. coast summer pelagic tours and is common inshore in n. NSW.

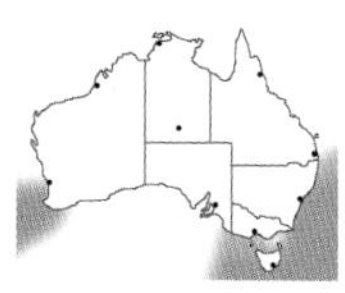

2 SOOTY SHEARWATER *Puffinus griseus* 47cm/18.5in

A large, all-dark shearwater with dark brown plumage above and a silvery-whitish panel across the centre of the underwing. It is very similar to Short-tailed Shearwater, but Sooty always has a longer bill and a more angular head. Sooty's underwing panel is almost always lighter than that of Short-tailed, but individual variability and viewing conditions make this character unreliable. Sooty's wings tend to taper more and have sharper tips. Its flight is fast, with rapid wing beats and long glides, but appears slower than that of similar Short-tailed Shearwater. Sooty Shearwater is uncommon from the Great Australian Bight (not seen from shore or on near-shore pelagic tours) to s. QLD, with most records during spring migration.

3 SHORT-TAILED SHEARWATER *Puffinus tenuirostris* 43cm/17in

This is a large, all-dark shearwater with dark brown plumage. It is very similar to Sooty Shearwater, but always has a shorter bill and a slightly shorter tail, resulting in a more compact look in flight. Its wings taper less and appear more blunt-ended. The feet often project behind the tail, but this can be seen only at close range. The rounded head, especially the forehead, give it a 'cuter' look. The underwing panel is almost always duskier and duller (less pronounced) than in Sooty Shearwater, but individual variability and viewing conditions make this character unreliable. Flight is fast, with rapid wing beats, and seems more frantic than that of Sooty Shearwater. Short-tailed Shearwater is very common from shore on the e. subtropical–temperate coast during spring migration and is found throughout summer around TAS.

1
1
2
2
3
3

PETRELS AND SHEARWATERS (PROCELLARIIDAE)

1 FLUTTERING SHEARWATER *Puffinus gavia* 30–36cm/12–14in

A medium-sized, pied shearwater with dark brown upperparts and face. The dark of the crown and face extends onto the neck sides in a dirty brown smudge. The underside is diffusely dirty white at the throat, becoming white from the upper breast to the vent. The underwing is variable, but usually has whiter central coverts, dirty armpits, and dark flight feathers forming a thick, dark trailing edge. A shorter head and bill give Fluttering Shearwater a more compact shape than Hutton's, and the narrow, pliable wings alternating rapid, deep wing beats with short glides give this species its name. It is found from Adelaide (SA) to Brisbane (QLD), often very close to shore. Though stragglers can be found at any time of the year, the species is most common in winter to mid-spring and is commonly seen from shore.

2 HUTTON'S SHEARWATER *Puffinus huttoni* 36–38cm/14–15in

A medium-sized, pied shearwater with dark brown upperparts and face. The dark of the crown and face extends onto the neck sides and throat in a more extensive fashion than in similar Fluttering Shearwater, giving Hutton's a hooded appearance. The underside is white from upper breast to vent. The underwing is much darker than in Fluttering Shearwater, with dark armpits extending onto the leading edge of the underwing, occasionally joining the collar. The remainder of the underwing is often dirty grey-brown, although occasionally white. Hutton's occasionally travels in mixed groups with Fluttering Shearwater, compared to which it appears slightly larger. It is present year-round in all of Australia and in winter is found far more commonly in the south-east.

3 LITTLE SHEARWATER *Puffinus assimilis* 25–30cm/10–12in

A small pied shearwater with blackish-blue upperparts and white underparts. The dark crown stops near the eye, and the white begins behind the eye, giving this species a distinctly 'capped' appearance compared to other pied shearwaters. The underwing is white, with a narrow black trailing edge. The species flies very low to the water, and its small size results in very rapid wing beats. It is smaller and faster than Hutton's and Fluttering Shearwaters. Little Shearwater is a rare visitor to far offshore waters of sw. and se. Australia and is only rarely encountered on pelagic boat tours.

4 COMMON DIVING PETREL *Pelecanoides urinatrix* 25–30cm/10–12in

A small, rotund, barrel of a bird that flies, auk-like, with extremely rapid wing beats on small stubby wings. Unlike the *Pterodroma* petrels of the family Procellariidae, it is in the diving petrels family, Pelecanoididae. The bill is short and broad, the head is rounded, and the tail is short and square. Upperparts are all sooty black, and underparts are whitish. This fairly common species breeds in Bass Strait and can be seen from Adelaide (SA) to s. NSW, though it is most easily observed around TAS.

5 KERMADEC PETREL *Pterodroma neglecta* 38cm/15in

A squat, compact, short-tailed petrel. Its flight is lazy, varying from deep wing beats and slow unhurried glides to effortless dynamic soaring with high arcs in strong winds. Plumage is highly variable, and pale, dark, and intermediate morphs occur. Consistent features on the underwing include primaries and primary coverts that are white with dark feather tips, forming large and small crescents juxtaposed on the end of the underwing, as on Providence Petrel. But at close range, Kermadec Petrel always has diagnostic pale shafts to the primaries, which show distinctly in the upperwing, differentiating it from all other Australian petrels. The dark morph can be confused with Providence Petrel but for the diagnostic upperwing pale primary shafts. The pale morph could be confused for a White-headed Petrel, but it lacks the latter's diagnostic black eyepatch and has a dark, not white tail. Kermadec Petrel's closest breeding area is Lord Howe Island, and it is an uncommon wintertime visitor to Australia's ec. coast, present from late summer to early spring.

1
1
2
2
3
3
4
4
light morph 5
dark morph 5

1 HERALD PETREL *Pterodroma arminjoniana* 36cm/14in

A medium-sized tropical petrel that flies low and forcefully over the water. Herald Petrel is extremely variable, with highly mutable plumage, and one of the trickiest seabirds to identify. It occurs in dark, pale, and several intermediate morphs, although the pale morph is most frequent in Australian waters. Pale morph has charcoal-grey upperparts and cap, but whitish belly, lores, and throat, and a variable dark throat band. The underwings are dark grey with a characteristic white leading edge at the shoulder and a thin median white strip that broadens into a large white patch at the base of the primaries. Pale morph differs from Kermadec Petrel by lacking pale shafts in the upperwing primaries, and from Soft-plumaged by being darker and browner, lacking the latter's distinctive eyepatch, and having more white in the underwing, particularly in the outer ends. Dark morph has all charcoal-brown upperparts, and differs from Providence Petrel by presence of white leading edge along underwing shoulders, greyer or whiter underwing, and lack of grizzled pattern around the base of the bill. Herald Petrel is a rare visitor from the Pacific tropical islands to coastal areas of e. Australia from n. QLD to NSW. It breeds in the outer Great Barrier Reef and is most often located on a reef trip or a pelagic trip off s. QLD.

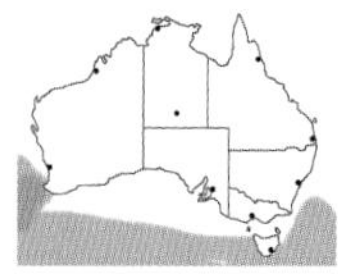

2 SOFT-PLUMAGED PETREL *Pterodroma mollis* 33cm/13in

A moderately small grey-blue and white *Pterodroma* petrel. The tail, mantle, and crown are grey-blue, and the colour creeps below the eye area, which has an indistinct white eyebrow and darker eyepatch. A dark brown M pattern (sometimes indistinct) appears across the back. The underparts are mostly white, with the exception of a complete though indistinct grey collar. The underwing's leading edge from shoulder to halfway along forewing is very indistinctly white, though this is difficult to see in the field. The complete underwing pattern (if seen well) in combination with the complete grey collar on white underparts is diagnostic. Soft-plumaged Petrel is an uncommon species throughout the s. oceans and is usually found far south of the mainland, though it appears to be most common (or least rare) off sw. Australia.

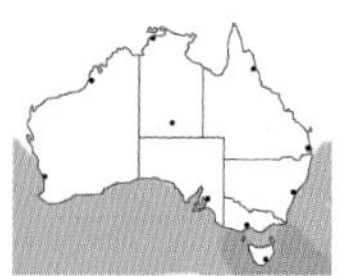

3 WHITE-HEADED PETREL *Pterodroma lessonii* 43cm/17in

A large, distinctive, chunky, white-bodied and dark-winged petrel with a diagnostic white head and distinctive deep black bill and black eyepatches. The dark grey upperwings with a indistinct M pattern grade into pearl on the rump and neck; the underparts are wholly white, including the tail, contrasting with the ashy underwings. White-headed Petrel is a fairly common species throughout the s. oceans around Australia, though it appears to be most common off VIC in winter.

4 KERGUELEN PETREL *Aphrodroma brevirostris* 36cm/14in

A small, almost all dark sooty-grey *Pterodroma* with a small bill, bulky head, tapering body and wings, and sharp-pointed tail. The upperparts are uniformly coloured, although the head often appears darker; the underwings have a paler greyish leading edge from shoulder to wrist and all greyish flight feathers that often 'flash' silver when the bird banks sharply. This is a solitary and strong, swift-like flyer that makes large looping arcs, raking glides, and short, sharp wing bursts. It is most similar to Great-winged Petrel, which is larger and more laborious in flight, has a bulkier bill, and appears browner and with a more grizzled chin and frons. Kerguelen Petrel is a rare winter visitor to Australia; irruptions are possible in some years.

5 GREAT-WINGED PETREL *Pterodroma macroptera* 43cm/17in

A large, almost all dark sooty-brown *Pterodroma* with a large, deep bill, chunky neck, and long wings. The face has a variable grizzled grey-and-white mask extending from in front of the eyes onto the frons and chin. In some individuals the patch may be residual or absent. Flight feathers are paler than coverts, but the wings seldom flash silver, as in Kerguelen Petrel. The New Zealand breeding taxon *gouldi* has a more extensive and stronger white face and is occasionally treated as a separate species. Great-winged Petrel is a strong, effortless flyer that carries its wings forward. A common species throughout the offshore s. oceans around Australia, it is very frequently seen on e. Australia pelagic tours during summer.

6 PROVIDENCE PETREL *Pterodroma solandri* 41cm/16in

A bulky petrel with a long wedged tail, broad wings, and robust chunky bill that give the bird a characteristic flight shape. It is larger and steadier in flight than other Australian *Pterodroma* petrels. Plumage is mostly grey-brown, darker on the head than the belly, and upperwings are all dark. Worn body feathers can sometimes appear whitish. Whitish scalloping on the muzzle (with black speckles interspersed) extends to the forehead, forming a variable dirty-white crescent around the front of the face. On the underwings, the primaries and primary coverts are white with dark feather tips, forming large and small crescents juxtaposed on the outer end of the underwing. This species breeds on Lord Howe and Norfolk Islands but disperses along the se. Australian coast between April and October; it is best seen from pelagic boat trips off NSW between July and October.

1
2
3
4
5
5
6
6

PETRELS AND SHEARWATERS (PROCELLARIIDAE)

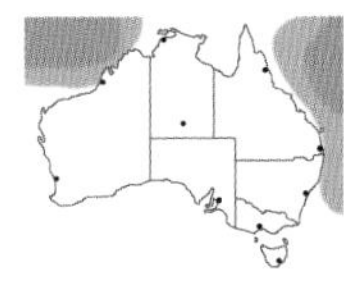

1 TAHITI PETREL *Pseudobulweria rostrata* 38cm/15in

A solid, sturdy, white-bellied dark petrel with a chunky bill. This slow- and low-flying bruiser seldom soars, interspersing strong, deep wing beats (resembling a skimmer in powered flight) with gliding. The entire head and upperparts are brown-black, as is the bill. The belly is white, and the margin between the dark-hooded head and the white breast is highly defined. The tail is dark, but the undertail coverts are white, creating a white wedge on the hindquarters. The underwing is dark, but a characteristic pale median bar runs the entire length of the wing where the flight feathers meet the underwing coverts. Tahiti Petrel breeds in the tropical Pacific but is recorded widely in Australian waters off n. WA and from QLD to Sydney (NSW), though scarce between December and May.

2 COOK'S PETREL *Pterodroma cookii* 28cm/11in

A small, blue-grey-backed petrel with a dark M across the wings and rump and an indistinctly dark eyepatch visible at close range. The tail is greyish and concolourous with the back and rump. Underparts, including the chin and belly, are mostly white. The underwing has a narrow black diagonal bar from the elbow to the inner wing; a narrow dark trailing edge and dark wing tips give the underwing a mild bordered appearance. Cook's Petrel differs from similar Black-winged in having a paler head and upperparts, with light scalloping on the covert feathers giving the upperwing and back a more frosted look. Cook's underwing is much more subtly margined, with a much narrower tapering dark bar, and its tail is all grey. It is a rare spring to summer visitor to waters off NSW, VIC, and TAS.

3 BLACK-WINGED PETREL *Pterodroma nigripennis* 30cm/12in

A small, blue-grey-backed petrel with a dark M on the back of the wings and rump, a grey hood that extends as a collar onto the sides of the neck, and a dark eyepatch that is distinct at close range. The tail has dark tips. The underparts are mostly white, including the chin and belly. On the underwing, a thick black diagonal bar runs from the elbow almost to the armpits of the inner wing; a dark trailing edge and dark wing tips give a strong bordered appearance to the underwing. Black-winged differs from similar Cook's Petrel in having a darker head, a more heavily margined underwing with a bolder dark bar, and a darker-edged tail. It is a very uncommon summer migrant to offshore e. Australian waters from s. QLD to e. VIC.

4 GOULD'S PETREL *Pterodroma leucoptera* 30cm/12in

A small, distinctive dark-capped petrel with a sooty crown extending diagonally past the front of the eye and forming a wedge on the side of the head that recurves on the neck and extends back around the nape forming a large black hood. Feathers above the bill are white, forming a narrow white band across the nasal bridge between the black bill and hood. On the upperparts, the back is greyish with feathers edged white, giving it a scaly appearance. A dark M cuts across the upperwings and rump. The grey rump and black tail give a strongly two-toned look to the hindquarters. The underparts are white, including the chin and belly. A narrow black diagonal bar runs from the elbow to the inner wing on the underwing, which has a narrow dark trailing edge and wing tips that give it a mild bordered appearance. Gould's Petrel can be told from White-necked by its smaller size, its hooded, not merely capped head, and its lack of a white collar; and from other smaller *Pterodroma* petrels by its distinctive black-capped appearance and strongly contrasting back and tail pattern. A rare and endangered breeding petrel from the se. Australian coast, Gould's is most commonly seen offshore of the c. and n. NSW coasts.

5 WHITE-NECKED PETREL *Pterodroma cervicalis* 41cm/16in

A large, long, sleek, and powerful petrel, similar in pattern to Gould's but easily told from it by its much larger size. It is another dark-capped petrel, with a sooty crown extending only past the eyes, a small white notch in front of the eyepatch, and a diagnostic broad white collar extending across the nape, separating the cap from the grey mantle. Feathers above the bill are white, forming a narrow white band between the black bill and the cap. On the upperparts, the back is greyish; an indistinct dark M cuts across the upperwings and rump. The underparts are white, including the chin and belly; a diagnostic small, grey thumb-shaped notch extends onto the hind neck below the wing. A narrow black diagonal bar runs from the elbow to the inner wing on the underwing, and narrow dark trailing edge and dark wing tips give the underwing a mild bordered appearance. White-necked Petrel can be told from Gould's by its larger size, its capped, not hooded head, and the diagnostic broad white collar. It is recorded as a vagrant around the Australian coastline, but is most common and likely to be seen off the e. coast during the spring and summer months.

1
2
3
3
4
4
5
5

1 LITTLE PENGUIN *Eudyptula minor* 43–48cm/17–19in

The smallest and one of the most distinctive of the penguins. The upperparts are all blue-grey in the adult, and the underparts are white. The bill is narrow and black, the eye pale, and the feet pink. Immatures have bluer upperparts; juveniles have downy, brown-grey plumes. Singles or small parties are seen at sea. The species commutes from breeding colonies to open ocean under cover of darkness. This species is found from s. WA eastward around the coast to s. NSW (rarely to n. NSW). It can be seen easily at colonies at Phillip Island (VIC) and Bruny Island and Bicheno (TAS). A small colony also exists in Sydney Harbour (NSW), but it is difficult to see the birds well.

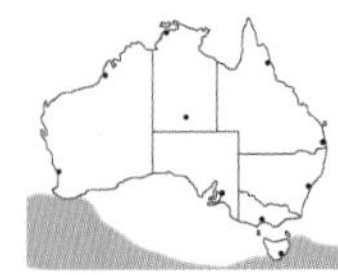

2 NORTHERN ROCKHOPPER PENGUIN *Eudyptes moseleyi* 44–60cm/17.5–23.5in

A small penguin with prominent yellow crests, which flare out widely behind each eye, and a heavyset, deep red bill. Similar in appearance to both Macaroni and Fiordland Penguins, Northern Rockhopper displays black crest feathers within the yellow crest, unlike the other two. It is considerably smaller than the Macaroni Penguin, which has a more saturated golden-yellow crest. Northern Rockhopper also has an area of naked skin around the base of the bill, lacking in the slightly larger Fiordland. Immature rockhopper is best separated from the Fiordland by the presence of naked skin around the base of the bill, and from Macaroni by its much smaller size. Northern Rockhopper Penguin is an uncommon, though regular, visitor to the south (mainly s. WA, se. Australia, and TAS) during the winter months (Jun–Aug).

3 MACARONI PENGUIN *Eudyptes chrysolophus* 65–76cm/25.5–30in

A large crested penguin with prominent golden-yellow head tufts. Much larger than the other similarly crested species (Northern Rockhopper and Fiordland), Macaroni is further differentiated from Fiordland by the presence of naked skin around the base of the bill. On Macaroni Penguin the thicker, golden-yellow crest does not begin as a defined line, or brow, above the eye (as it does on the other two species), but flares out from the front of the head and looks 'wilder' and unkempt. Macaroni Penguin is a rare, though annual, visitor to s. Australia, mainly to the south-east (incl. TAS).

4 FIORDLAND PENGUIN *Eudyptes pachyrhynchus* 55–60cm/21.5–23.5in

A black-and-white penguin with a stout, orange-red bill and a conspicuous tuft of yellow feathers running from the bill to the back of the head. Fiordland Penguin is very similar in appearance to two other species that annually occur in Australia, Northern Rockhopper and Macaroni, both of which also have similar bills and yellow tufts of feathers on the head. However, Fiordland often shows some white flecking below the eye (although only visible at close range) that is lacking in the other species. Furthermore, the yellow brow on Fiordland is thicker above the eye, and does not splay out wildly behind the eye, as on the other two, but droops down when it reaches the back of the head. It is also a considerably smaller bird than Macaroni Penguin. Fiordland Penguin is an uncommon visitor to offshore se. Australia and TAS, and is occasionally recorded in s. WA.

1
2
3
4

FRIGATEBIRDS (FREGATIDAE)

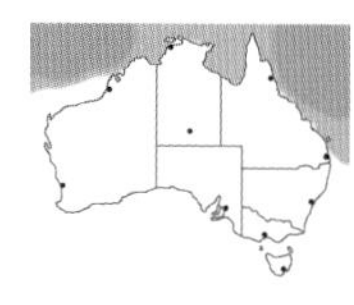

1 GREAT FRIGATEBIRD *Fregata minor* 86–100cm/34–39.5in

One of two very similar species of frigatebird in Australia. It is a distinctively shaped bird, with a long, hooked bill, narrow, sharply pointed wings, and a deeply forked tail, the combination of which gives it an unmistakeable silhouette. The male is all glossy black with a green sheen. Its red pouch, used in courtship, is always deflated when flying. Females lack the pouch, and have a white chin and breast but no white in underwing. Immatures have a mottled white head, chin, and breast, sometimes with an incomplete darker chest band. Great Frigatebird differs from Lesser in never having white armpits in any plumage. It is found at sea off n. Australia, though is less common across the top of Australia than Lesser Frigatebird; it is most likely to be seen on boat trips into deeper Australian waters (e.g., Great Barrier Reef cruises).

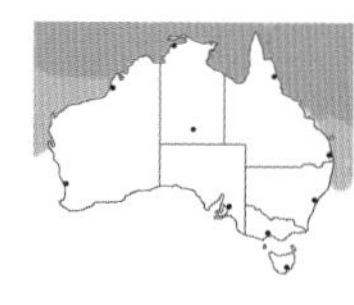

2 LESSER FRIGATEBIRD *Fregata ariel* 70–80cm/27.5–31.5in

Lesser Frigatebird male appears all glossy black with a green sheen but can be told from Great Frigatebird by the thin white line extending from the flanks onto the inner wing (armpit), which is completely absent in Great. The female Lesser is most easily distinguished by its dark throat, whereas the female Great has a white throat concolourous with its white chest patch. Lesser Frigatebird is found at sea off n. Australia, and is most likely to be seen either on boat trips into deeper Australian waters (e.g., Great Barrier Reef cruises), or when coming into roost around coastal ports such as Darwin (NT) and Weipa (QLD).

♂ 1
♀ 1
♂ 2
♀ 2

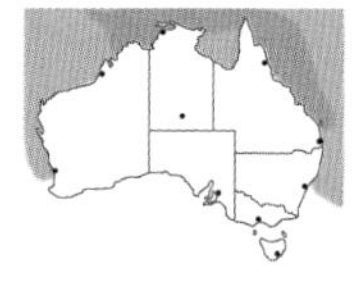

1 BROWN BOOBY *Sula leucogaster* 65–75cm/25.5–29.5in

Brown Booby is a brown seabird with a dark brown hood, chest, and upperparts, and contrasting white underparts and wing linings that make it a striking bird, even at some distance. It has a prominent, bone-coloured bill that is blue at the base in male birds (concolourous with the rest of the bill in females), and has a piercing yellow eye. Immatures are almost all dark brown, except for a slightly paler brown belly and underside, and have dull-coloured bills. Immature Masked Booby could be confused with Brown Booby, as it also possesses a brown hood, although from above Masked shows a contrasting white collar, lacking in Brown. Brown Booby is found throughout the tropical waters of Australia and breeds on Michaelmas Cay (Great Barrier Reef, QLD). It comes closer to shore than other boobies and can sometimes be seen off Darwin (NT) and Brisbane (QLD).

2 RED-FOOTED BOOBY *Sula sula* 65–80cm/25.5–31.5in

A variable booby with three colour morphs, all of which show striking red feet in adult plumage, which are usually also visible in flight. Plumage ranges from solid coffee-coloured brown (dark morph), to light brown (intermediate morph), to white (white morph). Dark morph could be confused with Brown Booby, although adult Brown displays a clean white belly. Immature Brown also displays a lighter belly in contrast with the rest of the underparts and upperparts, unlike the uniform colouration of dark Red-footed Booby. White morph Red-footed is similar to both Australasian Gannet and Masked Booby, which are also largely pied in colour. However, Red-footed Booby displays red feet, and has an all-white tail, while the other two species show extensive black in the tail. Immature Red-footed is uniformly dark brown and does not yet show red feet, but the uniform colouration differentiates this species from bi-coloured immature Brown Booby. Red-footed Booby is seen rarely around the tropical waters of Australia; it breeds on offshore cays and is rarely found on Michaelmas Cay reef trips out of Cairns (QLD).

3 MASKED BOOBY *Sula dactylatra* 75–85cm/29.5–33.5in

A striking, black-and-white booby with a prominent yellow bill. It is most likely to be confused with Australasian Gannet or white morph Red-footed Booby, which are also largely black-and-white in colour. Both Masked Booby and gannet have an all-white body except for a black mask, black outer and rear portion of wing (secondaries), apparent in flight, and black tail feathers. However, Masked Booby has an all-black tail, lacking the white sides the gannet tail shows, while Red-footed Booby has an all-white tail lacking black completely. Immature Masked Booby is brownish above with a brown hood, and white below. At this age it could easily be confused with adult Brown Booby, although young Masked shows a clear white collar, absent in Brown. Masked Booby is found throughout the tropical waters of Australia and breeds on offshore cays, though is nowhere easy to find. It is sometimes seen on Michaelmas Cay (QLD), which can be reached via a barrier reef cruise from Cairns.

ad. 1
imm. 1
dark morph 2
white morph 2
ad. 3
imm. 3

DARTER; CORMORANTS

1 AUSTRALASIAN DARTER *Anhinga novaehollandiae* 86–94cm/34–37in

Male darters are all glossy black with a bold white line that runs from the bill onto the upper neck. They have white streaking on the wings, which is more prominent when birds are drying themselves. Females have an off-white chin and underparts. Young birds are strikingly buff-coloured. In all plumages darters (family Anhingidae) are best identified by their distinctively sinuous necks and slender build. Australasian Darter is a common bird around freshwater areas in e. Australia and coastal w. Australia, and absent from the Great Sandy Desert, Great Victoria Desert, and the Nullarbor Plain.

2 GREAT CORMORANT *Phalacrocorax carbo* 80–85cm/31.5–33.5in

Cormorants are in the family Phalacrocoracidae. This is the largest cormorant, best identified by its size and generally all-blackish colouration, except for a prominent pale facial patch and pale band above the thighs (in adults). Breeding birds develop white plumes on the sides of the neck. Young birds are all brownish and lack the bold markings, and so are best identified by size alone. When large flocks are moving between feeding sites the birds often fly high in a V formation. Great Cormorant is found widely in freshwater and coastal environments throughout the continent, except for w. desert areas; they are often seen resting on piers or jetties or in dead trees.

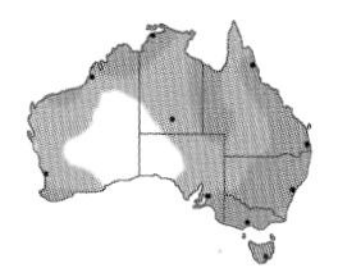

3 LITTLE BLACK CORMORANT *Phalacrocorax sulcirostris* 53–65cm/21–25.5in

The smallest Australian cormorant species, easily identified by its small size, its all-black colouration, and the lack of pale facial markings. Little Black Cormorant is a common and gregarious species, found nearly continent-wide in both freshwater and saltwater environments, and is most likely to be encountered in groups.

4 BLACK-FACED CORMORANT *Phalacrocorax fuscescens* 60–69cm/23.5–27in

A medium-sized, pied cormorant species. It is superficially similar in appearance to the larger Australian Pied Cormorant, with which it overlaps along the s. coast. Black-faced Cormorant can be differentiated by a black (not peach-coloured) face and a blackish (not bone-coloured) bill. It is confined to the s. coasts and TAS, and is a marine and coastal ground-nesting species that frequents harbours, docks, estuaries, and rocky coastlines. It is usually found in small groups.

5 LITTLE PIED CORMORANT *Microcarbo melanoleucos* 55–61cm/21.5–24in

A small black-and-white species that is the most abundant cormorant in Australia. Little Pied Cormorant is most readily identified by its small size, which can be striking when it is directly compared with the much larger Australian Pied or Black-faced. Further differences from the other two black-and-white species include Little Pied Cormorant's yellow bill, white above the eye, and its white thighs, compared with black thighs in Black-faced and Australian Pied. Little Pied is found throughout nearly all of the mainland and TAS on small and large bodies of water in both coastal saline habitats and freshwater environments. It is a colonial, tree-nesting cormorant species.

6 AUSTRALIAN PIED CORMORANT *Phalacrocorax varius* 65–85cm/25.5–33.5in

A large pied cormorant with a black back and all-white underparts except black thighs. Differs from Black-faced in having a peach, not black patch above the bill, and from Little Pied in having a longer and darker bill, larger overall size, and having black, not white thighs. Australian Pied is found throughout most of Australia, though it is more common in the south-east and most common in coastal areas.

♂ 1
♀ 1
♂ br. 2
♀ nbr. 2
3
4
5
6

PELICAN; STORK; CRANES

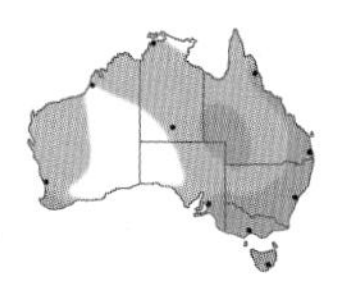

1 AUSTRALIAN PELICAN *Pelecanus conspicillatus* 160–180cm/63–71in

Australia's only pelican (family Pelecanidae) and its largest waterbird. It is mostly white, with contrasting black markings on the wings (most visible in flight), and carries a distinctive, instantly recognisable, large pink bill with a huge throat pouch for catching fish. Like many birds that live on this desert-like continent, which undergoes massive shifts in the availability of water through the years and seasons, Australian Pelican can be highly nomadic, travelling vast distances in search of water. Permanently found throughout most coastal areas, it moves inland during flood years to breed within vast colonies.

2 BLACK-NECKED STORK *Ephippiorhynchus asiaticus* 110–135cm/43.5–53in

This eye-catching bird is the only stork (family Ciconiidae) in Australia. Australian birders sometimes call it 'Jabiru', a misnomer, as that common name is given to *Jabiru mycteria*, a South American stork. A huge, long-legged, boldly patterned black-and-white bird with striking, bright red legs, Black-necked Stork can reach 1.3m (4ft) tall and has a monstrous 2m (6.5ft) wingspan. It is therefore unlikely to be confused with any other Australian bird; only the cranes are of similar shape and height, but they are not black-and-white in colour. Immature birds have the same general pattern as adults, but the black is replaced by a dirty grey-brown, and the white appears very dirty. Black-necked Storks are most often found around an assortment of freshwater wetlands, from wet pastures to large lakes and reservoirs, and more rarely on tidal flats along the coastlines. They are found mainly in tropical n. Australia and down the e. side of the mainland, and are much more abundant in the n. portion of their range. Black-necked Storks are especially abundant in the NT and are usually seen from boat cruises in Kakadu NP.

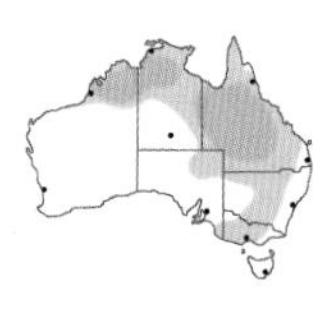

3 BROLGA *Grus rubicunda* 77–135cm/30.5–53in

A crane (family Gruidae), best distinguished from Sarus Crane by its dull greyish legs, by the red colouration that is limited to only the head (not extending down onto the neck), and by its distinctive head shape with the prominent dewlap on the underside of the chin (lacking in Sarus). In parts of n. QLD these birds can be seen flocking in the hundreds with Sarus Cranes. More widespread in Australia than Sarus Crane, Brolga is an Australasian endemic found extensively in tropical n. Australia and down the e. side of mainland Australia, although not present on TAS.

4 SARUS CRANE *Grus antigone* 150cm/59in

Sarus Crane is a massive bird, standing nearly 1.5m (5ft) tall and with a huge wingspan of almost 2.5m (8ft). Both Australian crane species, Sarus and Brolga, are remarkably similar in appearance: large pale grey birds with striking red heads. At large roost sites in QLD, where they flock in the hundreds, both species can occur side by side. They are best told apart by leg colour and examination of the head and neck. The legs of Sarus are a dull reddish-pink, and the red on the head extends down onto the upper neck (a crucial field mark). Furthermore, Sarus lacks the distinctive flap or dewlap on the chin that Brolga shows. As these are truly massive birds, these features can often be seen even at long range. Both Sarus Cranes and Brolgas are best looked for at communal roost sites where they gather in large numbers and draw attention to their presence with their frequent trumpeting calls. They are also frequently found feeding in agricultural fields, where they become red-stained, and are often easy to find courtesy of their large size and the open habitat.

1
1
2
2
3
3
4
right birds 4

HERONS AND BITTERNS (ARDEIDAE)

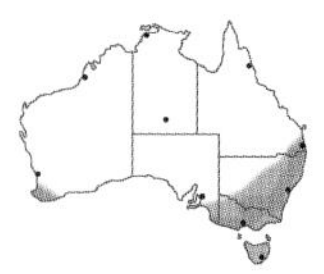

1 AUSTRALASIAN BITTERN *Botaurus poiciloptilus* 71–76cm/28–30in

A massive brown bittern, similar to American (*B. lentiginosus*) and Eurasian (*B. stellaris*) Bitterns. Its upperparts are chocolate brown mottled and vermiculated with fawn, producing extremely good camouflage. It has a buff throat, a chocolate moustachial streak, and nankeen underparts streaked with chocolate arrow marks. Rather than flushing when disturbed, bitterns tend to stand dead still with the bill held upward, and are very difficult to see on the ground when not feeding. Australasian Bittern is found in large marshlands, sometimes in irrigation ditches, of se. and sw. Australia. An uncommon and secretive species, it is rarely seen on the ground and best found at dusk when it flies over extensive reed beds.

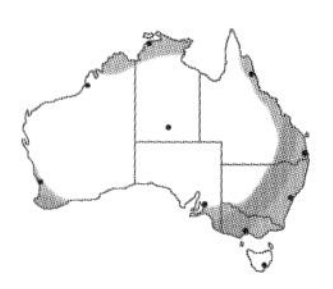

2 BLACK-BACKED BITTERN *Ixobrychus dubius* 25–33cm/10–13in

An attractive but very shy little bittern, with chestnut sides of breast, neck, and mantle. The back and cap are dark chocolate brown (black on males). It has a massive cream shoulder patch, scalloped with black and white on the very shoulder, and cream underparts streaked chocolate on flanks. The buff throat and centre of breast are offset by a black streak from chin to lower breast. It most resembles Striated Heron, which differs in having a more pronounced cap, pale facial skin (limited and dark on Black-backed), and no shoulder or flank markings. Found in e. Australia, ne. WA through the Top End of the NT, and sw. WA, this bird is nowhere common and always hard to find. Although it has some strongholds, it appears to be highly nomadic.

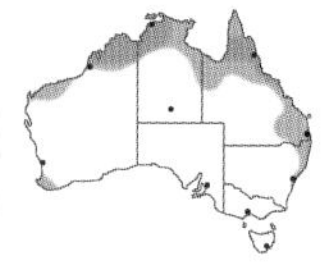

3 BLACK BITTERN *Dupetor flavicollis* 56–66cm/22–26in

A very impressive chocolate-brown to sooty-black bittern that sits dead still on the sides of creeks and is very easy to overlook until you are right in front of it. However, it does not freeze as often as Australasian Bittern, preferring to creep away. Black Bittern has a beautiful, golden-cream moustachial streak, which continues down the sides of the neck and breast to the edge of the armpits. The centre of the throat and breast is chocolate brown streaked with white. Black Bittern occurs throughout the e., n., and w. coastal regions of Australia. In the east and south-west it prefers overgrown streams in rainforests or wet sclerophyll forests. In the north, where it is far more common, it can be also found in swamp edges and billabongs in tropical savannahs; it is regularly seen on the Yellow Water boat trip in Kakadu NP.

4 STRIATED HERON *Butorides striata* 43cm/17in

A small heron with a prominent black cap. Its back is grey-brown, and its underparts range from pale rufous to grey. It has a thin, black-and-white chin stripe. Unlike some larger herons and egrets, Striated Heron often sits in a hunched posture while hunting, with the neck tucked in tight and the body crouched low. It is most likely to be seen at edges of mangrove stands or along the edge of a tidal channel at low tide It is confined to coastal areas of n. and e. Australia, where it can be difficult to detect as it sits quietly and motionless on the edges of mangroves and tidal areas, on the watch for crabs, molluscs, and small fish.

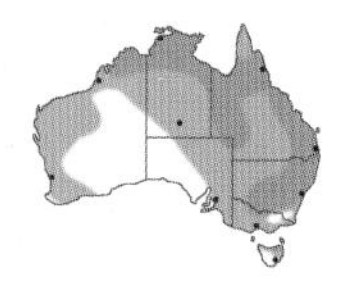

5 NANKEEN NIGHT HERON *Nycticorax caledonicus* 55–65cm/21.5–25.5in

A predominantly nocturnal heron, mostly seen while roosting during the day. The adult is strikingly marked, with richly coloured, nankeen-rufous upperparts, a bright white underside, and a neat black cap. In breeding plumage, in common with many other herons and egrets, it grows delicate white breeding plumes that protrude from the nape. Immature looks similar to Black Bittern but is paler and streaked all over and has a shorter, thicker bill. Nankeen Night Heron is found in a variety of wetlands (tidal and freshwater) over all of e. and n. Australia, even occurring in Australia's biggest cities, and also patchily in w. WA, although it is largely absent from desert areas.

1
1
2
2
3
3
4
4
br. 5
imm. 5

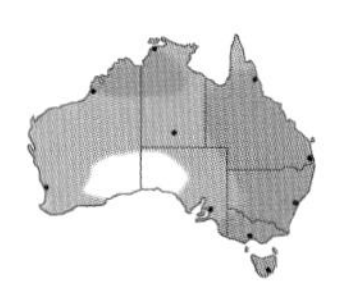

1 WHITE-NECKED HERON *Ardea pacifica* 75–105cm/29.5–41.5in

A large heron with a dark blackish-blue body and a long, all-white neck and head. The grey legs and huge size help to distinguish this heron from considerably smaller young Pied Herons, which also possess an all-white head and neck. White-necked Heron's dark wings have a small but conspicuous white window, visible on the leading edge of the wing when the bird is in flight and showing as a distinct shoulder patch when it is perched. This is a common and widespread heron throughout the continent, except for the interior of se. WA and w. SA. Rarely found in tidal areas, it usually favours small wetlands or marshy areas; it is found on flooded fields, small farm dams, and many other small pools and flooded areas.

2 GREAT-BILLED HERON *Ardea sumatrana* 100–110cm/39.5–43.5in

The largest heron in Australia, standing over 1m (3.3ft) tall. It flies with its neck folded into an S shape. It is a dark heron with a dusky-grey head, neck, and body; dull greyish legs and bill; and a dull whitish breast. The sheer size and the overall dull colouration, lacking white on the head or neck, differentiate this massive waterbird from any other heron. Young birds are all brown and therefore could be confused with the similarly coloured immature Black-necked Stork. However, storks have a much thicker bill with a markedly different shape, and they fly with their necks extended. Great-billed Heron is found in forest-lined creeks, mangroves, and tidal areas, usually where there is plenty of surrounding cover. This large, shy heron is usually encountered quietly feeding alone along the banks of creeks or large rivers. Great-billed Heron is an uncommon bird of coastal tropical Australia from near Broome (nw. WA) to tropical QLD.

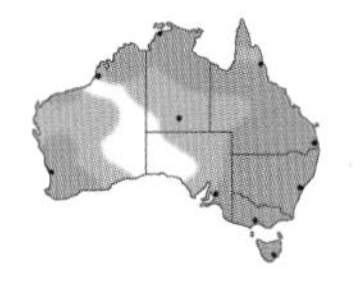

3 WHITE-FACED HERON *Egretta novaehollandiae* 66–67cm/26–26.5in

A common and familiar, light ashy-grey heron with a bold white face, piercing pale eye, and yellow legs. The overall pale colouration is quite unlike that of any other species. Although young birds lack the white face, they are all pale grey like adults and therefore equally distinctive. White-faced is less wary and more approachable than many other herons. It is a common and widespread heron in a variety of wetland habitats nearly throughout Australia (including TAS). It occurs in both freshwater and saline environments, and can be found in many habitats, including tidal flats, small farm dams, salt marshes, flooded pastures, harbours, and beaches.

4 PIED HERON *Egretta picata* 43–55cm/17–21.5in

A striking little black-and-white heron. Pied Heron has a dusky-grey body, a white neck, and a dusky-grey cap. It also has bright yellow legs and bill and a piercing yellow eye. Although it is superficially similar to the much larger White-necked Heron, the adult Pied Heron can be separated from that species by its grey cap. Immature Pied Herons, while lacking the grey cap of the adult, are much smaller, have yellow legs, and lack the bold white shoulder of White-necked Heron. Pied Heron is a distinctly gregarious species relative to other herons, and is rarely found alone. It is a northern species confined to the tropics of ne. WA, n. NT, and n. QLD, where it occurs in the coastal zone. Pied Herons are found in both tidal saline environments and freshwater wetlands, and are usually encountered in flocks.

1
1
ad. 2
imm. 2
3
3
ad. 4
imm. 4

HERONS AND BITTERNS (ARDEIDAE)

1 PACIFIC REEF EGRET *Egretta sacra* 66–71cm/26–28in

Short legs and an oversized bill give this bird a distinctive shape. Two morphs exist: The white morph, more common in the tropics, has a yellow bill and tawny legs. The grey morph, more common in temperate areas, is dark grey with bone legs, a bone-grey bill, and a very small white patch on the throat, visible at close range. Pacific Reef Egret is found on most coasts of Australia, though very rare or absent from SA to VIC and TAS. It is most common from Broome (nw. WA) around to Brisbane (QLD), and found mainly on coastal beaches and offshore islands, including cays of the Great Barrier Reef.

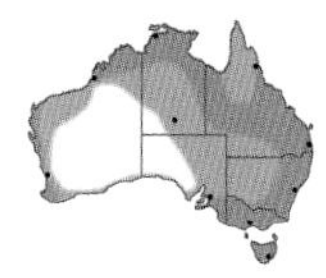

2 GREAT EGRET *Ardea alba* 84–104cm/33–41in

Great Egret is the largest egret species. When it is in breeding plumage, the dark legs are flushed with reddish on the upper section (from the knees up), the bill is blackish, and the lores are pale green. When it is not breeding, the bill and facial skin are yellow. The bare facial skin patch always extends beyond the eye in the Great Egret. The fine breeding plumes grow out from the back and extend down beyond the tail, but these egrets never possess breast plumes. Great Egret is a common and widespread species over almost all of Australia, found on both saltwater and freshwater wetlands, and often found well inland.

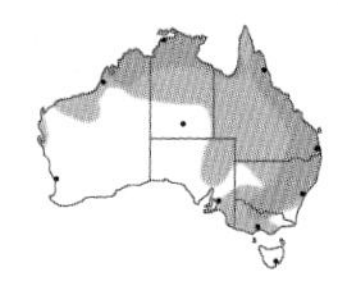

3 INTERMEDIATE EGRET *Egretta intermedia* 55–70cm/21.5–27.5in

Intermediate in size between smaller Little and larger Great Egrets. In breeding plumage Intermediate Egret has an orange bill, green lores, and dark legs washed reddish from the knees up. It also grows fine breeding plumes both from the breast (unlike Great) and the back. When not breeding, Intermediate has blackish legs and yellow bill and facial skin. An important feature, the facial skin patch does not extend beyond the back of the eye, as it does in Great. Both Great and Intermediate heron have a similar structure, often holding their necks kinked in the shape of a question mark. Intermediate is a common egret over n. WA and much of the NT, where it often outnumbers both Great and Little Egrets. It is less common in the south-east, and very rare in TAS.

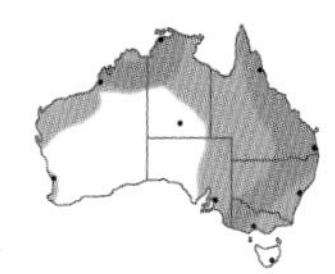

4 LITTLE EGRET *Egretta garzetta* 55–65cm/21.5–25.5in

A small heron with black legs and bill and yellow facial skin. The facial skin flushes red or orange-red at the height of the breeding season. When in breeding plumage, Little Egret grows fine breeding plumes that extend from both the breast and the back, and also grows a few plumes from the nape. When fishing, this egret can be quite acrobatic, quickly and suddenly dashing after fish while raising its wings in the process, which makes for a distinctive behavioural feature, quite different from the slow, deliberate feeding style of both Intermediate and Great Egrets. Little Egret is found in coastal and sub-coastal regions from sw. WA around the north and east, south to Eyre Peninsula (SA).

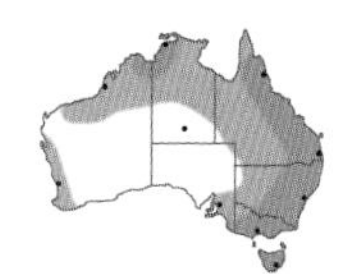

5 EASTERN CATTLE EGRET *Bubulcus coromandus* 70cm/27.5in

A small egret that is colourful in breeding plumage, turning rich orange on the head, neck, breast, and back. At this time it also has reddish legs and the bill flushes red. It grows short, spiky breeding plumes from the back of the head and the breast that are not long and fine, like the breeding plumes of other egrets. When not breeding, Western Cattle Egret has an entirely white body, dark legs, and yellow bill and facial skin. A common and spreading species, found around the coastal zone of almost all of mainland Australia, it is absent from much of the interior and the coasts of SA and s. WA. Western Cattle Egret, which has adapted well to expanding farmlands, is one of the most widespread birds on earth.

grey morph 1
white morph 1
2
3
4
br. 5

IBISES AND SPOONBILLS (THRESKIORNITHIDAE)

1 GLOSSY IBIS *Plegadis falcinellus* 65–75cm/25.5–29.5in

Glossy Ibis is distinct from the other Australian ibis species in being wholly dark in colouration, with almost no white at all. The Glossy Ibis generally appears all black in colour, although is actually deep reddish brown and glossed with purple-green on the back and wings, which can usually be discerned only in strong sunlight. It is most often found in single-species flocks probing in shallow wetlands for frogs and fish or foraging in pastures for insects. It occurs throughout most of tropical and e. Australia, is absent from most of the inland south and west, and is generally uncommon. Glossy is never seen around human habitations, unlike Australia's other two ibis species, which can be regularly found around towns and farmlands.

2 AUSTRALIAN WHITE IBIS *Threskiornis moluccus* 65–75cm/25.5–29.5in

Also known as Australian Ibis. It has an almost wholly white body with a contrasting black neck and head, and shows striking black wing tips in flight. A common and familiar white ibis, it is found in a range of wetlands and also regularly in town parks of the east, where the birds are often highly visible, tame, and approachable. They are most often seen, as are most ibis, in flocks, sometimes mixing with Straw-necked Ibis, and are easy to find as they feed in open habitats, including parks and gardens. They are found throughout n. and e. Australia, where they are most abundant, although only patchily in coastal and n. WA. Absent from the most arid parts of the continent.

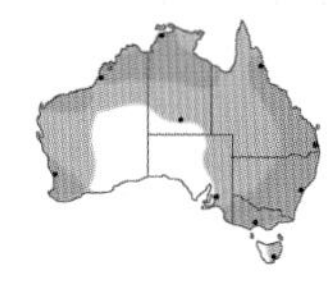

3 STRAW-NECKED IBIS *Threskiornis spinicollis* 60–70cm/23.5–27.5in

A common ibis species that, like Australian White Ibis, also turns up regularly in town parks and gardens. Unlike Australian White, with which it sometimes flocks, Straw-necked Ibis is largely dark in colouration, with white mainly on the underside and the sides of the neck. At close range the strange, straw-yellow plumes on the neck that give it its name are visible. Like Australian White Ibis, and unlike Glossy, Straw-necked is also quite approachable and often tame around towns. It is found in a wide variety of freshwater wetlands and fields over much of Australia, except for the dry interior of the west where no ibis occur.

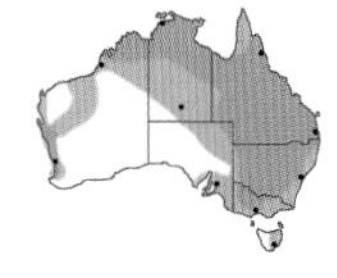

4 ROYAL SPOONBILL *Platalea regia* 75–80cm/29.5–31.5in

The massive, spoon-shaped bill is hard to miss, even in flight. Royal Spoonbill is all white in non-breeding plumage, with a black bill, black legs, and black skin on the crown, face, and chin; it has a small yellow eyebrow. In breeding plumage it develops long plumes from the head and a golden wash on the breast. It is found on a variety of wetlands and in coastal areas in e. and n. Australia, and in the n. section and w. coast of WA. It is generally absent from w. TAS.

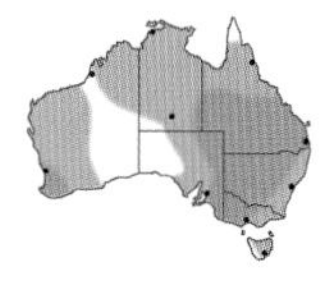

5 YELLOW-BILLED SPOONBILL *Platalea flavipes* 81–91cm/32–36cm

An all-white spoonbill with bone-coloured legs and bill. The white face is bordered by a very thin black line. In breeding plumage it develops golden plumes on the breast. Yellow-billed Spoonbill is rarely coastal, more common inland, and is often found on smaller wetlands and dams than Royal Spoonbill. It is found across much of n. and e. Australia and in the wetter parts of w. WA; it is a rare visitor to TAS.

1
2
3
4
5

OSPREY; KITES, HAWKS, AND EAGLES

1 EASTERN OSPREY *Pandion cristatus* 50–65cm/19.5–25.5in

A distinctive fish-eating hawk occurring along most Australian coastlines. Unlike the other species shown on this page (all in family Accipitridae), it is in the osprey family, Pandionidae. From underneath Eastern Osprey appears largely white, with a dark line through the eye and faint, scattered dark markings on the underside of its white wings. Perched it is very distinctive, with a chocolate-brown back, white underparts, white crown, and brown face, which gives it a masked appearance. Ospreys often perch prominently and construct huge, highly visible nests of sticks in open places, such as on the tops of power poles. Eastern Osprey is becoming an increasingly common sight in coastal e. Australia, though is still rare in VIC and TAS. It is generally coastal, although it can be found along large rivers and around substantial inland wetlands.

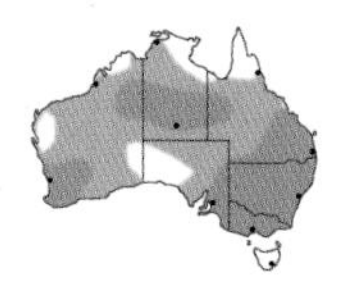

2 AUSTRALIAN KITE *Elanus axillaris* 34cm/13.5in

Also known as Black-shouldered Kite. A distinctive small, slender-winged, black-and-white kite that hunts by hovering for prey such as rodents and grasshoppers. It is a mainly white bird with pale ashy-grey upperparts and a bold black shoulder patch, visible as a distinct bar in flight. A widespread and common bird found throughout the mainland in open woodlands, farmlands, and grasslands with scattered trees, it is most likely to be noticed hovering for prey or perched on an exposed branch or wire.

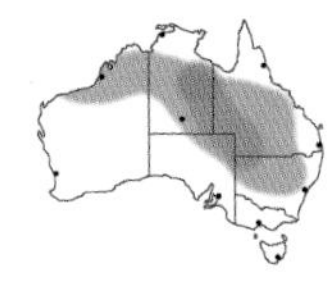

3 LETTER-WINGED KITE *Elanus scriptus* 33–38cm/13–15in

A small, mainly white kite with a prominent black M or W on the underwing, very obvious in flight. When perched it looks similar to Australian Kite, but Letter-winged Kite's face has a more owl-like appearance. A rare bird, it can turn up almost anywhere in Australia after a population boom in its main areas of grasslands from e. NT through to sw. QLD. This crepuscular and nocturnal species is best seen at dawn or dusk, feeding or searching thickets on inland watercourses for roosting birds.

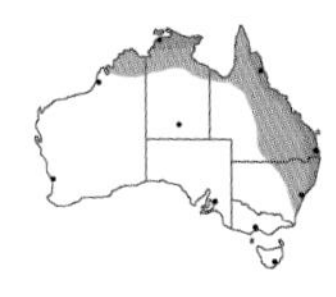

4 PACIFIC BAZA *Aviceda subcristata* 38–44cm/15–17.5in

Pacific Baza, a member of the family Accipitridae, has broad, rounded paddle-like wings similar to those of Hook-billed Kite (*Chondrohierax uncinatus*) of the Americas. The wings have pale secondaries giving the appearance of a large white patch, heavily barred primaries, and a rufous leading edge. When perched, Pacific Baza looks like no other Australian raptor, with its prominent crest, grey head, grey-brown back, and rufous underparts that are heavily barred with white. This bizarre hawk is found in rainforests, wet sclerophyll forests, monsoon forests, and tropical savannahs of far n. WA, the Top End (n. NT), and tropical e. QLD down to n. NSW, though nowhere common. It is usually seen soaring quietly above rainforest canopy or in spring when doing noisy, butterfly-like, rising and falling, fluttering and displays.

1
1
2
2
3
3
4
4

1 BLACK KITE *Milvus migrans* 44–55cm/17.5–21.5in

Black Kite is often confused with Whistling Kite but is darker overall, having an all-dark body, tail and underwings except for a small window on the outer wing, and it has a deeply forked tail. When perched it is a darker-appearing bird than Whistling, with an obvious dark smudge around the eye. Black Kite is widespread over mainland Australia, absent only from se. WA, w. SA, e. NSW, e. VIC, and TAS. It is found in open grasslands, farmlands, timbered watercourses, and other open habitats. The consummate scavenger, it regularly gathers in flocks around slaughterhouses and rubbish dumps and frequently attends bush fires.

1

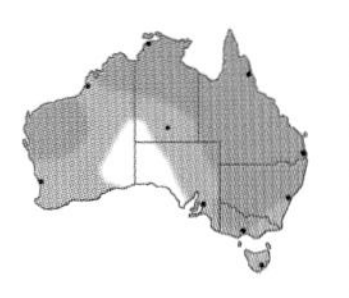

2 WHISTLING KITE *Haliastur sphenurus* 50–60cm/19.5–23.5in

Similar to Black Kite, although with a paler body and pale wing linings on the underwing, visible in flight. Whistling also shows a prominent pale vertical wedge on the underside of the secondaries in flight, which in combination with the pale underwings creates a distinct pale M shape on the underwing. It has an unmarked rounded tail, quite unlike Black's forked tail, an important distinguishing field mark. Whistling Kite is very widespread over the mainland, absent only from the most arid areas of the w. interior, and does occur on TAS, where Black Kite is absent. Whistling Kite frequents open woodlands, plains, open forests, timbered watercourses, and even tidal flats, and is usually encountered singly or in pairs, flocking less often than Black Kite. Whistling has a closer affinity with watercourses than Black Kite.

3 BRAHMINY KITE *Haliastur indus* 44–50cm/17.5–19.5in

A striking kite, with a rufous body and clean white hood and upper breast. In flight the rufous upperwings contrast noticeably with the black wing tips. Immature is much harder to distinguish from Whistling Kite, as it has a similar M along the underwing; but Brahminy Kite immature holds its wings flatter, and the M is whiter, the wing tips blacker, and the tail rounder than on Whistling Kite. Brahminy Kite is strictly coastal and found from the c. WA coast eastward around the top of Australia to n. NSW. Its strong affinity to coastal areas is reflected in its old name, Red-backed Sea Eagle. It occurs around tidal flats, mangroves, beaches, estuaries, harbours, and coastal forests. Brahminy Kites are most likely to be seen soaring alone or in pairs high above the coastline.

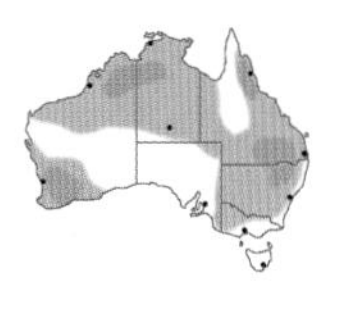

4 SQUARE-TAILED KITE *Lophoictinia isura* 51–56cm/20–22in

At first glance, Square-tailed Kite in flight gives the impression of a Black Kite, with the tail appearing notched when the bird is turning, but this species is distinguished by the strongly upswept wings, much like those of harriers, and glaring white patches in the otherwise barred underwings. When seen well, Square-tailed Kites appear remarkably graceful. When perched, they are rufous with an off-white-streaked face and an indistinct fawn-grey band across chocolate-brown wings. This raptor has a widespread but patchy distribution through most of Australia, and is sometimes found in rainforests but more often in mixed open woodlands and gallery forests. Not an easy bird to target, it is best looked for around regular nesting territories such as Mt Molloy or Daisy Hill (QLD).

1
1
2
2
3
3
4
4

1 BROWN GOSHAWK *Accipiter fasciatus* 41–56cm/16–22in

Brown Goshawk is very similar to Collared Sparrowhawk, and the two are best separated using a combination of size, shape, and general impression. Unfortunately, this requires experience of both species before it can be done confidently. Both are nearly identical in plumage, with adults having grey-brown wings, head, and tail; finely barred rufous underparts; and a rufous collar. Young birds are brown above and pale below, with a complex pattern of streaks and bars. Brown Goshawk is a larger and bulkier bird, with a large head and, if seen well, a 'fierce'-looking eye with a pronounced brow. In flight the tail is long and rounded. This bird will remind North American birders of Cooper's Hawk (*A. cooperii*) and the larger Northern Goshawk (*A. gentilis*), which also occurs in Europe. Brown Goshawk can be found throughout Australia, almost anywhere there are trees, and is often seen soaring in mid-morning.

1

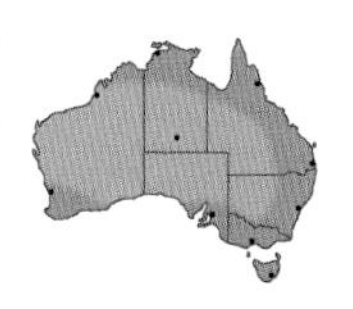

2 COLLARED SPARROWHAWK *Accipiter cirrocephalus* 30–38cm/12–15in

Identical in plumage to Brown Goshawk (don't be fooled by the name—both have a collar), Collared Sparrowhawk is best separated by comparing size and shape. The sparrowhawk is much smaller than the goshawk and has a finer build. Be careful though: As with most raptors, the female is larger than the male, so while males will be obviously smaller than goshawks, large female sparrowhawks may be nearly as large as male goshawks, particularly in n. Australia, where the subspecies of Brown Goshawk is quite small. Sparrowhawks also have a smaller head, and the eye appears 'wide open' rather than 'fierce', like that of the goshawk. In flight the Collared Sparrowhawk's tail is square; this feature can be obvious, but beware of birds moulting their tail feathers. North American birders will be reminded of Sharp-shinned Hawk (*A. striatus*), while European birders will recall Eurasian Sparrowhawk (*A. nisus*). The species is found throughout Australia, in almost any treed habitat.

3 GREY GOSHAWK *Accipiter novaehollandiae* 41–56cm/16–22in

This beautiful raptor has two distinct colour variations. The most common is the grey morph, which is pale grey above and white below with fine grey bars on the breast. The white morph is a stunning bird, pure white all over. Both morphs have a large, bright yellow cere (which contrasts with the black bill) and yellow legs. They are closely associated with dense forests, usually seen flying across openings in the forest or soaring above the canopy, where their pale colouration makes them easy to identify. The species is found across n. and e. Australia and is generally difficult to see. A good strategy is to find a vantage point with a view over a large area of forest canopy and watch for them soaring around mid-morning.

4 RED GOSHAWK *Erythrotriorchis radiatus* 48–61cm/19–24in

As one of Australia's rarest birds, this species has achieved near-mythical status. On the few occasions it is seen, it is usually flying powerfully through the forest canopy or sitting quietly on a screened perch. It is a large raptor with a reddish body and pale head, both heavily streaked, and a long barred tail. It is restricted to tropical woodlands in n. and e. Australia, often near large watercourses. It is very difficult to see reliably away from the nest. Birds are occasionally seen in Kakadu NP (NT), and also on Cape York Peninsula in QLD and in the Kimberley in n. WA.

1
1
imm. 1
imm. 2
ad. 2
imm. 2
grey morph 3
white morph 3
4
4

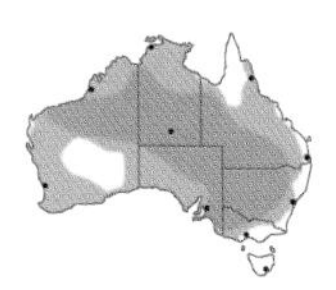

1 SPOTTED HARRIER *Circus assimilis* 53–61cm/21–24in

Adult Spotted Harrier is a beautiful raptor with smoky-blue upperparts and chestnut underparts covered in fine white spots. Young birds are brown overall and variably mottled and streaked. Unlike Swamp Harrier, Spotted never has a white rump. These harriers are usually seen quartering low over grasslands, such as cereal crops, on broad, upswept wings held in a deep V. They can be found across Australia, and are most populous in drier inland areas, where they can be fairly common on open plains. The species can be irruptive, and after good seasons young birds may locally outnumber adults.

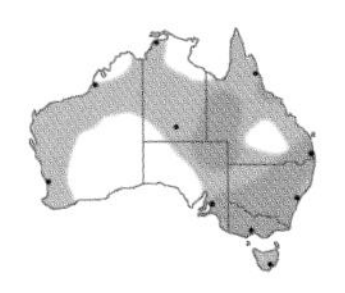

2 SWAMP HARRIER *Circus approximans* 51–61cm/20–24in

Swamp Harrier is usually seen flying low over swampy areas with its broad wings in a deep upswept V. Its plumage is quite variable, but at all ages and in all plumages it has a white rump, usually quite an obvious feature. It is the only raptor in Australia with a white rump, so if you see this, identification is easy. Young birds are dark brown, but as they get older they become paler, and adult birds are generally brown above with pale and streaked underparts. Swamp Harrier is found across much of mainland Australia, almost exclusively on swamps, and is quite common in se. Australia. For some reason, in TAS it is found in a wider range of habitats, including open woodland areas.

1
1
1
imm. 1
2
2
2
2

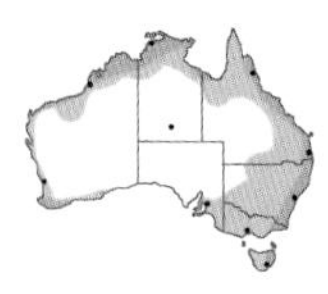

1 WHITE-BELLIED SEA EAGLE *Haliaeetus leucogaster* 75–85cm/29.5–33.5in

A massive bird with very broad wings held in a deep V shape in flight and a relatively short, wedge-shaped tail. It is uniform grey on the upperside of the wings and back, while the head, neck, and underside of the body are bright white; the underwings are also white, with a broad black border on the back edge. Younger birds are brown above, fawn below, and have large white patches on their primaries, visible in flight, which distinguishes them from Wedge-tailed Eagles. White-bellied Sea Eagles are most often seen gliding high over the sea or along coastlines, singly or in pairs, or sitting up on a high, prominent perch around wetlands and in coastal areas. They are found on all the coastlines of Australia, including TAS, and occur some way inland from the coast in the east, where they can be found around larger wetlands and dams.

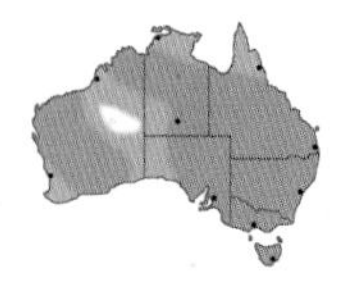

2 WEDGE-TAILED EAGLE *Aquila audax* 85–107cm/33.5–42in

The largest raptor in Australia, identified by a combination of its massive size, all-dark colouration, and distinctive long, diamond-shaped tail. Wedge-tailed Eagle occurs widely across Australia and TAS in a range of habitats, although is perhaps most abundant inland within open country. It is often seen feeding on road kill along an Outback road, where one or two may attend the carcass much in the manner of a vulture.

2

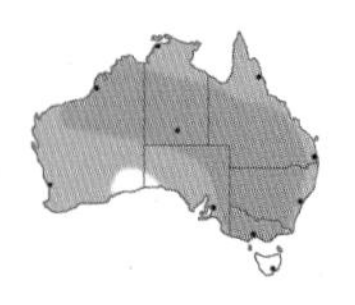

3 LITTLE EAGLE *Hieraaetus morphnoides* 46–56cm/18–22in

A small eagle, more likely to be confused with kites than other eagles. Pale morph birds give first impression of Booted Eagle (*H. pennatus*) of Europe. The species has a square tail, a very strong cream M pattern on the underwing, and a band on the upperwing. The less common dark morph is dark rufous with a paler patch at the base of the primaries. Little Eagle holds its wings flat in flight, which distinguishes it from Black, Whistling, and Square-tailed Kites. It can be found in most habitat types on mainland Australia except thick, closed wet sclerophyll forests and rainforests. It tends to be more common away from the coastal areas.

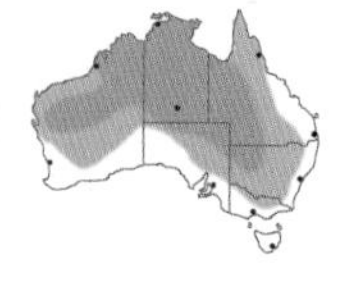

4 BLACK-BREASTED BUZZARD *Hamirostra melanosternon* 51–61cm/20–24in

A stunning large hawk. When seen at a distance in its rocking, harrier-like flight on upswept wings, the chocolate-brown adult, with a black breast and back, appears all black with absolutely beaming white patches at the base of the primaries. When seen well, the adults are unlikely to be confused with any other bird. Young birds are paler, with less distinct primary patches and pale secondaries, and therefore can be confused with Square-tailed Kite or Red Goshawk; but Black-breasted Buzzard is best distinguished by an absolute lack of barring in any plumage. The species is found throughout much of tropical and inland Australia; it is far more common in sub-coastal savannah areas than in the truly arid terrains.

1
1
2
2
pale morph 3
pale morph 3
4
4

FALCONS (FALCONIDAE)

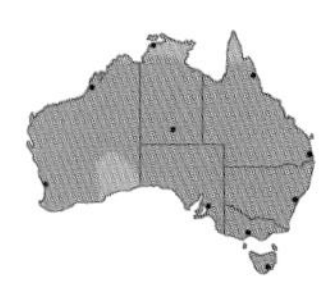

1 NANKEEN KESTREL *Falco cenchroides* 30–36cm/12–14in

Also called Australian Kestrel. Nankeen Kestrel has a characteristic hovering hunting style. Males and females are quite different: Both are rufous birds, but the male has a contrasting dove-grey head and tail that stand out in flight especially. Both sexes have a strong dark subterminal tail band and contrasting black tips to the upperwings, which are striking features in flight. The rufous colouration and hovering behaviour make this falcon easily recognisable. It is common and widespread throughout the mainland and TAS in open country.

2 AUSTRALIAN HOBBY *Falco longipennis* 30–36cm/12–14in

An agile, narrow-winged, long-tailed falcon that hunts at high speed, scything through the air to swoop on prey. Its typical falcon shape, narrow, sharply pointed wings, dark grey upperpart colouration, and hunting style make it readily confused with Peregrine Falcon, which has broader wings, a shorter tail, and dull buff underparts. The rufous underside of Australian Hobby distinguishes it from all other falcons. Usually observed in flight, either alone or in pairs, when hunting on the wing over open areas, this bird is widespread over open wooded habitats and scrubby areas and even wanders into urban environments over most of mainland Australia and TAS.

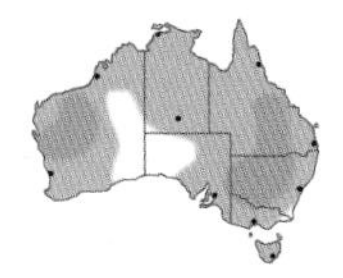

3 PEREGRINE FALCON *Falco peregrinus* 35–50cm/14–19.5in

A large, dark-backed falcon, much like Australian Hobby on the upperparts, Peregrine is distinguished by heavily barred or streaked underparts that are dull buff in colouration, not rich rufous. It is also more heavily built than the hobby, with broader, shorter wings. It is found sparsely throughout Australia and TAS, occupying many habitats, although most common where rocky outcrops provide vital nesting areas. In some towns and cities in Australia Peregrines hunt pigeons, and in tidal areas they haunt large flocks of shorebirds.

4 BROWN FALCON *Falco berigora* 39–50cm/15.5–19.5in

A thickset brown falcon that is broader-winged than all other Australian falcons. It is variable: Some birds are almost entirely dark brown; others are rufous above, similar to Nankeen Kestrel; and some are dark brown with a pale buff head. In most plumages it can be told from the other falcons by its dark brown upperparts and a double moustachial stripe. The rufous morph lacks the obvious dark subterminal band on the uppertail Nankeen Kestrel possesses, and has dark 'trousers', while the kestrel's are white. Although Brown Falcons can hunt by hovering, like kestrels, they do this for only short periods, and are always more powerfully built than the slender kestrel. They also regularly soar, which can then give them a hawk-like appearance, and usually hunt by gliding down and capturing their quarry on the ground. Brown Falcon is widespread throughout the mainland and TAS in a range of habitats that includes open woodlands, farmlands, and grasslands.

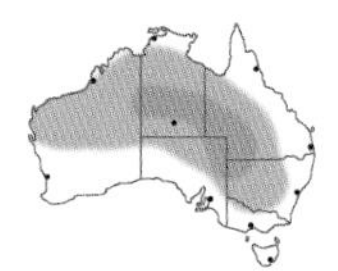

5 GREY FALCON *Falco hypoleucos* 33–43cm/13–17in

Grey Falcon is medium grey above and very pale grey below, and has a grey moustachial streak and a yellow cere, eye ring, and legs. In flight the sooty black wing tips contrast with the whitish underwings. It is distinguished from Grey Goshawk when perched by the yellow eye ring and the lack of barring on the breast, and in flight by its slimmer, pointed wings. A rare, nomadic, and therefore hard-to-pin-down falcon of inland Australia, Grey Falcon prefers timbered watercourses in otherwise pretty desolate gibber plains or grasslands.

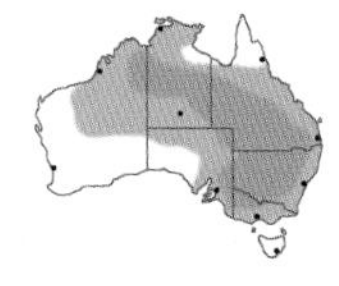

6 BLACK FALCON *Falco subniger* 46–56cm/18–22in

A large, powerful-looking chocolate-brown falcon with subtle pale streaking on breast. In flight, the broad-based, pointed wings give an almost triangular impression. Dark morph Brown Falcon appears similar but is more hawk-like and less acrobatic in flight, and it bears a double moustachial stripe, which Black Falcon lacks. This is an open-country falcon of mainland Australia with a very patchy distribution. Regular locations include the Cunamulla region of QLD and the Round Hill region of NSW. These birds follow locust swarms in farming districts.

♀ 1
♂ 1
2
2
3
3
4
4
5
5
6
6

1 CHESTNUT RAIL *Eulabeornis castaneoventris* 46–51cm/18–20in

This large rail is found in large areas of tidal mangroves across the far north of Australia, where it can sometimes be seen feeding at the edge of an ebbing tide. It has a chestnut body, and its wings are chestnut in the e. part of its range (QLD and NT) and olive in the w. portion (WA). The head is grey, and the powerful legs and bill are yellow. Its raucous braying call is often heard, but the bird is usually quite shy. A mangrove species from the coastlines of the Kimberley (WA) to the Gulf Country (Gulf of Carpentaria shoreline, NT and QLD), it comes out of the mangroves at dropping tides to feed on exposed mudflats. It can be found in mangroves around Darwin (NT), such as those at the Buffalo Creek boat dock, near Lee Point.

2 RED-NECKED CRAKE *Rallina tricolor* 28cm/11in

This crake is found in rainforests, where it prefers thick undergrowth, often near water. It has a chestnut head and breast, dark brown wings and body, and a yellow bill. Young birds are dark all over. It is not uncommon within its range but is shy, as are most crakes, and difficult to see in the gloomy rainforest, though it has become habituated in a few locations. This species is found mainly near streams or bogs in rainforests of far ne. QLD.

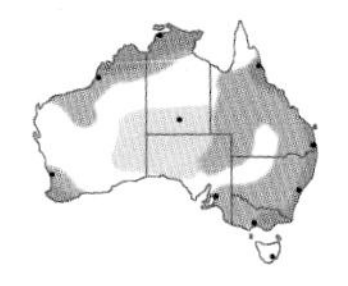

3 BUFF-BANDED RAIL *Gallirallus philippensis* 30–33cm/12–13in

The most commonly seen rail, this bird is found nearly anywhere there is water and/or dense damp vegetation. It usually skulks out of sight, but in the early morning or late afternoon will come to edges to forage on lawns, mudflats, and other open areas. It is beautifully patterned, with brown wings, black-and-white-barred underparts, a grey throat, and a rufous head with a white eyebrow. Across the breast is a single orange band. Found in all but the driest parts of mainland Australia and absent from Cape York Peninsula, it lives in both rainforest fringes and monsoon forests, as well as open marsh-type habitats. This is one of the quickest rails to become habituated in areas where people abound and can become quite tame.

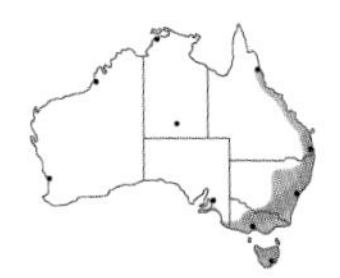

4 LEWIN'S RAIL *Lewinia pectoralis* 22–23cm/8.5–9in

This very secretive rail can be extremely difficult to see. It has a dumpy body, a long neck, and a long red bill. The wings are brownish with white speckling, the back is brown with chocolate streaking, the breast is greyish brown, and the belly is finely barred with black and white. The head and top of the neck are rufous. The common call sounds as if two rocks are being knocked together and often broadcasts the bird's presence. Uncommon, secretive, and nomadic, this is a difficult species to locate. In its e. range it is found in many habitat types from wet heaths and cumbungi- (bulrush-) lined rivers to marshlands, and is usually spotted near dusk when it ventures into more open areas to feed.

5 PALE-VENTED BUSH-HEN *Amaurornis moluccana* 25cm/10in

Because of the name 'bush-hen', birders often expect this species to look like a moorhen or coot, but it is more like a rail. Quite small and slender, it has a greyish breast, olive wings and head, a yellow-green bill, and slender yellow-green legs. It is very difficult to see well, usually staying well hidden in dense vegetation, and is very shy and hard to beckon out. It is most often seen in early mornings on roadside edges or during spring and summer when it is vocal and more likely to be seen moving about. Found across the top of ne. WA and the Top End (NT) and down the e. coast of QLD to n. NSW, this bird inhabits rainforest edges, tall grass, and cane fields near thick forests.

1 WHITE-BROWED CRAKE *Porzana cinerea* 19cm/7.5in

This is the most easily seen of the small crakes, as it regularly forages in the open, often walking across lily pads. It has a grey body, brown wings, and two distinctive white lines, one above and one below the eye. Found on wetlands, particularly those with floating vegetation, it is a common but generally shy small crake of n. Australia, mainly the Top End (NT) and Cape York Peninsula (QLD). This species has a broader habitat range than the other small crakes and can be found in suburban ponds, mangroves, and forest edges near water. It can often be seen from the Yellow Water cruise in Kakadu NP (NT), and the Cattana Wetlands near Cairns (QLD) and Knuckey Lagoon in Darwin (NT) are also good places to see it.

1

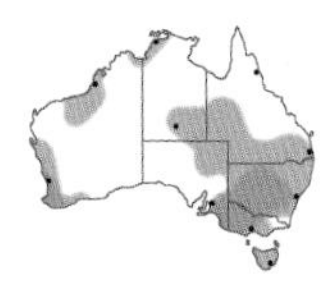

2 SPOTLESS CRAKE *Porzana tabuensis* 19cm/7.5in

The most easily identified small crake, it is also very shy and thus the most difficult to see. It has plain olive-brown wings, a sooty-grey body, a black-and-white-barred undertail, red legs, and a red eye. The red legs contrast with the dark body, making identification easy if the bird is seen well. It is quite secretive, rarely foraging far from cover. Spotless Crake is a common species throughout most wet regions of se. and sw. Australia and TAS, and also has a population in the Top End (NT). In n. Australia, it is far less confiding than White-browed Crake.

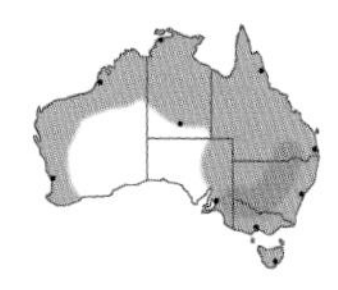

3 BAILLON'S CRAKE *Porzana pusilla* 15cm/6in

Baillon's Crake is pale brown above with dark streaks, and has a pale grey breast and face. The belly is barred black and white, and so is the undertail, which is often seen as the bird flicks it up and down. Although shy, this bird does occasionally come out in the open and can be seen with patience. Look for it at wetlands when water is receding and leaving a muddy edge where the bird can forage while remaining close to cover. It is found throughout most wet regions of Australia and TAS, though it is far more common south of the tropics. This species prefers reed- and grass-dominated wetlands, and can easily be seen running in and out of the reed clumps.

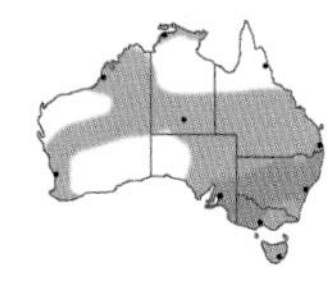

4 AUSTRALIAN CRAKE *Porzana fluminea* 20cm/8in

This small crake is very similar in both plumage and behaviour to Baillon's Crake, but is slightly stockier and has a darker grey breast, darker brown upperparts, and a white undertail (not barred). Although remaining shy, it can be quite easily seen if there are exposed muddy edges for it to feed on. It is the most common small crake in se. and sw. Australia, occurring in the same habitats as Baillon's and Spotless Crakes, with which it associates at marsh edges.

1
2
3
4

RAILS, CRAKES, AND COOTS (RALLIDAE)

1 PURPLE SWAMPHEN *Porphyrio porphyrio* 36–39cm/14–15.5in

A large, deep purple gallinule with a shocking scarlet bill and frontal shield, duller pinkish-red legs, and bright white undertail coverts. The combination of purple colouration and bright red bill makes the adult unmistakeable. Immatures are duller versions of the adults, but still usually show the red bill and some purple colouration. Purple Swamphen is found in wetlands all over e. Australia, though it is absent from much of the west, except sw. and n. WA.

2 DUSKY MOORHEN *Gallinula tenebrosa* 36–39cm/14–15.5in

A small, blackish-grey gallinule with a bright red shield, a red bill that is tipped yellow, and bright white undertail coverts. Immatures are brown and lack the bright bill colouration, and can be readily confused with Eurasian Coot; however, the immature moorhen displays whitish undertail coverts. Dusky Moorhen regularly forages out of the water on land, which coots rarely do, preferring to remain in the water except when nesting. Found around a variety of freshwater wetlands, including urban ponds and lakes, Dusky Moorhen is a common and abundant gallinule over e. Australia but is absent from large sectors of the west, where it occurs only in sw. WA.

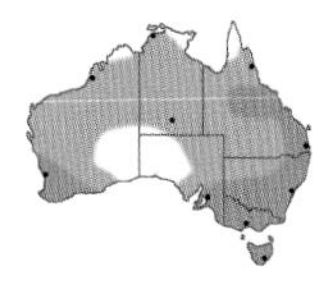

3 EURASIAN COOT *Fulica atra* 36–39cm/14–15.5in

A slate-grey gallinule with a conspicuous ivory-white bill and shield, most often observed on the water, where it frequently dives for food. Adults are readily differentiated from other gallinules by the white bill and shield, while the browner immatures also show pale bills and lack the white undertail coverts of some of the other gallinules (e.g., Purple Swamphen and Dusky Moorhen). Eurasian Coot is found commonly throughout mainland Australia (including TAS), absent only from interior WA and w. SA. It occurs on many types of wetlands with abundant aquatic vegetation, such as freshwater lakes, reservoirs, lakes in town parks, and sewage ponds, as well as brackish lagoons and saline offshore waters. Most often seen in large rafts on large lakes, it rarely leaves the water to graze.

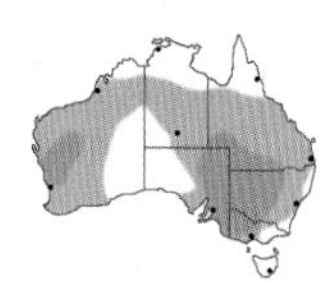

4 BLACK-TAILED NATIVEHEN *Tribonyx ventralis* 30–38cm/12–15in

A distinctive gallinule with a uniquely greenish-blue bill bearing a prominent red area on the lower mandible, bright red legs, and an all-dark body (deep blue below, brownish above). It holds its black tail in a distinctive fashion: cocked and fanned. It also shows some white flank markings. The most similar member of the family Rallidae, Tasmanian Nativehen, is confined to TAS, where Black-tailed does not occur. Its unique bill colour, white flank markings, and red legs differentiate Black-tailed Nativehen from all other Rallidae. It is mainly found inland, away from the coasts, and there it is common, though highly irruptive and nomadic. It is absent from the Top End of the NT, tropical areas of n. QLD, the Great Sandy and Great Victoria Deserts, and the Nullarbor Plain. In years with good rains, the whole population moves to c. Australia, where its numbers increase rapidly.

5 TASMANIAN NATIVEHEN *Tribonyx mortierii* 43–51cm/17–20in

A huge, flightless gallinule found only in TAS. As it is unable to fly, it runs, dives, or swims to escape danger. This conspicuous nativehen is dark-bodied with a bold white flank stripe, its most prominent ID feature. It also has black undertail coverts that are often exposed, as it cocks its tail when grazing. Moorhens and swamphens both show white undertail coverts. The bill is dull greenish, and the legs are grey. Tasmanian Nativehen is locally common on TAS over much of the island, although absent from the south-west (an area dominated by rainforests, which it avoids). Nativehens feed mostly by grazing around the edges of marshes, in marshy paddocks, and along grassy margins of farms and woodlands (usually near water), where small parties are normally conspicuous and easily found.

1
1
2
3
4
5

1 AUSTRALIAN BUSTARD *Ardeotis australis* 80–119cm/31.5–47in

A large, long-legged, heavyset terrestrial bird (family Otididae) with a brown back, a grey face, neck, and underparts, and a black chest band. The male performs a bizarre display, during which he inflates a large gular (upper throat) sac that droops down spectacularly from the neck and touches the ground, fans out his breast feathers, and raises the tails over his back while strutting around slowly and making a number of strange grunts, croaks, and booms. Australian Bustard is found throughout inland Australia in grasslands, tropical savannahs, pasturelands, and open woodlands.

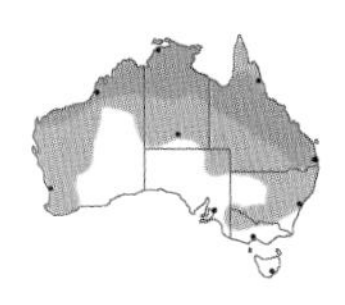

2 BUSH STONE-CURLEW *Burhinus grallarius* 53–58cm/21–23in

Also known as Bush Thick-knee; a member of the family Burhinidae. A long-legged bird with well-camouflaging plumage—mottled grey-brown above, off-white and heavily streaked grey-brown below—a prominent white eye stripe, and a white bar on wings. Its cryptic plumage makes it easy to overlook; it is also most active at night, so the species can appear rarer than it is at times, as it rests in shaded areas of the leaf litter during the day. Bush Stone-Curlew is found in a range of dry areas, including rainforest edges, woodlands, pastures, parks, and gardens, throughout most of Australia except the extreme south and desert areas.

3 BEACH STONE-CURLEW *Esacus magnirostris* 53–56cm/21–22in

Also known as Beach Thick-knee; a member of the family Burhinidae. A heavyset stone-curlew with a massive bill. It has striking plumage: pale brown underparts, black-and-white striping on the wings, and a piercing yellow eye. It walks slowly and deliberately in a heron-like fashion. Despite its huge size, it can be hard to find, as it is a shy bird that often roosts by day in dense mangroves; pairs can occupy territories that cover long stretches of coastline. Beach Stone-Curlew is found very locally in coastal areas of n. Australia from n. WA around to n. NSW.

1
2

3

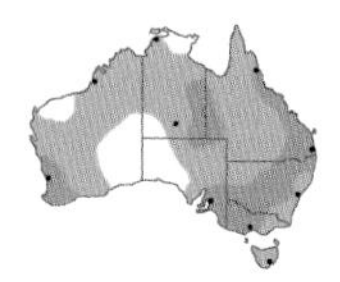

1 WHITE-HEADED STILT *Himantopus leucocephalus* 33–37cm/13–14.5in

Stilts and avocets make up the family Recurvirostridae. This stilt has a long white neck with a bold black patch on the nape, an all-blackish back, and the incredibly long, bright bubble-gum pink legs that are the defining feature of stilts. It is found in a range of wetland habitats throughout the continent, absent only from the driest part of the interior, and is most often encountered in flocks.

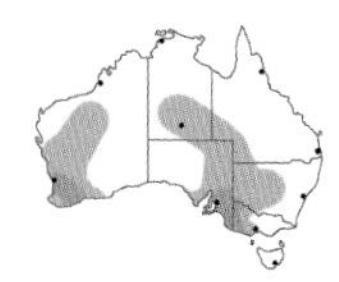

2 BANDED STILT *Cladorhynchus leucocephalus* 38–43cm/15–17in

This species has an all-white head, nape, and upper breast, which contrast with the black wings and the pale red legs. In breeding plumage, adults have a spectacular broad chestnut breast band that blends into a black belly stripe. Non-breeding adults and immature birds have an immaculate white breast and belly. The black bill is very thin and straight. A desert breeding species, Banded Stilt nests on saline lakes and disperses to the s. coastal areas in drier years. It is difficult to locate, though some birds are usually present near Adelaide (SA) and in s. WA.

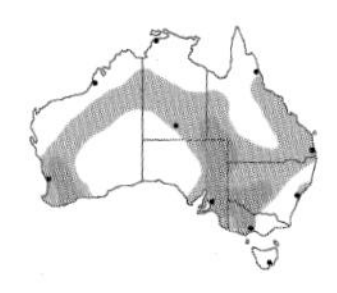

3 RED-NECKED AVOCET *Recurvirostra novaehollandiae* 39–48cm/15.5–19in

An extremely distinctive and elegant shorebird with a very long neck and legs and a distinctive upturned bill, which lends it a unique outline. Red-necked Avocet is an immaculate bird with a neatly pied body, a crisp chestnut head and neck, and grey-blue legs. It employs a distinctive feeding action, wading into deeper waters and sweeping the oddly upturned bill from side to side, foraging on aquatic insects and crustaceans. It can forage while swimming too, almost in the manner of a duck, by floating on the water, with its long legs hidden below, and picking up aquatic prey from the surface. Red-necked Avocet occurs in an array of wetland environments, both in freshwater and saline areas, from inland swamps and large lakes to small dams and coastal salt pans.

4 PIED OYSTERCATCHER *Haematopus longirostris* 42–50cm/16.5–19.5in

Oystercatchers are in the Haematopodidae family. Pied is a very distinctive shorebird with all-black upperparts except for a white rump and a white wing bar visible only in flight. The black head and breast are sharply demarked from the pure white lower breast and belly. It has red legs and a thick, straight crimson bill; the red eye just adds to the beauty of this species. It occurs on sandy and muddy coastlines and estuaries, while Sooty Oystercatcher prefers rocky shorelines. Pied Oystercatchers are usually encountered in pairs or small groups, and like many oystercatchers are very vocal, their piercing, piping calls heard most when they are disturbed from their well-hidden nesting sites, which are nothing more than a shallow scrape on the ground. They are generally more approachable than Sooty Oystercatchers.

5 SOOTY OYSTERCATCHER *Haematopus fuliginosus* 39–52cm/15.5–20.5in

Sooty Oystercatcher is an entirely black-plumaged shorebird with a bright orange beak and legs. It generally prefers rocky shorelines, which the only other Australian oystercatcher, Pied, generally avoids. However, in some areas the two species can be seen foraging together on the same beaches. Unlike Pied, Sooty never shows any white in its plumage. It feeds mainly on molluscs, although it will also capture crabs and feed on starfish, small fish, and even sea urchins. Sooty is generally more wary than Pied Oystercatcher and less often allows close approach.

ad. 1
imm. 1
br. 2
3
4
5

PLOVERS (CHARADRIIDAE)

1 PACIFIC GOLDEN PLOVER *Pluvialis fulva* 25cm/10in

A short-billed and long-legged shorebird. In breeding plumage, which some birds attain as early as March, it is mottled golden and black above and has striking black underparts from throat to undertail and a broad white supercilium that forms a border around the face and continues down the sides of the neck and breast. In flight, the armpits and inner underwing are plain grey. Similar Grey Plover lacks golden colouring and has a larger bill and striking black armpits. In non-breeding plumage, the two are much more confusing, though Pacific Golden Plover is browner, faintly speckled chocolate, with a faint pale eyebrow. The feet of Pacific Golden Plover extend just beyond tail in flight, differing from Grey Plover, but this is not easy to see at a distance. This plover feeds out of the water with a run, stop, run, stop action. A common summer migrant found on most mudflats and sand beaches around the Australian coast, it arrives in August, leaving in April.

1 br.

2 GREY PLOVER *Pluvialis squatarola* 30cm/12in

Also known as Black-bellied Plover. A large, heavy-billed plover, in non-breeding plumage a grey wader, mottled grey and black above, off-white below, with a black bill and legs and a large black eye. If seen in flight, it shows distinctive black armpits, which makes identification straightforward, as well as a clear, white, square rump and a white wing bar; its feet do not extend past the tail. In breeding plumage, which is rarely seen in Australia, it is a stunning bird with mottled white (looks silver) and black upperparts that are separated from the jet-black underparts by a white border. It lacks the golden colouring of Pacific Golden Plover. Very rarely found away from coast, it displays the typical plover run, stop, run, stop feeding movement. An uncommon Holarctic migrant appearing in the austral summer, Grey Plover may possibly be found on any mudflat or sandy beach.

br. 1
br. 1
nbr. 1
nbr. 1
partial br. 2
partial br. 2
nbr. 2
nbr. 2

1 DOUBLE-BANDED PLOVER *Charadrius bicinctus* 18cm/7in

In non-breeding plumage, when most likely to be encountered, a small brown-and-white plover resembling Lesser Sand Plover. It has a white forehead, a long white supercilium extending over and behind the large black eye, and a faint upper breast band, and while it lacks a lower breast band it shows patches over sides of breast. Male in breeding plumage has white forehead and supercilium, black lores, a narrow black upper breast band, and a distinctive, broad chestnut-red lower breast band, striking against the white underparts. Breeding female has the same pattern, but colours are subdued. All plumages show a white wing bar in flight. The call is a loud *pit, pit*, given as a rapid trill. Double-banded Plover is a non-breeding winter migrant from New Zealand to coastal se. and sw. Australia, often found on broad sandy beaches though also on salt pans and sandy fields. It arrives in February and leaves in August.

2 LESSER SAND PLOVER *Charadrius mongolus* 20cm/8in

A small, delicate-looking, short-billed plover. In non-breeding plumage, it is brown and white, with breast band, very similar to Greater Sand Plover. Identification requires careful attention to the structure. Lesser Sand Plover has a small, fine bill compared to the heavy bill of Greater. The head of Lesser is rounded, giving the bird a softer look, and the legs are proportionally shorter and thinner than those of Greater. In breeding plumage, which can be seen in some individuals during their whole winter visit, Lesser Sand Plover is a striking bird, with an orange-rufous breast and crown, white throat and forehead bordered black, black lores, and black around the eye. Some birds show a black forehead. Lesser Sand Plover is a common Palaearctic migrant to the n. and e. coasts of Australia in the austral summer, occurring in fewer numbers around the entire coastline and sometimes at inland sites. It mainly feeds on mudflats, but roosts on sandy spits, on breakwaters, and in mangroves. It is easily seen on the Cairns Esplanade.

3 GREATER SAND PLOVER *Charadrius leschenaultii* 24cm/9.5in

Very similar to Lesser Sand Plover but a larger, chunkier bird with a heavy bill, stout legs, and an angular head shape that makes it look less 'friendly' than Lesser. Pay careful attention to the structure, and if possible compare it directly to Lesser Sand Plover: Greater's non-breeding plumage is identical to Lesser's non-breeding, but if the two are seen together the size and structural differences are clear. In breeding plumage, which is more rarely encountered in Australia but seen in some individuals during their visit, Greater has an orange breast, paler than that of Lesser; an orange crown; a blackish face (not jet black); and a mottled blackish forehead less cleanly delineated than the black or white forehead of Lesser. Greater Sand Plover is a common Palaearctic migrant in the austral summer to the nw. and n. coast of Australia, occurring in fewer numbers around the entire coastline. It mainly feeds on mudflats, but roosts on sandy spits, breakwaters, and in mangroves. It is common around Broome (WA) and Darwin (NT).

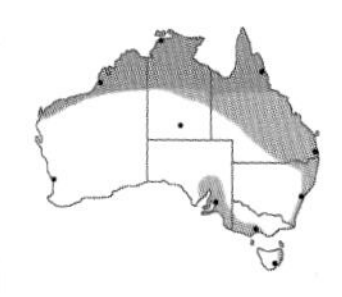

4 ORIENTAL PLOVER *Charadrius veredus* 23cm/9in

A delicate, long-legged, long-winged plover with a long bill and an upright stance. Its non-breeding plumage, most likely to be seen in Australia, is similar to that of the sand plovers: pale brown upperparts, white underparts, black bill, and yellow legs. It also shows an indistinct mottled-brown breast band. In flight, the underwing is uniform brown and the feet extend beyond the tail, an easily seen feature that distinguishes this bird from Grey Plover. The underwing is mostly white on the rather similar Caspian Plover (*C. asiaticus*), a rare vagrant. In breeding plumage, Oriental Plover's head, neck, and face are almost white, and the crown and ear coverts are brownish; the breast is chestnut-orange, bordered with black below, and the rest of the underparts are clean white. The breeding female is less colourful and lacks the black below. Oriental Plover is an uncommon Palaearctic migrant during the austral summer to the n. savannahs and grasslands of tropical Australia. It shows a preference for dried lake beds and remote gravel or grass airstrips; a very regular site is the Timber Creek Airport in the NT.

♂ br. 1
nbr. 1
1
br. 2
nbr. 2
nbr. 2
br. 3
nbr. 3
partial br. 3
br. 4
nbr. 4
nbr. 4

PLOVERS (CHARADRIIDAE)

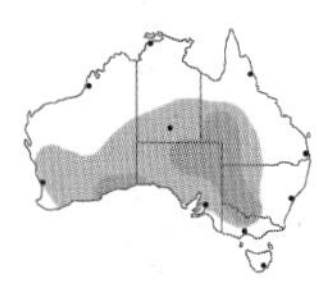

1 INLAND DOTTEREL *Peltohyas australis* 20–22cm/8–8.5in

A striking, nocturnal dotterel of the dry inland plains with distinctive plumage characterised by a unique, bold black Y on the chest, set against a rusty-orange background, best viewed head-on. It also shows a striking thick black vertical line from the top of the head down through the eye. Upperparts are dark buff streaked with dull brown. Non-breeding plumage is just a slightly paler version of breeding plumage. The bill is slight and short. In flight, the bird shows a clear white wing bar and white trailing edge to the inner wing. Inland Dotterel is a nocturnal shorebird of the desolate gibber plains of c. Australia, where it occurs at low densities, which makes it difficult to track down. It also occurs in some human-modified, treeless plains in sw. NSW, where it can sometimes be spotlighted at night during organised searches for Plains-wanderer.

2 RED-KNEED DOTTEREL *Erythrogonys cinctus* 17–19cm/6.5–7.5in

Like the larger Banded Lapwing, this dotterel has a greyish back and a black head, though this species lacks the white blaze behind the eye, which also differentiates it from Black-fronted Dotterel. Red-kneed Dotterel's most striking feature is the contrasting white throat, bordered by a black chest below; the black extends down the flank in a bold stripe that gradually fades into dark chestnut at the rear. It also has a blood-red bill with a black tip, in common with Hooded Dotterel and Black-fronted Dotterel. However, Hooded Dotterel has a complete black hood with a dark throat, which differentiates it from this species. In flight Red-kneed Dotterel also has a bold, broad white hind wing. This handsome little wader is most likely to be found in small numbers around the edges of dams, swamps, billabongs, and sewage ponds well inland from the coast, although like many of Australia's waterbirds it is nomadic and wanders widely in response to water conditions.

3 BLACK-FRONTED DOTTEREL *Elseyornis melanops* 16–18cm/6.25–7in

This tiny, boldly marked shorebird has a brown cap, a bold white line over the eye, a black line through the eye, a bold white throat bordered by a black Y-shaped chest band underneath, and all-white underparts. The upperparts are grey-brown with a distinct chestnut area on the shoulders. It also has a reddish bill with an obvious dark tip, in common with other birds on this page. Black-fronted Dotterel is distinguished by its brown cap (not black), bold white supercilium (lacking in the all black-headed Hooded Dotterel), broad chest band but no markings along the flank (unlike Red-kneed Dotterel), and bold chestnut patches on the shoulders. In flight it shows a white band across the centre of the wing. Immatures have a very washed-out version of the same pattern with the black areas of the adult appearing mid-brown. This dotterel is most often encountered in small groups around the edges of freshwater wetlands. It is found across much of Australia around small dams, large freshwater wetlands, and occasionally also in tidal areas.

4 RED-CAPPED PLOVER *Charadrius ruficapillus* 14–16cm/5.5–6.25in

A small, delicate plover. Males show a rufous head and nape, bordered on the upper forehead by a thick black border, which contrasts with the clean white forehead below. There is a bold, black loral line linking the eye with the bill on an otherwise bright white face. The underparts are clean white below but show bold black patches on the sides of the upper breast. The legs and bill are black. Non-breeding birds and females, less boldly coloured in the same overall pattern, lack the bright rufous head and nape and are generally dull grey-brown, sometimes with a light chestnut wash. In flight this plover shows a slim white wing bar and a dark rump and tail with white sides. Red-capped Plover is a resident species found nearly throughout Australia, on freshwater and saline lakes and in tidal areas. It is generally more common in the coastal zone, on sand or mudflats.

5 HOODED DOTTEREL *Thinornis cucullatus* 19–23cm/7.5–9in

This handsome greyish shorebird has a contrasting all-black hood bordered by a clean white collar on the nape and is bright white below. It has a reddish bill with a black tip. Hooded Dotterel is easily identified by its all-black head and throat; the other plovers and dotterels with large amounts of black on the head also have large sections of white on the throat and therefore do not have a hooded appearance. This is a very particular shorebird, found only locally in coastal areas of the extreme south of the e. mainland, in sub-coastal ephemeral lakes of sw. Australia, and on TAS. It is the most wary of the plovers and generally occurs only in relatively undisturbed areas, usually found in pairs along sandy beaches.

1
2
3
imm. 3
partial br. 4
5

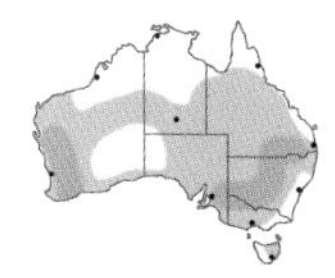

1 BANDED LAPWING *Vanellus tricolor* 25–29cm/10–11.5in

Lapwings are in the plover family, Charadriidae. A scarce, localised lapwing of inland grasslands and agricultural lands, Banded Lapwing has a greyish back and largely white underparts; it differs from Masked Lapwing in having more black in its plumage, including an almost wholly black face except for a white flash behind the eye and a broad black band across the chest, which lends the bird its name. In flight Banded Lapwing displays a bold white band across the upperwing, whereas Masked has plain upperwings. The Banded has a yellow bill and red lores but lacks any extravagant facial wattles. Most commonly found in the south of mainland Australia, Banded Lapwing is highly nomadic in response to water levels and therefore changing vegetation, so is quite unpredictable in its appearances. It is a scarce species that is not encountered frequently.

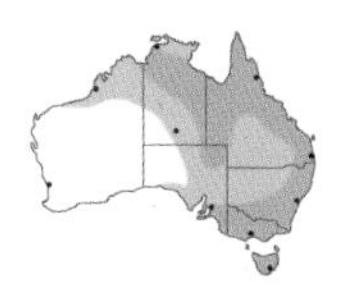

2 MASKED LAPWING *Vanellus miles* 30–37cm/12–14.5in

A common, easily recognised shorebird that is often tame and approachable where it occurs in urban areas. It is a large, distinctively patterned plover with a black cap, grey-brown back, gleaming white underparts, long reddish legs, and a yellow bill. However, its most noticeable feature is the bright canary-yellow facial wattle that hangs down from each side of the head. Northern birds have a more extensive wattle, and southern birds have a black collar. In flight this lapwing has plain upperwings lacking any bold markings. Masked Lapwing is found widely on the e. half of the continent, where it is commonly found in pairs on lawns in gardens, towns, and airports, making it a familiar sight to Australians.

3 COMB-CRESTED JACANA *Irediparra gallinacea* 20–27cm/8–10.5in

Comb-crested Jacana (family Jacanidae) is unlikely to be confused with any other bird due to its distinctive toes and bold colouration. Adults are black-backed and have a whitish underside, a black band across the chest, and a striking vermilion-red shield or comb on the forehead that glows in strong sunlight and is its most striking feature, besides its extremely long toes. Immatures are much more subdued, lacking the black breast band and the red comb of the adult. Jacanas are locally common on large lagoons and natural freshwater wetlands, where groups of them are usually found conspicuously walking on floating mats of vegetation, often well away from the margins of the wetland. They are especially striking in flight, when their long legs and toes trail well beyond the tail, creating a truly unique appearance. They are found in tropical n. Australia and down the e. side of the coast into n. NSW, and are especially common around wetlands in Kakadu NP (NT).

1
1
s. 2
n. 2
3
imm. 3

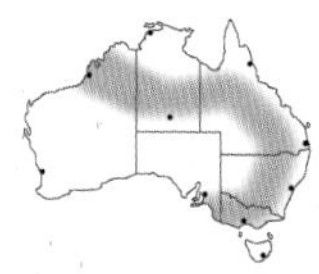

1 AUSTRALIAN PAINTED-SNIPE *Rostratula australis* 24cm/9.5in

A very strange-looking wetland bird. Unlike other snipes (which are in Scolopacidae), it is in the family Rostratulidae. Male and female have different plumages. Adult female, brighter than male, has a red-brown breast with a black border separating it from white underparts; green upperparts with a buff V down the back and a red-brown collar bordered white; and a black head with a yellow crown stripe and a white patch around and extending behind the eye. Smaller and less colourful, the adult male has olive-brown upperparts, white underparts, and white eye markings as in female, but has a white collar at the base of the neck and a white V down the back. In both sexes the bill is thick, long, and slightly decurved, and the legs are strong and pale green. Australian Painted-Snipe is found extremely sporadically across n. and e. Australia, though most common (occurring at least irregularly) in the south-east. It seems to prefer very shallow edges of well-vegetated wetlands. Closely related species in Asia have a preference for overgrown drainage channels, and this may also be the case with this species.

2 LATHAM'S SNIPE *Gallinago hardwickii* 29cm/11.5in

Also known as Japanese Snipe; a member of the family Scolopacidae. A cryptically coloured buff, brown, and white bird with a long, straight bill. It is almost identical to Swinhoe's Snipe, and very good views are needed to separate the two in the field, though even with great views identification often is just not possible. In Latham's Snipe the broad buff loral line and supercilium are wider than the blackish eye stripe. On its upperparts, there is an obvious V formed by the edge of the mantle and the lower border of the upper scapulars. The bird looks long-tailed on the ground, and the tail clearly projects beyond the closed wings. The basal half of the bill is greenish-brown, the lower bill blackish. The legs are dull green. In flight, the toes do not usually project beyond the tail, as they can do in Swinhoe's Snipe. Latham's has fewer (16–18) tail feathers (vs. 20–24 in Swinhoe's), but this is rarely observable in the field. Latham's Snipe is a common Palaearctic migrant in the austral summer to n. and especially e. Australia. It is more common in coastal and semi-coastal regions but is found through most of NSW and VIC, possibly as just a passage migrant. Its habitat choices range from reeds in extensive deep marshes to boggy dairy cow pastures and thick roadside ditches.

2

3 SWINHOE'S SNIPE *Gallinago megala* 28cm/11in

Extremely similar to Latham's Snipe (see above). Plumage differences are very subtle, and colouration of bare parts is similar. Swinhoe's has thickish, yellowish-green legs, and its bill is greyish green on the basal half. Latham's has fewer (16–18) tail feathers (vs. 20–24 in Swinhoe's), but this is rarely observable in the field. Swinhoe's is a rare Palaearctic migrant in the austral summer to the n. coast from Cape York Peninsula across to the tropical coastal regions of WA. Swinhoe's frequents drier (though still marshy) wetland edges than Latham's and sometimes ventures out onto exposed mudflats in semi-coastal locations.

♂ 1
♀ 1
2
2
3
3

1 BAR-TAILED GODWIT *Limosa lapponica* 41cm/16in

A large, long-legged, long-billed shorebird, rather similar to Black-tailed Godwit but very different in flight, showing uniform wings with no wing bar, a barred tail, and a white rump that extends onto the back. On the ground Bar-tailed Godwit appears smaller and more compact, with shorter legs; its bill is slightly upturned; and it shows a stronger supercilium. In breeding plumage, it is brick red on the entire underparts, with the colour more extensive than in Black-tailed. When feeding, Bar-tailed usually probes with its bill close to its feet, employing a distinctive head-down sewing-machine feeding action. It is less likely to call than Black-tailed Godwit, giving a harsh *kirrucc*. Bar-tailed Godwit is a very common Palaearctic migrant in the austral summer to most Australian coasts and inland as a passage migrant. A few can be found on most coastal mudflats in e. Australia, as well as sandy beaches, cays, and estuaries.

2 BLACK-TAILED GODWIT *Limosa limosa* 41cm/16in

A large, long-legged, long-billed shorebird. Very distinctive in flight (reminiscent of Pied Oystercatcher), showing a striking white wing bar, white rump and tail, and black terminal tail band. In non-breeding plumage it is grey-brown above, white below. In breeding plumage the upper breast, neck, and head are brick red, and the mantle is mottled brick red and black. The bill has a pink base; the legs are black. Black-tailed Godwit feeds by probing ahead with its long, straight bill. It often calls in flight, an urgent *wicka, wicka, wicka*. A common Palaearctic migrant in the austral summer to n. Australian coasts, it also occurs in lesser numbers inland as a passage migrant to the se. and sw. coasts. It is found mainly on coastal mudflats but also on sandy cays and inland shallow lakes.

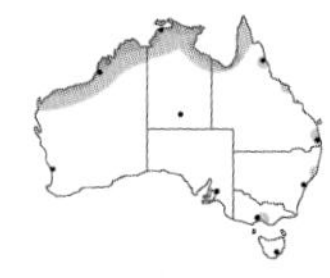

3 ASIAN DOWITCHER *Limnodromus semipalmatus* 36cm/14in

This rare shorebird is most likely to be confused with Bar-tailed Godwit but is smaller and has a straight, all-dark bill, compared to the slightly upturned, pink-based bill of the godwit. Given a good view, one can see that Asian Dowitcher's bill is swollen at the tip, very unusual among shorebirds. In flight, the dowitcher lacks extensive white on the back (shown by Bar-tailed Godwit), having pale colouration confined to the rump and uppertail, and the underwing is white. Asian Dowitcher often feeds in deep water up to its belly, unlike Bar-tailed Godwit, which prefers mudflats. A rare Palaearctic migrant in the austral summer to n. and especially nw. Australian coasts. It is most likely found on tidal mudflats, but in Asia it regularly visits rice fields and salt fields, so should also be expected in these habitats in Australia.

nbr. 1
nbr. 1
1
br. 2
nbr. 2
bottom bird 2
3
3
3

1 LITTLE CURLEW *Numenius minutus* 30cm/12in

This really is a very small curlew, about the size of a Pacific Golden Plover, so once its size is accurately judged, it is easily identified. It has a short, thin neck, a flat crown, and a very slightly decurved bill. The head is distinctive with a pale cream central stripe and dark brown lateral stripes; a long buff supercilium extends and flares out just behind the eye. In flight it shows a uniform brown rump. The bill is relatively short for a curlew, down-curved at tip, dark brown with a pinkish base, the pink more extensive on the lower mandible. The legs are dull bluish grey. Little Curlew is an uncommon summer migrant to n. Australia, though is recorded around much of Australia. Not limited to the coastal zone, it can be found in marshlands, bare areas, and short grasslands throughout the tropical savannah zones.

2 WHIMBREL *Numenius phaeopus* 43cm/17in

A medium-sized, short-billed curlew with a distinctive head pattern. Whimbrel looks dark and compact, with a flat crown, a flat back, and a shortish bill with a downward curve near the tip. It has a narrow, pale central crown stripe, most obvious when seen head-on, banded by dark lateral stripes; and a long, broad, prominent supercilium. The bill is blackish brown with an indistinct pale pinkish base. The legs are bluish grey. In flight it shows a white rump extending up the back in an inverted V. It is very vocal, giving a far-carrying and often repeated rippling titter, usually in flight. Whimbrel is a common Palaearctic migrant in the austral summer to the n. coasts, less common to the se. and sw. coasts. It tends to stick more closely to coastal zones than Little Curlew.

3 FAR EASTERN CURLEW *Numenius madagascariensis* 64cm/25in

A large shorebird with a staggeringly huge bill. It has the longest bill of all the shorebirds, and is easily identified by this feature, given a good view. The head and neck are warm buff-brown with a very faint supercilium, which gives the bird an open-faced, rather round-headed appearance in contrast to Whimbrel's dark, flat head with a strong supercilium. Far Eastern Curlew's extremely long bill, even longer in females, is strongly decurved and dark brown with a pale pink base to lower mandible. In flight, it shows a mottled brown rump concolourous with the wings and tail, in contrast to Whimbrel's white rump and back. Its legs are long and bluish grey, proportionately much longer than Whimbrel's. Far Eastern Curlew is a common Palaearctic migrant in the austral summer to n. and e. Australian coasts. It occurs on mudflats, beaches, salt marshes, and mangroves.

1
1
2
2
3
3

SANDPIPERS AND SNIPES (SCOLOPACIDAE)

1 GREY-TAILED TATTLER *Tringa brevipes* 27cm/10.5in

A medium-sized sandpiper with a stout, straight bill and yellowish legs. Extremely similar to Wandering Tattler. In breeding plumage, which is rarely seen in Australia, it shows a distinctive, long supercilium—the pair of which meet on the nape—bordered by a black loral line. The upper breast is barred grey; the underparts are white. In non-breeding plumage, when it is most like Wandering Tattler, it lacks barring and has a grey wash to the breast, neck, flanks, and white underparts. In flight it appears virtually uniform grey above, the rump concolourous with the tail, back, and wings, which lack wing bars. The strong-looking bill is straight and blackish. Legs are stout, shortish, and yellow. The diagnostic call is *tu-weeep*, rising on the second note. Grey-tailed Tattler seldom associates with other shorebirds. A common Palaearctic migrant in the austral summer to n. Australian coasts, it also occurs in lesser numbers inland as a passage migrant to the se. and sw. coasts. It is found mainly on beaches, spits, rocky reefs, and mudflats.

2 WANDERING TATTLER *Tringa incana* 28cm/11in

A medium-sized sandpiper with a stout, straight bill and yellow legs. Extremely similar to Grey-tailed Tattler in plumage, best separated by call. In breeding plumage, Wandering Tattler's underparts have heavier barring than those of Grey-tailed, the barring extending onto the undertail coverts, which are white on Grey-tailed. The bill is blackish, its base tinged yellow. Legs are stout, shortish, and yellow. Wander Tattler seldom associates with other shorebirds. It has a diagnostic flight call, made up of six to ten notes of rippling call, accelerating but falling away in volume towards end of call. A Holarctic migrant in the austral summer to the ne. and e. Australian coasts, Wandering Tattler is much less common than Grey-tailed and far more habitat specific, in Australia much preferring rocky habitats compared with Grey-tailed Tattlers.

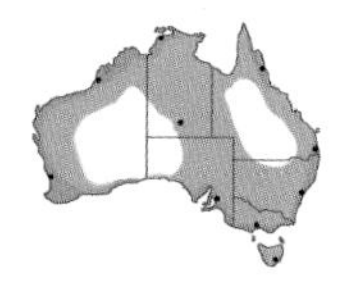

3 COMMON GREENSHANK *Tringa nebularia* 33cm/13in

A large, long-legged, long-billed 'shank' that moves quickly and elegantly through shallow water in much the same way yellowlegs (*T. flavipes* and *melanoleuca*) do in North America. In non-breeding plumage Common Greenshank is a strikingly pale and uniform shorebird, clean white below and pale grey above. In breeding plumage, it has a mid-grey back heavily mottled with dark grey; grey on the crown, face, and hind neck, all streaked with white; and black scalloping on a white breast. In flight, it always shows the white of the tail and rump—very distinctive against the plain greyish wings. The bill is long, stout, slightly upcurved, and grey with a blackish tip. Legs are long and dull green. Common Greenshank often feeds by chasing prey through water, showing an active and alert posture. It has a distinctive ringing *tu-tu-tu* call. It is a common Palaearctic migrant in the austral summer to most of Australia. It is mainly found on coasts, but can occur on any lake edge, shallow wetland, mudflat, or salt pan.

4 MARSH SANDPIPER *Tringa stagnatilis* 24cm/9.5in

A small and very delicate-looking shank, considerably smaller than the rather similarly plumaged Common Greenshank. Proportionately long-billed and long-legged, Marsh Sandpiper shows an active feeding style, often in water. Size is the key to its identification; once the small size has been correctly assessed, no other shorebird shows the combination of a fine needle-like bill, clean white underparts, greyish upperparts, and yellowish-green legs. In breeding plumage, the upperparts are mottled, the face is black, and the neck is speckled. In flight, it shows a white tail and rump, with the white extending onto the lower back. When feeding, Marsh Sandpiper often tips forward, holding the tail and rear end well above horizontal. It is an uncommon Palaearctic migrant in the austral summer to most of Australia. It is found mainly on coastal mudflats and marshes but occurs inland on shallow wetlands, mudflats, or salt pans.

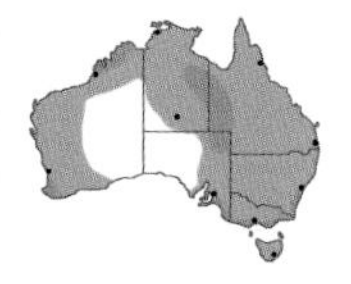

5 WOOD SANDPIPER *Tringa glareola* 20cm/8in

A small, compact sandpiper with a striking supercilium, thin white eye ring, yellow legs, and a stout, straight bill. In breeding plumage, the crown and hind neck are dark brown streaked with white, and the upperparts are dark brown with buff spotting. The underparts are white with heavy streaking on the breast and neck. In non-breeding plumage, the upperparts are darker with less spotting. Immature has bright spotting with golden and black markings. When the bird is perched, the wings extend just beyond the tail. In flight, it shows a square white rump and a barred tail. The bill is blackish with a yellowish-green base. Legs are dull yellow. Wood Sandpiper is a fairly common Palaearctic migrant in the austral summer to most of n. Australia, in lower numbers to s. Australia. It is found mainly on the edges of tropical wetlands, such as the Yellow Water in Kakadu NP, though also occurs on coastal mudflats and marshes.

nbr. 1
2
3
nbr. 3
4
4
5
nbr. 5

SANDPIPERS AND SNIPES (SCOLOPACIDAE)

1 RUDDY TURNSTONE *Arenaria interpres* 24cm/9.5in

A very distinctive, small, short-billed, short-legged shorebird that lives up to its name. Its piebald plumage and distinctive feeding behaviour—pushing through the tide line turning over weeds and small stones—make this a real character among shorebirds. In breeding plumage the upperparts are chestnut-orange and black, and the underparts clean white. The head and neck are a complex black and white pattern with a black mask through the eye and a black collar and upper breast surrounding the white centre of the breast. Non-breeding plumage is much duller, showing just hints of the chestnut-orange head and the brownish-black breast pattern. In flight, it shows a strong white wing bar, a white patch on the back, and white uppertail coverts, resembling a diminutive Pied Oystercatcher at quick glance. The bill is small, stout, and black. The legs are short, stocky, and orange. Ruddy Turnstone is a very common Holarctic migrant in the austral summer to the n. coasts and a common migrant to the s. coasts. It prefers beaches, coral cays, and rock spits, though also occurs on mudflats.

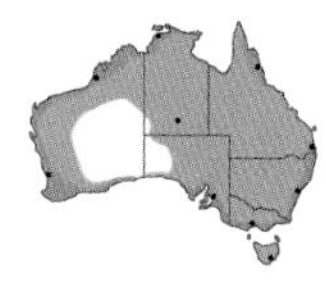

2 COMMON SANDPIPER *Actitis hypoleucos* 20cm/8in

A small brown-and-white wader found at the water's edge. It flies fast and low with distinctively bowed wings almost touching the water surface. Its plumage is olive brown above, clean white below; the breast is washed greyish, and a distinctive white point shows between the folded wing and the greyish breast, obvious even at long range. The head shows a narrow white eye ring and an indistinct whitish supercilium. There is very little difference in plumages; the white point on the side is consistent in all. In flight it shows a white wing bar and white edges to the tail. On the ground it has a horizontal carriage and is almost constantly bobbing the tail and rear end, even at rest. The bill is short, straight, and dark brown. The legs are dull greenish, occasionally yellowish. Common Sandpiper is a common Palaearctic migrant in the austral summer to the n., e., and sw. coasts, an uncommon migrant to the s. coast and TAS, and passes through much of interior Australia. It occurs on a very wide variety of wetland and coastline habitats, natural and man-made.

3 TEREK SANDPIPER *Xenus cinereus* 24cm/9.5in

The only small shorebird possessing a diagnostic upturned bill, blackish with a pale orange base. Its plumage is olive brown above and white below; perhaps the only confusion species is Common Sandpiper, but a clear view of the bill will distinguish the two. In breeding plumage, Terek Sandpiper has a black line on the scapulars. In flight it shows an eye-catching white trailing edge to the inner wing and a dark outer wing that contrasts with a pale panel at mid-wing. The legs are usually dull greenish yellow but in breeding plumage can be quite bright orange. Terek Sandpiper is a common Palaearctic migrant in the austral summer to the n. coast mudflats, an uncommon migrant to s. coast mudflats.

4 SANDERLING *Calidris alba* 22cm/8.5in

A small, compact shorebird with a straight black bill and black legs that often runs along the shoreline like a wind-up toy. Sanderling's breeding plumage resembles that of several other shorebirds (e.g., Red-necked Stint and some vagrant species), so great care should be taken. In breeding plumage Sanderling's head, neck, and upper breast are mottled rufous and grey, streaked with dark brown, and can look very rufous in fresh plumage—and note, the intensity and extent of the rufous is very variable. Upperparts are mottled rufous, black, and white. Underparts are clean white. It lacks the supercilium shown by many other small shorebirds. In non-breeding plumage Sanderling is strikingly white below; has pale grey upperparts and a pale grey wash on the sides of the upper breast; and its beady black eye stands out in a pale face with a grey-streaked cap. Sanderling is a common Holarctic migrant in the austral summer to all coasts. It prefers beaches, though also occurs on mudflats.

1
partial br. 1
2
2
3
3
br. 4
nbr. 4

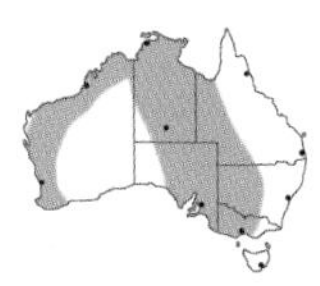

1 LONG-TOED STINT *Calidris subminuta* 14cm/5.5in

A tiny, long-legged wader whose main identifying features are its dull greenish-yellow legs, a black, straight, pointed bill, and a crouched feeding posture, which give it the overall impression of a very small Sharp-tailed Sandpiper. Once Long-toed Stint's tiny size and leg colour are seen, only extreme vagrants need to be eliminated. Immature bird shows a bright rufous cap with dark streaking, a prominent supercilium, and a very striking white V on the mantle. Adult non-breeding is dull grey-brown above; its structure and bare-part colouration are the best features: dull greenish-yellow legs (black in all other stints) and fine, black, pointed bill. Long-toed Stint is a rare Palaearctic migrant in the austral summer to freshwater wetlands; it is much more likely found in WA but is also possible through the mainland to the south-east.

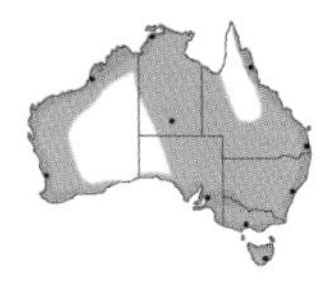

2 RED-NECKED STINT *Calidris ruficollis* 15cm/6in

A tiny, short-legged wader (a 'peep'), with black legs and a short black bill. The size and leg colour are distinctive in Australia. In breeding plumage, this stint has a bright rufous neck bordered by dark streaks; the rufous extends onto the head but this feature is variable. The back has rufous feathers with black centres. In non-breeding plumage it is grey-brown above and white below with an indistinct grey half-collar. In all adult plumages it shows a strong supercilium. Immature birds show a white V on the mantle, mottled brown-and-black upperparts, and clean white underparts. Red-necked Stint is a very common Palaearctic migrant in the austral summer to almost all of Australia, especially the north-west and south-east. It occurs on a wide variety of wetland and coastal habitats.

2 br.

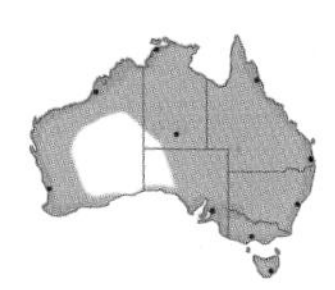

3 SHARP-TAILED SANDPIPER *Calidris acuminata* 19cm/7.5in

A medium-sized wader with pale legs and a prominent cap. Breeding plumage shows a rufous cap, streaked brown, and a white supercilium. Upperparts are buff with darker centres to the feathers; the breast is heavily marked with buff, extending down the flanks, and the border of breast is nowhere near as clearly defined as on a Pectoral Sandpiper. In the duller non-breeding plumage, the breast is grey-washed and has dark streaks. Immature birds have a bright rufous cap streaked blackish, a strong white supercilium, a warm buff breast with dark streaks, grading into a white belly, and on the upperparts the feathers are dark-centred with bright buff and white edges. In all plumages, the bill is blackish brown with a pale base to the lower mandible, and the legs are dull grey-green to yellowish. Sharp-tailed Sandpiper is a very common Palaearctic migrant in the austral summer to almost all of Australia, especially the south-west and south-east. It occurs on a wide variety of wetland and coastal habitats.

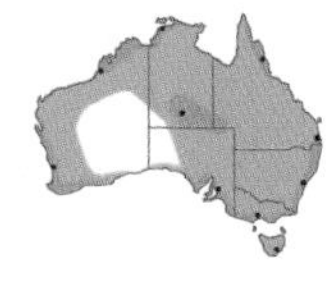

4 PECTORAL SANDPIPER *Calidris melanotos* 30cm/12in

Unlikely to occur in breeding plumage in Australia. In non-breeding plumage, this bird has a chestnut cap streaked with black, a brown back with black streaking, a very heavily streaked breast, and a distinctly demarcated pectoral band. On the upperparts, the feathers are blackish, edged in warm buff. The transition between the breast streaking and the whitish underparts is much more defined than on Sharp-tailed Sandpiper. Immature is a bright version of adult, with a bright white V on the mantle, distinctive breast markings, and bright yellow legs. The bird's bill is shortish, decurved, and dark brown with a dull yellowish base. Pectoral Sandpiper is a very uncommon Holarctic migrant in the austral summer to nearly all of Australia. It occurs on coastal mudflats but has a preference for freshwater marshes, mudflats, and wet grasslands.

1
1
nbr. 2
2
nbr. 3
3
4
4

1 RED KNOT *Calidris canutus* 25cm/10in

A beautiful shorebird in full breeding plumage—brick red below and black and rufous above—which is most often seen in n. Australia in September and March during migration. Non-breeding adult is a plain grey bird, darker grey above with blackish primaries, and paler grey below with faint and indistinct grey-brown arrow marks on the breast. The structure is distinctive: It is a pot-bellied, rounded bird with a straight, strong black bill and rather short black legs. Red Knot forms huge flocks. It is a common Holarctic migrant in the austral summer to the n., w., and e. Australian coasts and also occurs in lesser numbers inland as a passage migrant to the se. and sw. coasts.

2 GREAT KNOT *Calidris tenuirostris* 27cm/10.5in

In breeding plumage Great Knot has bold thick black markings from the throat downward and dark upperparts with a distinctive contrasting chestnut patch on the shoulder. In non-breeding plumage it is an indistinct grey bird with strong arrow marks down the flanks, bolder and more extensive than those of the similar and closely related Red Knot. The bills of the two species are similar in shape, but Great has a longer bill that is often slightly decurved, as well as a larger body size, both useful ID features. In flight it has a bright white rump, unlike the duller barred rump of Red Knot. Great Knot is a non-breeding migrant to Australia between September and March, when it is common along coastlines. Large flocks of knots tend to favour tidal flats and estuaries and are most likely to be encountered within big mixed-species flocks of shorebirds.

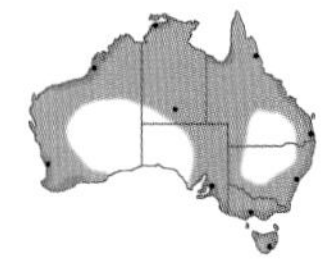

3 CURLEW SANDPIPER *Calidris ferruginea* 20cm/8in

This sandpiper's long legs, long decurved bill, square white rump, and attenuated rear end are consistent features in all plumages. The breeding plumage is brick red with the upperparts heavily marked black and the face showing a white eye ring and a whitish mark around the base of the bill. The non-breeding bird is grey above, white below, and shows a strong white supercilium. Immature is warm buff on upper breast, white below, and has a strong supercilium. Curlew Sandpiper is a common but decreasing austral summer migrant to mainly coastal Australia. It occurs on a very wide variety of wetland and coastline habitats, natural and man-made.

4 BROAD-BILLED SANDPIPER *Limicola falcinellus* 18cm/7in

A small, stint-like shorebird with a striking head pattern and a very distinctive long, tube-like bill with a deep base and a decurved tip. In breeding plumage it has a very obvious striped crown with a forked supercilium, a back black bordered in warm buff, heavy streaking on the breast and extending down the flanks, and a white belly. The non-breeding bird wears a faded version of the breeding plumage with whiter underparts, a more open head pattern, and broader white stripes over the crown. Immature is similar to non-breeding adult; its underparts are lightly streaked. The legs are short and black. Broad-billed Sandpiper is a rare Palaearctic migrant in the austral summer to coastal mudflats around n. Australia.

5 RED-NECKED PHALAROPE *Phalaropus lobatus* 19cm/7.5in

A small, delicate shorebird with a needle-fine bill usually found swimming on water. It will spin in circles and is often confiding and allows close approach. In breeding plumage the more colourful female has long orange-red patches running down both sides of the neck and meeting on the breast. The head is dark blackish grey with a striking white chin and a white spot above eye. Beautiful gold stripes mark the back; the flanks are mottled, and the belly is white. Male is a dull version of female, with much less orange-red on the neck. Non-breeding birds are striking black and white, with all-white underparts, a white face with a black patch extending behind the eye and down onto the upper neck, a black cap, and a white forehead. The upperparts are black with dull buff, almost whitish stripes. The bill is needle-like and jet black. The legs are dark grey, and the toes are lobed. Red-necked Phalarope is a regular Holarctic migrant to the nw. Australian coast. It spends much of its time at sea but also occurs in saline lakes, estuaries, and bays.

nbr. 1
br. 1
nbr. 2
br. 2
br. 3
nbr. 3
nbr. 4
nbr. 5

BUTTONQUAIL (TURNICIDAE)

1 LITTLE BUTTONQUAIL *Turnix velox* 13–16cm/5–6.25in

The smallest buttonquail, it is also probably the easiest to identify in flight. Buttonquail are shy birds usually found in grassy habitats, with specific habitat preferences differing slightly among the species. Females are larger and generally have brighter and more distinct plumages than males. The female Little Buttonquail is bright rufous or cinnamon above with pale streaking, and has a white belly and flanks. The male is plainer brown. In flying birds look for the white flanks, which are usually obvious, combined with the small size and rufous colouration. This bird is very widespread and irruptive throughout the inland, sometimes turning up in large numbers. It can be found in a variety of habitats but shows a general preference for grassy areas.

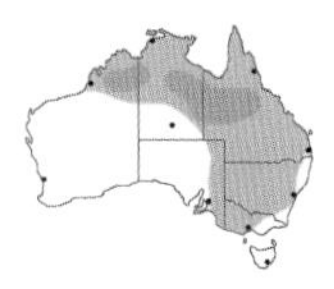

2 RED-CHESTED BUTTONQUAIL *Turnix pyrrhothorax* 13–16cm/5–6.25in

This buttonquail is perhaps the most difficult to identify in flight, being relatively plain with no very obvious field marks. It is small and has fairly plain grey-brown upperparts finely patterned with pale streaks and bars. The female has rich orange-brown underparts, which are diagnostic. The male is duller, with paler underparts. Red-chested Buttonquail is most likely to be confused with Little Buttonquail; look for the red flanks and less bright upperparts in Red-chested. Like Little Buttonquail, it is found in a variety of habitats, usually with a grassy understorey.

3 RED-BACKED BUTTONQUAIL *Turnix maculosus* 13–16cm/5–6.25in

Habitat is a good indicator when trying to identify this species; unlike other buttonquail it prefers damp grassy areas, often close to water. The female is quite distinctive, showing dark chestnut shoulders and breast and dark grey-brown wings. The male lacks the dark chestnut shoulders and breast, showing just a faint rufous collar. Both have a fine yellowish bill that is quite different from the bills of all other buttonquail. The chestnut back of the female can be seen in flight, and both sexes have a buff panel in the wing that contrasts with the dark back. The species is found in n. and e. Australia, usually close to the coast.

♂ 1
♀ 1
♂ 2
♀ 2
♀ 3
♂ 3

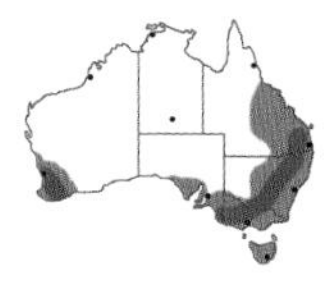

1 PAINTED BUTTONQUAIL *Turnix varius* 18–20cm/7–8in

This species occurs in open woodlands, often with a complex but open understorey of grass tussocks and leaf litter. It generally avoids densely grassed areas. It is often found on stony ridges and can also be found in mallee or more open woodlands. The female is a large buttonquail with a chestnut-coloured back intricately patterned with black bars and white streaks. Her breast and face are grey with white streaks and spots. The male is similarly patterned but duller. This is the buttonquail most likely to be encountered in woodlands in se. Australia, where it is fairly common. It is also found in far sw. WA.

1 ♀

2 CHESTNUT-BACKED BUTTONQUAIL *Turnix castanotus* 18–23cm/7–9in

This species replaces Painted Buttonquail in the tropical woodlands of the n. NT to n. WA. Its plumage is similar to but slightly duller than that of Painted Buttonquail. It is usually found on stony hillsides and ridges or around the tops of escarpments where there is a sparse cover of grass. The hills around Copperfield Dam near Pine Creek in the NT can be a good place to find it.

3 BUFF-BREASTED BUTTONQUAIL *Turnix olivii* 18–23cm/7–9in

This truly enigmatic bird is very rarely seen, and there are no known photographs of wild birds. It is a large buttonquail, and those who have seen it say the large size of the female is obvious. It has pale cinnamon-buff upperparts that are weakly patterned and a plain buff breast. If seen on the ground, the large pale bill may be noted. It occurs in very sparse grass at the base of stony slopes and even in this open habitat is very difficult to see and very difficult to flush. Once flushed it is rarely refound. Most recent sightings have come from the Lake Mitchell area of the Atherton Tableland (QLD).

4 BLACK-BREASTED BUTTONQUAIL *Turnix melanogaster* 18–20cm/7–8in

The first clue to this species' presence, and indeed several other buttonquail, is the 'platelets' it leaves when foraging. The birds dig in the leaf litter while turning in a circle, leaving a characteristic circular depression. The female of this species is stunning, with a black head and breast covered in white spots. The wings are brown and intricately patterned with black bars and white streaks. The male has a paler head and face. Within its restricted distribution the bird is found in dry rainforests or vine scrub and also lantana thickets. It is fairly easy to see at Inskip Point in se. QLD.

♀ 1
♂ 1
♀ 2
♀ 2
♀ 4

PRATINCOLES; PLAINS-WANDERER

1 AUSTRALIAN PRATINCOLE *Stiltia isabella* 23–24cm/9–9.5in

Pratincoles are in the family Glareolidae. A medium-sized, lanky wader with sandy-coloured plumage and very long wings. It is often seen scurrying around on open plains or flying gracefully on its long, pointed, black-tipped wings, looking almost like a tern. In summer, when the bird is in breeding plumage, it has a beautiful red bill and chestnut flanks, while in non-breeding plumage it is duller. Although generally found across n. Australia in the winter and migrating to s. inland Australia in summer, the bird can be quite unpredictable in its movements. If conditions are good, inland birds will congregate in certain areas to breed, but they may not appear in those places the following year. In winter it can often be seen on the floodplains in Kakadu NP (NT), while in summer it is often found on the Hay Plains (NSW).

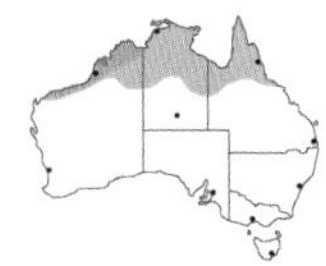

2 ORIENTAL PRATINCOLE *Glareola maldivarum* 23–24cm/9–9.5in

Another member of the Glareolidae, this bird appears each summer on the plains of nw. Australia, often in huge flocks of thousands and on one occasion, in 2004, some 2.5 million birds. Although it is seen in large numbers only in the north-west of Australia, for example around Broome (WA), birds do wander across the country and occasionally turn up much farther afield. The species is a possibility almost anywhere in Australia during summer. Oriental Pratincole is similar in shape and habits to Australian Pratincole, and has very long, pointed wings. It is generally brown above and has a white belly and rufous underwings. It has a pale throat encircled by a black 'necklace', which is pronounced in breeding plumage but less distinct in the more commonly seen non-breeding plumage.

3 PLAINS-WANDERER *Pedionomus torquatus* 15–18cm/6–7in

Plains-wanderer, the only member of the family Pedionomidae, is very small. The male in particular is just the size of a small quail, while the female is only slightly larger. The male is pale brown with a fine pattern of black scallops and crescents on the upperparts and head. The female is similar but has a complete black collar covered in white spots and below this a rufous crescent on the upper breast. Both sexes have a yellow bill and longish yellow legs. Plains-wanderer is very rarely seen during the day, tending to freeze when disturbed, which makes it very difficult to find. At night it is less wary, and birders can see it by spotlighting. On the rare occasions it is flushed, it flies feebly with its legs dangling, like a small rail. The species is very difficult to see independently, as most known populations occur on private property around Hay (NSW) or in national parks such as Terrick Terrick (VIC) and Oolambeyan (NSW).

3 ♀

br. 1
1
br. 2
2
♀ 3
♂ 3

SKUAS (STERCORARIIDAE)

1 BROWN SKUA *Stercorarius antarcticus* 64–66cm/25–26in

This bird breeds mostly on sub-Antarctic islands and is a regular winter visitor to waters off s. Australia. It is rarely seen close to the coast, preferring to remain far offshore. First impressions are always of a large, stocky, and powerfully built bird with broad, pointed wings, a short tail, and a large stout bill. Adults are dark brown with some fine pale streaking and in flight have obvious pale patches in the outer wing. Young birds are similar but more uniform brown without streaking. Brown Skua is most likely to be seen on pelagic birding trips off s. Australia.

2 SOUTH POLAR SKUA *Stercorarius maccormicki* 55cm/22in

Also known as South Polar Jaeger. Breeding mostly along the coastline of Antarctica, this species migrates north during winter and is rarely seen off the Australian coast. Like Brown Skua, it is seldom close to shore and is most likely to be seen from pelagic birding trips far offshore. It is also large and powerfully built but is always slightly smaller and less bulky than Brown Skua. It has two morphs. The pale morph, most likely to occur in Australia, has a pale buff head and body, quite different from Brown Skua. The dark morph is similar to Brown Skua but usually has a pale buff collar.

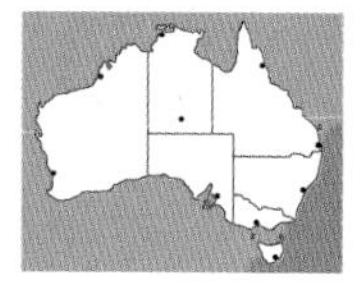

3 POMARINE SKUA *Stercorarius pomarinus* 47–51cm/18.5–20in

Also known as Pomarine Jaeger. This species is the largest and most powerful of the 'jaegers' found off Australian coasts during the summer. It is most likely to be confused with Parasitic Jaeger but is larger, has broader wings and a bulkier bill, and to a practiced eye appears more 'barrel-chested' than Parasitic. It has a powerful, steady flight and is most likely to be seen far offshore over the edge of the continental shelf. Dark morph birds are generally dark brown all over and have a dark cap, while pale morph birds have dark wings and a dark cap but a pale breast. In breeding plumage the birds have two tail streamers, which some birds may still have when they arrive in Australian waters in early summer. These are broad and twisted in Pomarine, fine and pointed in Parasitic. Immature birds can be confusing; they are dark all over but covered in fine pale bars.

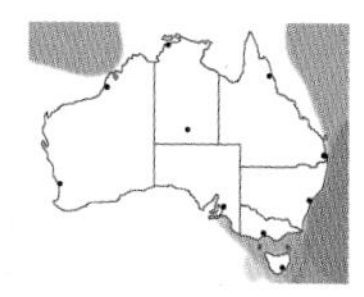

4 LONG-TAILED JAEGER *Stercorarius longicaudus* 41–47cm/16–18.5in

The smallest and most slender of the 'jaegers', this species is the easiest to identify but is also the rarest. It is much slimmer and less powerful than Pomarine Skua and Parasitic Jaeger, the other two species that occur off Australia in summer, and is a graceful flyer. It can also be separated on plumage alone. In both dark and light morphs, Long-tailed has greyer wings and lacks pale patches in the outer wings, which both Pomarine Skua and Parasitic Jaeger have. Like Pomarine it is likely to be seen far offshore in pelagic waters, although it tends to be more restricted in distribution and is most often seen in waters off se. Australia.

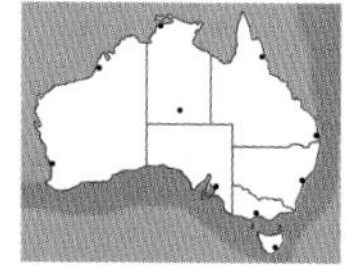

5 PARASITIC JAEGER *Stercorarius parasiticus* 41–47cm/16–18.5in

In all plumages this bird is very similar to Pomarine Skua, and body shape is the best clue to identification. Parasitic is smaller and more slender than Pomarine Skua, and has narrower wings. Its flight is more graceful and has been described as 'falcon-like', while Pomarine's is steadier, recalling a Pacific Gull, for example. Parasitic Jaeger is more likely to be found inshore than Pomarine Skua, often occurring close to the coastline. It can be seen following fishing trawlers or harassing other seabirds such as Silver Gulls for food.

1
pale morph 2
pale morph 3
4
light morph 5
dark morph 5

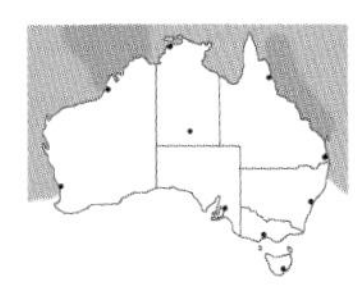

1 BROWN NODDY *Anous stolidus* 39–44cm/15.5–17.5in

Brown Noddy is one of the more distinctive terns, all chocolate brown, including the underparts, except for an off-white cap. The main confusion species in its range is Black Noddy, which is smaller and sooty coloured and has a more distinct, cleaner cap. Brown Noddy occurs across tropical Australian waters from s. WA around to Brisbane (QLD). It breeds on islands off QLD and WA, but it may be seen anywhere between these points, as it is a widely wandering oceanic bird. It is most likely to be seen from a boat trip to the sandy cays of the Great Barrier Reef.

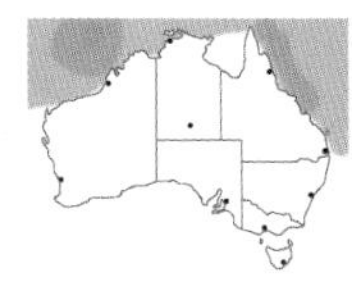

2 BLACK NODDY *Anous minutus* 34–37cm/13.5–14.5in

A sooty-black tern with a prominent, well-delineated white cap. It is smaller than Brown Noddy, and the two are best told apart by size when they are sitting together. Widespread in ne. waters, though less common than Brown Noddy, Black Noddy also has a patchy distribution in WA waters. It is most likely to be seen from a boat trip to the sandy cays of the Great Barrier Reef.

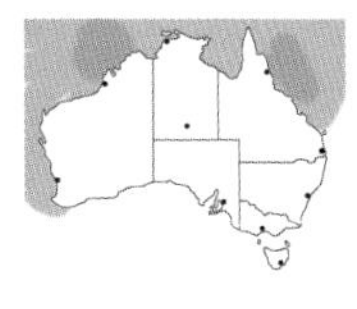

3 BRIDLED TERN *Onychoprion anaethetus* 30–32cm/12–12.5in

Bridled Tern is greyish chocolate brown (not jet black) above and white below, with a black cap and nape. It is told from the very similar Sooty Tern by the white line running from the base of the bill, over the eye, and beyond the eye. In Sooty this eye line is reduced, not extending behind the eye. Bridled Tern is found in the tropical seas of n. Australia, from sw. WA eastward around the coast, rarely to n. NSW. A pelagic species, it is rarely seen from the mainland. It breeds on offshore islands in n. Australia, and is regularly seen from boat trips to Michaelmas Cay.

3

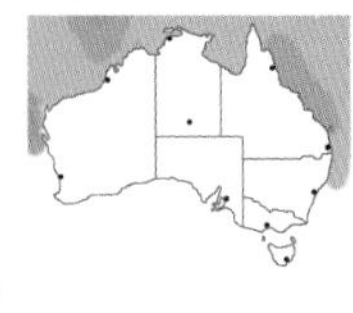

4 SOOTY TERN *Onychoprion fuscatus* 33–36cm/13–14in

Jet black above and white below, Sooty Tern is very similar to Bridled Tern. Sooty differs in having a black back, not greyish chocolate brown, and a shorter white line on the face that runs from the bill over the eye but does not extend past the eye, as it does in Bridled. In island colonies, juvenile Sooty Terns and Brown Noddies could be confused, as young Sooties are all dark brown but show bold speckling all over the upperparts, while juvenile noddies are all uniform brown. Large colonies of Sooty Terns nest on offshore cays and islands off n. QLD and n. WA; they are most easily seen from cruises out to Michaelmas Cay in the Great Barrier Reef.

1
1
2
2
3
3
4
4

GULLS AND TERNS (LARIDAE)

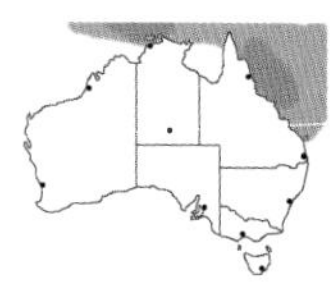

1 BLACK-NAPED TERN *Sterna sumatrana* 30–32cm/12–12.5in

Black-naped is a small and dainty, crisply marked tern of the north and north-east. Its most conspicuous features are the dazzlingly white overall appearance and its head pattern: all white on the crown and with a bold black mask running behind the eyes and around the nape. Otherwise it is silvery on the upperparts and gleaming white on the underside. It has a long and deeply forked tail, unlike Little Tern, which is also significantly smaller. This coastal tern is patchily distributed from n. NT eastward to n. QLD and south down the coast to s. QLD. As it is generally not close inshore, it is best seen from boat trips out to sandy cays on the Great Barrier Reef.

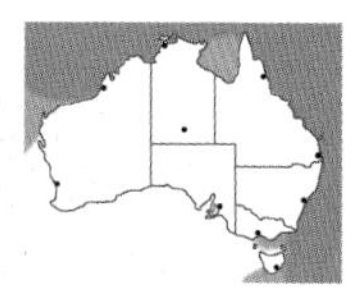

2 COMMON TERN *Sterna hirundo* 33–38cm/13–15in

There are several smaller pale terns that occur in Australian waters, and separating them can be a challenge even for experienced birders. It can be frustrating, but don't be put off if at first they all look the same. Common Terns breed in the Northern Hemisphere before migrating southward for the austral summer. Some birds stay around Australia over winter, but most are present from about October to April. Common Tern is fairly small and usually seen in non-breeding plumage, with a grey back and wings and a white body. It has a mottled black cap, black bill, and black or dull red legs. When resting on the ground it usually shows a dark bar on the shoulder, and the tip of the tail just reaches the wing tips. Similar terns have longer tails that go past the wing tips. In flight it has dark outer tips to the wings and a dark trailing edge on the underwing; other similar terns have paler wing tips, so this is a good feature to become familiar with. Common Tern can be seen around the coast in summer, particularly in e. and se. Australia.

3 ROSEATE TERN *Sterna dougallii* 34–38cm/13.5–15in

This bird is very similar to Common Tern but is smaller and slimmer and has a longer, finer bill and paler grey upperparts. It also has a longer tail and comparatively paler wing tips than Common Tern. Roseate Tern breeds on many offshore islands, from near Perth (WA) around the n. coast to s. QLD, so it may be seen in breeding plumage, when it has a solid black cap, a bright red bill, and often a pale pink wash to the underparts, hence the name. It is found mainly across n. Australia and tends to be found well offshore or on small offshore islands; it is rarely seen from the mainland. There are usually a few Roseate Terns on Michaelmas Cay off Cairns (QLD).

4 WHITE-FRONTED TERN *Sterna striata* 41cm/16in

This tern breeds mainly in New Zealand during the austral summer, though a few birds breed off ne. TAS. In winter, some birds visit the coasts of se. Australia, so the species is usually seen only between March and September (unlike Common Tern, which is a summer visitor). White-fronted is larger than Common Tern, with a longer tail and broader wings. It also has a steadier flight; Common Tern tends to bob around more when flying. If seen in breeding plumage, White-fronted Tern has a white band between the bill and the dark cap, while the dark cap of Common Tern in breeding plumage touches the bill.

1
1
2
2
3
3
imm. br. 4
4

GULLS AND TERNS (LARIDAE)

1 FAIRY TERN *Sternula nereis* 20–25cm/8–10in

Rather similar to Little Tern, Fairy Tern is also grey above and white below and has a black cap. It lacks the black lores and bill tip shown by Little Tern in breeding plumage, and in non-breeding plumage Fairy retains yellow on the base of bill, which is otherwise dusky. A mainly s. coastal tern, from e. VIC around to Perth (WA), it also occurs in lesser numbers north to the Kimberley (WA) on the w. coast.

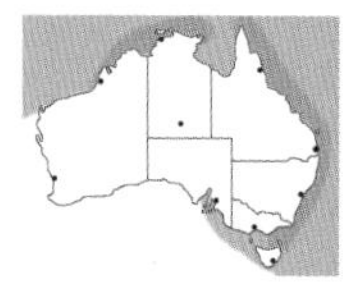

2 LITTLE TERN *Sternula albifrons* 20–25cm/8–10in

Little is tiny, the smallest tern in the world. It is mid-grey above, white below, and has a shallowly forked tail. In breeding plumage it has a yellow bill, tipped black, a black cap and large white forehead, and a bold black line connecting the bill to the eye. It is most likely to be mistaken for the similarly patterned Fairy Tern in breeding plumage, when Fairy lacks the black bill tip and never has a complete dark bar linking the bill and the eye. In non-breeding Little Tern develops a white crown, loses the black lores, and has a black bill, whereas Fairy retains some yellow in the bill and more black on the crown. Little Tern is fairly common along n. and e. coastlines but is absent from the coasts of s. and w. WA, w. SA, and w. TAS.

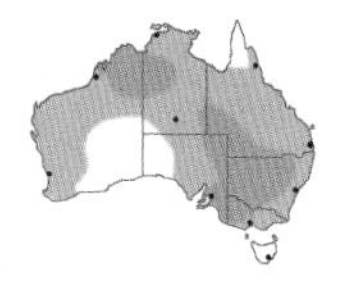

3 WHISKERED TERN *Chlidonias hybrida* 24–25cm/9.5–10in

A small, dark tern of marshes. In breeding plumage it has a blackish breast and underbelly and bold white cheeks, as well as a dull deep red bill, a solid black cap, and silvery-grey upperparts. In non-breeding plumage, the bill is dark, and the cap is incomplete, with white on the forehead, and the underparts are white, all making it less distinctive. The short tail has a very shallow fork. This is the only tern likely to be found in large groups hawking insects over marshes, as most others terns are fish-eaters. It is found widely across the continent, absent only from the sandy deserts of c. Australia, in various freshwater wetlands, sewage ponds, well-irrigated fields, and also sometimes coastal estuaries and brackish lagoons.

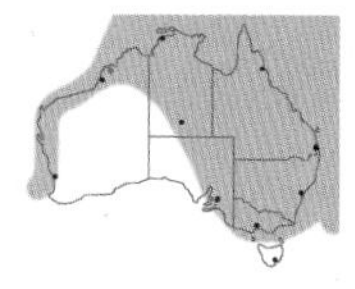

4 WHITE-WINGED TERN *Chlidonias leucopterus* 23–24cm/9–9.5in

Also known as White-winged Black Tern. A rare bird, very much like Whiskered Tern. In breeding plumage it has a jet-black hood, nape, throat, and breast; the lower back is slate grey and the belly is white. The striking feature is the white upperwing, visible when the bird is perched and in flight. Most birds seen in Australia are in non-breeding plumage, which is very similar to that of Whiskered Tern. Non-breeding White-winged Tern has all-white underparts, a black crown contiguous to a black ear patch, and upperwings tipped black. Whiskered lacks the black wing tips and black ear patch. White-winged Tern is a very uncommon summer migrant to the w., n., and e. coasts of Australia. Not limited to the coastline, this bird may turn up on freshwater bodies such as lakes or sewage ponds.

1
imm. 1
nbr. 2
br. 2
br. 3
nbr. 3
nbr. 4
br. 4

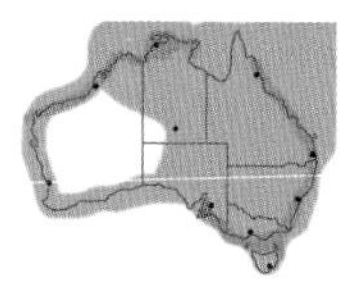

1 CASPIAN TERN *Hydroprogne caspia* 47–53cm/18.5–21in

Caspian Tern sports a massive carrot-coloured bill with a dark smudge at the tip. In body and bill size it is much larger than any other tern on earth. Its bill is notably heavy and thick-based, quite unlike the longer, more slender bill structure of the crested terns. Caspian Terns have a square-shaped head with just a short, stunted crest evident at the nape, unlike the shaggy crest displayed by the crested terns. In flight, the feathers on the underside of the outer wing (the primaries) are washed with black, which is quite contrasting and conspicuous relative to other species. Caspian is common and widespread on the coasts all around mainland Australia and TAS, and although it is mainly a coastal species, it does turn up inland, mainly in the e. half of the continent, where it can frequent freshwater lakes.

2 GREATER CRESTED TERN *Thalasseus bergii* 39–50cm/15.5–19.5in

Australia's second-largest tern, Greater Crested has a slender, long, uniform yellow (not orange) bill that aids in identification. This species never shows a dark smudge near the bill tip, as Caspian Tern does. Both crested terns have a shaggy black crest extending from the nape, which gives them quite a different head shape from the block-headed Caspian. The crested terns have a solid black cap in breeding plumage and have an incomplete black cap with a prominent white forehead in non-breeding. The extent of the white on the forehead in non-breeding birds is usually less in Greater (relative to Lesser), and Greater possesses more black on the head in this plumage. Crested terns are best separated by relative size (Greater is larger) and bill colour (yellow in Greater and orange in Lesser). Greater Crested Tern is a common, strictly coastal bird throughout the coasts and pelagic areas of Australia, often encountered fishing just offshore.

3 LESSER CRESTED TERN *Thalasseus bengalensis* 38–43cm/15–17in

Lesser Crested Tern has a long, slim orange bill with no dark markings. It is notably smaller than the huge Caspian and also smaller than Greater Crested. In all plumages the bill colour and relative body size are the best ID features. In non-breeding plumage Lesser sports a more extensive white forehead with less black on the head relative to Greater Crested. Like Greater Crested this species also has a conspicuous shaggy black crest that gives it a very different head shape from Caspian. In n. Australia, Lesser Crested Tern can frequently be seen close inshore, while on the e. coastline it tends to be more pelagic, not often seen unless one takes a boat trip out into deeper waters. It occurs in the tropical seas of n. Australia around to s. QLD.

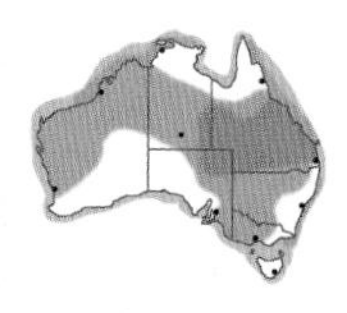

4 GULL-BILLED TERN *Gelochelidon nilotica* 38–43cm/15–17in

Gull-billed Tern has a black bill, stubbier than that of most terns, pale grey upperwings with fine black tips, and a white rump and tail. In breeding plumage it has a black cap and nape. In non-breeding plumage the black is reduced to an ear patch. Its broad-based wings and short tail give it a less graceful appearance in flight than most terns. Widespread and fairly common over most of continental Australia, it can occur on both saline and fresh waters and is present year-round in most large tropical wetlands.

1
1
2
imm. 2
3
4
nbr. 4

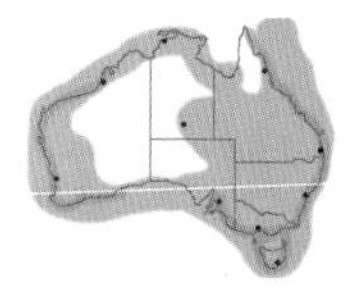

1 SILVER GULL *Chroicocephalus novaehollandiae* 43–44cm/17–17.5in

The smallest and most abundant of the three regular gulls in Australia, Silver Gull is distinguished from the others by its much smaller size; slim, bright red bill; and paler, silvery-grey upperparts. The larger Pacific and Kelp Gulls both display blackish-grey upperparts when adults and exhibit heavy, largely yellow bills with red only near the tip. Silver Gull is not confined to coasts, occurring in a variety of habitats, including tidal flats, city dumps, and lakes, throughout mainland Australia and TAS, and is the only gull found in inland Australia. It is a familiar bird to Australians and like many gulls is highly gregarious, most often seen in flocks; some birds are regular scavengers in urban areas, frequently squabbling noisily for food scraps.

1

2 PACIFIC GULL *Larus pacificus* 50–67cm/19.5–26.5in

The largest of the Australian gulls, Pacific is a real brute—large-bodied with a thickset, heavy bill—and is often identified by its bill shape and body size alone. It has a truly massive bill with a very pronounced bump (the gonys) on the underside, compared with the bill of Kelp Gull. When Pacific Gull is in breeding plumage its bill is deep orange-yellow with a large amount of red at the tip on both the upper and lower mandibles. Pacific has a dark tail band at all ages, unlike adult Kelp, which has a pure white tail. Immature Kelp and Pacific Gulls are confusingly patterned, with variable amounts of brown, and the structure of the bill and overall body size are the best clues to identification. Pacific Gull is confined to coasts on the s. mainland and TAS. A coastal species, it is likely to be seen in small flocks, either perched on rocks along the shore or on the tidal flats, or even resting on boats around docks.

3 KELP GULL *Larus dominicanus* 50–62cm/19.5–24.5in

A large, dark-backed gull. Although it is similar to Pacific Gull, Kelp Gull is smaller in body size and has a much slimmer bill with a less pronounced bump on the underside. In adults its bill is a paler yellow than that of Pacific, with noticeably less red and that confined to the lower mandible, rather than on both sections as in Pacific. The adult bird also lacks the obvious dark tail band of Pacific, making it readily identifiable in flight at this age. Immatures are an ID challenge, as in both species they are essentially variably patterned brown birds; they are then best told apart by the overall size and the structure of the bill, as outlined above. Kelp Gull is a coastal species, rarely straying far from shorelines of the s. mainland and TAS. It often occupies the same harbours and coasts as Pacific Gull; when the two can be compared side by side they can appear structurally very different. Kelp Gull is often encountered in small groups along s. shorelines.

imm. 1
1
2
2
3
imm. 3

1 PALM COCKATOO *Probosciger aterrimus* 61–64cm/24–25in

A massive black cockatoo with a conspicuous red facial-skin patch and an almost ridiculously long, shaggy crest. The head shape, facial pattern, and lack of colouring other than black on the body or tail make this bird very easy to identify. It is a fairly common and noisy cockatoo of n. Cape York Peninsula (QLD). It prefers rainforests and rainforest edges but also occurs in vine forests, heavy riverine environments, mangroves, and adjacent tropical savannahs dominated by *Eucalyptus tetrodonta*.

2 YELLOW-TAILED BLACK COCKATOO *Calyptorhynchus funereus* 56–65cm/22–25.5in

A huge black cockatoo with yellow ear patches, yellow tail panels, and a laboured, almost pterodactyl-like flight. In the s. part of its range it does not overlap with any similar species, and it is the only black cockatoo with yellow markings on the tail. This cockatoo occurs in pines, wet and dry sclerophyll forests, temperate rainforests, heathlands, scrublands, and farmlands. It is usually noticed when groups passing over in flight draw attention with their loud squeals. It is a bird of the coastal zone from s. QLD south through e. NSW into VIC and TAS (it is the only black cockatoo in TAS). It is quite common in Royal NP south of Sydney (NSW).

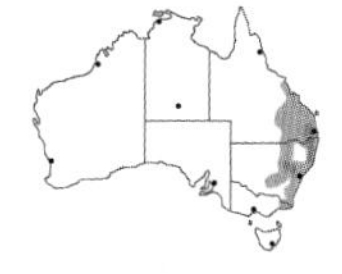

3 GLOSSY BLACK COCKATOO *Calyptorhynchus lathami* 47–51cm/18.5–20in

A relatively small, dirty-black cockatoo with a rounded head and oversized bill. Males have bright red panels in tail. Females have orange panels with black barring on the tail and irregular patches of yellow around the neck and head. This is an uncommon species of se. QLD, e. NSW, and far e. VIC. In some locations it is found quite a distance inland. The primary habitat requirement is *Casuarina* (she-oak) trees, its only food source, and it can be found in most wooded habitats that also support casuarinas. It is nowhere common but is occasionally recorded on the slopes of the Lamington Plateau and other ranges west of Brisbane (QLD), and also in Sydney (NSW). There is an isolated population on Kangaroo Island near Adelaide (SA).

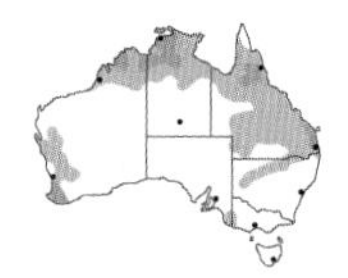

4 RED-TAILED BLACK COCKATOO *Calyptorhynchus banksii* 55–60cm/21.5–23.5in

Red-tailed is a huge, noisy black cockatoo with a prominent crest. Males are solid black with a scarlet tail panel. Females are browner with narrow barring on the underparts, fine pale speckling on the head and wings, and an orange-, yellow-, and black-banded tail panel. The species is widespread across tropical n. Australia, where it is the only black cockatoo, and has isolated populations in sw. WA and se. SA–sw. VIC. It occurs in open forests and woodlands and is most often found when large flocks gather to roost in the evenings or visit waterholes.

5 SHORT-BILLED BLACK COCKATOO *Calyptorhynchus latirostris* 53–56cm/21–22in

Also known as Carnaby's Black Cockatoo. Short-billed is a large black cockatoo that displays a pale ear patch and white panels in the tail. Its black body feathers are fringed with white. It is one of two very similar black cockatoos in WA, which differ in their bill shapes: Short-billed has a shorter and broader bill, adapted for cracking open the seeds of *Hakea*, *Banksia*, and *Dryandra* species, as well as introduced pines; while Long-billed has a longer, slimmer bill. Short-billed is confined to sw. WA, within a wider range than Long-billed, in pine plantations, open woodlands, Karri forests, sand-plain woodlands, and mallee.

6 LONG-BILLED BLACK COCKATOO *Calyptorhynchus baudinii* 53–56cm/21–22in

A large black cockatoo with a white ear patch and tail panels. It differs from the very similar Short-billed in having an obviously longer, narrower bill. The species is confined to far sw. WA, mostly south-west of a line from Perth to Albany in the Jarrah forests, where its primary food source is the nut of the Marri tree (*Corymbia calophylla*). It is also found in Karri forests and farmlands. It is fairly common in the far south-west and often seen around Augusta and Albany.

1
2
♀ 3
♂ 4
5
6

1 GANG-GANG COCKATOO *Callocephalon fimbriatum* 33–36cm/13–14in

A comical-looking grey cockatoo with a bright red head and crest in the male (these areas are grey in females). The body is ashy grey in colour and lightly barred with pale markings all over. If seen well, this tufted-topped cockatoo cannot be mistaken for any other. An inconspicuous but confiding cockatoo, it is best found by the soft 'creaky door' calls it gives when feeding or passing over in flight. It prefers wet sclerophyll forests, mountain woodlands, and the eucalypt groves within alpine heath, and is uncommon in the sub-coastal mountain belt from Sydney (NSW) to Canberra (ACT) to w. VIC and just into SA. It is occasionally seen around Barren Grounds NR (NSW) and is fairly common around Canberra.

2 GALAH *Eolophus roseicapilla* 36cm/14in

A fuchsia-pink cockatoo with a powdery-grey back and upperwings, pink underwings, and a solid creamy-white cap. It is one of only two pink cockatoos in Australia; the other, Major Mitchell's Cockatoo, is mostly pale pink and white in colour rather than grey like Galah. Found throughout the mainland in open country with sparse trees and often frequenting town parks, it can be found feeding in flocks on the ground, sometimes with other cockatoos such as Sulphur-crested or corellas. Galahs are usually tame and approachable, and are therefore familiar to many Australians.

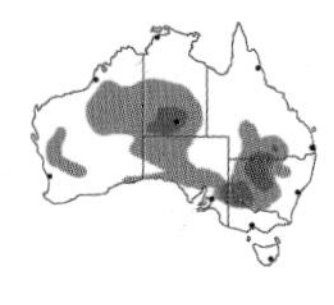

3 MAJOR MITCHELL'S COCKATOO *Lophochroa leadbeateri* 39cm/15.5in

Also known as Pink Cockatoo. A dazzling bird with light salmon-pink underparts (which vary greatly in intensity depending on light conditions), a white back, and a unique, remarkable red-and-yellow crest. A wide-ranging though uncommon species of interior Australia, it has a distinct preference for mulga regions and is relatively abundant in the mallee belt of the south-east. It is most often encountered either feeding quietly in trees or on the ground or in noisy flight overhead.

♂ 1
♀ 1
2
3

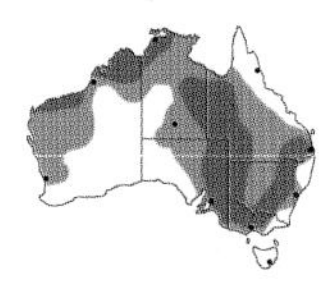

1 LITTLE CORELLA *Cacatua sanguinea* 36–39cm/14–15.5in

Little Corella is the ubiquitous white cockatoo of the Australian grain fields. It is found patchily across much of the mainland, absent from the driest parts of inland w. Australia as well as the far sw. corner of WA. Like all corellas it has a hidden crest that can be raised but is usually not seen. The main common confusion species is the larger Sulphur-crested Cockatoo, which has a large yellow crest and lacks the blue eye skin that this corella possesses. Little Corella favours a range of open habitats from agricultural lands to scrubby country over most of nw., n., and e. Australia.

1

2 LONG-BILLED CORELLA *Cacatua tenuirostris* 38–41cm/15–16in

This is probably the most distinctive of the corellas; most birds show obvious bright red feathers in front of the eye and on the upper throat, and a very long upper mandible that makes the bill look long. Like the other corellas it also has a patch of blue bare skin around the eye. The long bill, obvious even at a distance, in combination with the red face and throat make it easy to separate from Little Corella. Long-billed has quite a restricted natural distribution in rural w. VIC but like Little Corella has established feral populations in many coastal areas.

3 WESTERN CORELLA *Cacatua pastinator* 48cm/19in

This corella has two subspecies, both with very restricted distributions in sw. WA. In appearance it is intermediate between Long-billed and Little Corellas, with quite a long bill and some faint red or orange feathers around the eye and upper throat. Only Little Corella occurs in the same areas, and the best feature to use for separation is the bill: Western Corella has a much longer bill than Little. Like the other corellas, Western is found in open wooded areas, including farmlands and around towns. The best places to see it are around the town of Rocky Gully south-east of Perth (subspecies *pastinator*), or in the rural areas north-east of Perth (subspecies *derbyi*).

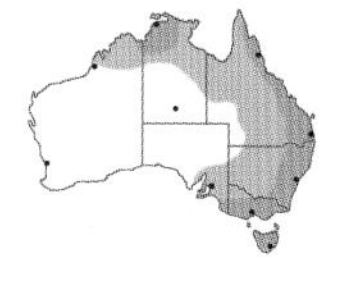

4 SULPHUR-CRESTED COCKATOO *Cacatua galerita* 48–55cm/19–21.5in

The familiar large white 'cocky' of e. Australia. It is all white with a sulphur-yellow crest and blackish bill. In flight the underwings and undertail are lightly washed with yellow. These boisterous and highly vocal birds are hard to ignore where they occur, as the noisy flocks draw attention wherever they are. The species is common in a coastal arc from n. WA through n. QLD and down e. Australia to s. NSW, SA, and TAS. It is a generalist, adapting to many habitats, including urban areas of e. Australia.

1
2
3
4

PARROTS (PSITTACIDAE)

1 RED-COLLARED LORIKEET *Trichoglossus rubritorquis* 25–30cm/10–12in

A long-tailed bright green parrot with a deep blue hood and a large dark blue belly patch, very similar to Rainbow Lorikeet. Red-collared differs in having an orange breast patch (not red and yellow), and a deep orange nape (not lime). Found from the Kimberley (WA), across the Top End (NT) to w. Cape York Peninsula (QLD), it occurs in tropical savannahs, melaleuca woodlands, and monsoon vine forests. It is a noisy and common bird in Darwin (NT), familiar to many, as groups regularly visit blooming shrubs within town parks and gardens.

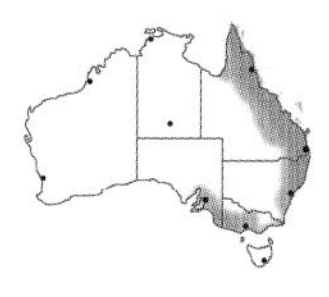

2 RAINBOW LORIKEET *Trichoglossus moluccanus* 26–30cm/10.25–12in

A stunning, long-tailed bright green parrot with a deep blue hood, red and yellow breast, lime nape, and a large dark blue belly patch. Rainbow Lorikeets inhabit tropical woodlands, open forests, and heaths. They are most often detected in noisy groups as they fly overhead, when they reveal even more colours in their vivid red underwings, long pale lime-green wing bars, and long bright green tails. The species occurs in the e. coastal zone in an arc southward from n. Cape York Peninsula to extreme se. SA, inhabiting most habitats but especially eucalypt forests and woodlands. It is quite common in Cairns and Brisbane (QLD), Sydney (NSW), Melbourne (VIC), and Adelaide (SA).

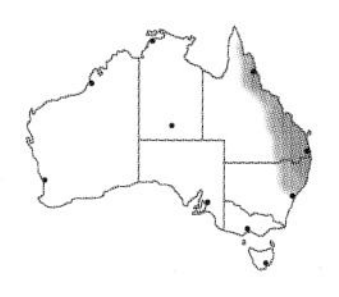

3 SCALY-BREASTED LORIKEET *Trichoglossus chlorolepidotus* 23cm/9in

At a distance, a medium-sized uniform green lorikeet. A close view reveals the underparts and upper back to be scaled golden yellow, and in flight the red underwing is very conspicuous. Scaly-breasted Lorikeet has a broad distribution from Cape York Peninsula (QLD) along the semi-coastal regions to s. NSW, where it is moderately common. It becomes much more common in coastal wet and dry sclerophyll forests, and towns and suburbs from subtropical QLD down to Sydney (NSW). It is quite common in Cairns, Brisbane, and Sydney.

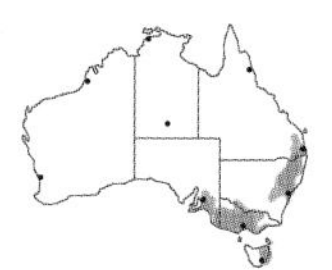

4 MUSK LORIKEET *Glossopsitta concinna* 20–22cm/8–8.5in

A small, nectar-feeding lime-green parrot with a relatively short tail and bright red facial markings. The pattern of the scarlet face patches is the key to identification: The forehead is bright red, and the broad crimson cheek patch extends behind the eye. Musk Lorikeet is conspicuous around blossoming eucalypts, where it occurs in noisy gatherings that can be detected from the racket produced. These tiny, hyperactive parrots regularly dart in and out of the trees at high speed, calling and moving often. The species is found throughout se. Australia, and is perhaps most common in coastal areas but is quite nomadic. It is quite common in e. TAS, and fairly easily seen around Hobart.

5 LITTLE LORIKEET *Glossopsitta pusilla* 16cm/6.25in

A tiny, short-tailed, seemingly all-green lorikeet. If seen at close range when perched, the red throat, lores, and forehead, which give the bird a red-faced appearance, can be viewed. The flight is very direct, less wavering than that of other lorikeets. It often occurs with other lorikeets, but its tiny size and colouration make it difficult to spot in the canopy. It also flies more readily than others when approached and rarely returns to a tree after leaving. Little Lorikeet is found from Cairns (QLD) to Adelaide (SA) in open eucalypt woodlands and dry sclerophyll forests, though is nowhere common. It is a blossom nomad but can usually be found in dry woodlands west of Brisbane (QLD) and throughout the Capertee Valley west of Sydney (NSW).

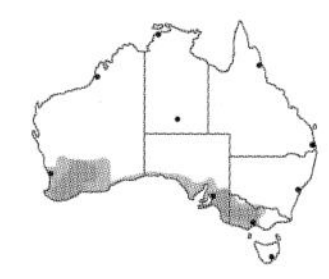

6 PURPLE-CROWNED LORIKEET *Glossopsitta porphyrocephala* 16cm/6.25in

The only lorikeet over much of s. Australia. When feeding it appears all green, as the purple crown, red frons, and the orange ear patch are all hard to see at a distance. Rather similar to the Little Lorikeet, it differs by having red not green underwings (seen in flight), a purple crown, and a bluish wash to the breast. Throughout its range Purple-crowned Lorikeet is found in a variety of wooded habitats, wherever there are flowering eucalypts, including mallee, dry sclerophyll forests, and open woodlands. It is fairly common across far s. Australia and can often be found in suburban Adelaide (SA) and even the outskirts of Melbourne (VIC).

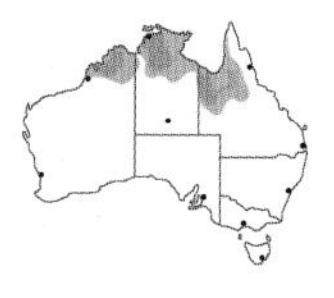

7 VARIED LORIKEET *Psitteuteles versicolor* 18–20cm/7–8in

A stunning, small, short-tailed lorikeet of tropical n. Australia with a conspicuous bright red cap, white eye ring, and yellowish ear. The upperparts are leaf green, underparts are a paler lime green with a pale red breast band. The body is covered all over in fine yellow streaks. The species has a broad, somewhat patchy distribution throughout the tropical savannahs of n. Australia, from the Kimberley (WA) right across to QLD. It is a blossom nomad and moves around quite a bit within its distribution but can often be found in the NT from Darwin to Katherine.

1 DOUBLE-EYED FIG PARROT *Cyclopsitta diophthalma* 13–15cm/5–6in

The smallest Australian parrot, Double-eyed Fig Parrot has a grass-green body with yellow sides to the breast, apparent with good views. Among its three subspecies, the most likely to be seen is that occurring around Cairns (*macleayana*), the male of which has a red frons, underbrow, moustache, and cheeks, and a bluish eye ring. In the female the red on the face is replaced by a pale peach to yellow. The Cape York Peninsula population (*marshalli*) has a more uniform red facial pattern. In the nearly extinct southern subspecies (*coxeni*), the red is replaced by a blue wash. This fig parrot looks like a small, round-tailed lorikeet, most closely resembling Little Lorikeet, which has a more uniform, single red facial patch. The three separate populations are based around Australia's three rainforest regions: Iron Range, on Cape York Peninsula (QLD), the Cairns region (QLD), and the QLD–NSW border area. The southernmost population (Coxen's) is extremely rare, with almost no reliable recent records. The northernmost subspecies is regular in Iron Range NP, and subspecies *macleayana* is a common bird even around Cairns city.

2 ECLECTUS PARROT *Eclectus roratus* 41–43cm/16–17in

A massive, short-tailed parrot. The much brighter female is bright crimson with a blue breast, belly, and upper back. The male is a uniform grass green with an orange bill, tipped yellow, and red flanks, which are difficult to see when the bird is perched but are contiguous with red underwings and very obvious in flight. Limited to the rainforests of Iron Range NP on Cape York Peninsula (QLD), it is common and easily seen there.

3 RED-CHEEKED PARROT *Geoffroyus geoffroyi* 22–25cm/8.5–10in

Chunky, short-tailed rainforest parrot of the far north-east. Because it is illustrated beside the huge Eclectus Parrot in many field guides, the expectation is often that this is a small parrot, but it is actually quite large. Both sexes are bright green and display blue underwings in flight. The male has a red face and purple crown; the female has a brown face and crown. The species is found mainly in the rainforests of Iron Range NP on Cape York Peninsula (QLD), where it is common and easily seen, but it also occurs sporadically in monsoon forest patches westward to near Weipa.

4 AUSTRALIAN KING PARROT *Alisterus scapularis* 39–44cm/15.5–17.5in

A large, striking parrot with a long, square tail. Males and females are markedly different: The male is bright scarlet except for a forest-green mantle, tail, and wings and a slim, lighter green shoulder stripe. The female is nearly all dull green, with a red belly and undertail. Despite their colourful appearance, these parrots can be quite inconspicuous as they forage within trees in rainforests, eucalypt forests, or densely vegetated parklands, usually giving themselves away when they burst out from cover in flight or through their high-pitched piping calls. The species is found in wet e. coastal regions from Cairns (QLD) to Melbourne (VIC). In some areas, most notably Lamington NP (QLD) and Pebbly Beach (NSW), a number of wild individuals have been habituated around feeding stations and can be seen at absurdly close quarters.

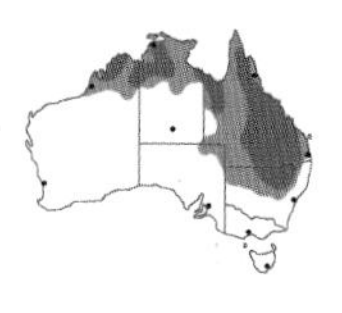

5 RED-WINGED PARROT *Aprosmictus erythropterus* 30–32cm/12–12.5in

A stunning green parrot with large scarlet panels in the wing, a contrasting black back (in males), and an orange bill. This dashing parrot is striking in appearance both perched and in flight. It flies with slow, deliberate wing beats and displays a zigzag flight pattern, which allows the observer to savour its most conspicuous feature, the scarlet wings, and also note the square-shaped tail. Unlike many other Australian parrots, Red-winged does not usually forage on the ground but feeds on blossoms within trees. It is a widespread and common bird of the tropical savannahs of n. Australia but is also widespread throughout the drier habitats of inland QLD and n. NSW.

♂ macleayana 1
♀ macleayana 1
2
♂ 3
♀ 3
♂ 4
♀ 4
♂ 5
♀ 5

1 SUPERB PARROT *Polytelis swainsonii* 38cm/15in

A very long-tailed bright green bird. The male has a yellow crown, face, and throat separated from the chest by a crimson band. The female has a bluish wash to the face and cheeks. Flight is very swift and direct, and the long tail is very obvious. The bird feeds on the ground, often on roadsides, and in trees in blossom. It is fairly common in the wheat-belt region of sc. NSW and n. VIC. It is usually associated with areas of Red Gum and Yellow Box trees (both eucalypts) near watercourses but is also found in grazing lands and croplands. It is quite common in the area around Griffith, Hay, and Deniliquin in s. NSW.

2 REGENT PARROT *Polytelis anthopeplus* 38cm/15in

A stunning lime-green and yellow parrot. The male has a green back, yellowish underparts, a black tail, and a very large golden patch on the shoulder. The female is much duller, with a darker green head, back, and breast and a much-reduced golden shoulder patch. The species is both a ground and tree feeder. It is a locally common bird of Red Gum habitat, mallee, and surrounding croplands in the Victorian mallee belt, the Riverland of SA, and sw. WA. In the west it is more common, less tied to these habitats, and can also be common in open eucalypt woodlands. Good places to find it are Gluepot Reserve (SA), Wyperfeld NP (VIC), and Stirling Range NP (WA).

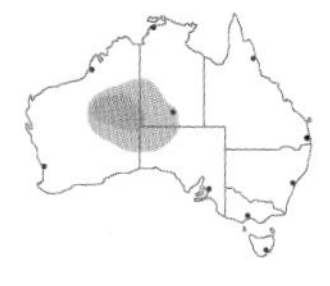

3 PRINCESS PARROT *Polytelis alexandrae* 38–44cm/15–17.5in

A highly sought, extremely long-tailed parrot. The male has an olive-green back, large lime-green shoulders, a peach-washed throat, a powder-blue crown, and lemon-washed underparts. Females and immatures are slightly duller and have shorter tails. Princess Parrot has a distinctively lazy flight compared to Regent and Superb Parrots. It is a rare, irruptive species with a large possible range and a much smaller core range in the Great Sandy Desert (accessible via the Canning Stock Route), where the main habitats are arid open woodlands of desert oak, mulga, or eucalypts with a spinifex understorey.

4 COCKATIEL *Nymphicus hollandicus* 29–32cm/11.5–12.5in

A familiar native bird commonly found in the cage-bird trade. Although in the cockatoo family (Cacatuidae), it appears more like a parrot, as it is much smaller and longer-tailed than most cockatoos. Mainly grey, it has large, conspicuous white panels in the wing (which are visible even at some distance), a lemon-yellow face and long, conspicuous crest, and a prominent chestnut cheek patch. Females lack yellow on the head. It is a common inland species of semi-arid open country with scattered trees across much of the continent but absent from the most arid desert areas and coastal regions. Cockatiels are usually encountered in small groups either foraging on the ground or perched in open dead trees, where their distinctive shape and striking combination of colours make them unmistakeable.

♂ 1
♀ 1
♂ 2
♀ 2
3
3
♂ 4
♀ 4

1 CRIMSON ROSELLA *Platycercus elegans* 36–38cm/14–15in

Crimson Rosella is a bird of e. coastal Australia, from n. QLD down to w. VIC and e. SA, where it inhabits rainforests, woodlands, eucalypt forests, and gardens. It is a striking bird, largely vivid scarlet in colour, with a royal-blue cheek patch, blue tail, and blue flashes in its wings. It is a common bird throughout its range, most likely encountered feeding quietly on the ground in groups, which when disturbed flush up with a flurry of red and a sudden outbreak of noisy calls, making them hard to miss once in flight. Orange and yellow forms (these colours replacing the scarlet areas on the body) are found in Adelaide (SA) and inland s. NSW respectively. They are easy to see in the ranges west of Brisbane and Sydney, such as the Lamington Plateau (QLD) and the Blue Mtns (NSW).

2 GREEN ROSELLA *Platycercus caledonicus* 30–36cm/12–14in

A parrot restricted to TAS, where it overlaps with just one other rosella species, the markedly different Eastern Rosella, which has an all-red head. On occasion these two rosella species can be found feeding together within the same flocks. Green Rosella possesses black-scaled green upperparts, a green tail, bright yellow underparts and head, a solid blue cheek patch, and a small bright red forehead. In combination these features are not found on any other parrot species on TAS. This Tasmanian endemic, common even in downtown Hobart, is found in a range of habitat types from dense forests to farmlands to eucalypt woodlands and neighbouring clearings. Like all the rosellas, it is most frequently observed in small groups.

3 EASTERN ROSELLA *Platycercus eximius* 28–33cm/11–13in

Eastern Rosella is found from se. QLD south to VIC, TAS, and se. SA. Its boldest feature is the bright scarlet head, which contrasts with a clean white cheek patch and gaudy yellowish underparts. It is yellow on the back with bold black scales down the mantle and has a bright green rump and a blue tail. This myriad of colours makes it unlikely to be confused with any other bird. It is common in parks, open woodlands, farmlands, and gardens, and like all rosellas is often encountered feeding quietly in small parties on the ground.

4 NORTHERN ROSELLA *Platycercus venustus* 30–32cm/12–12.5in

A parrot found in n. WA and the n. NT that does not overlap with any other rosella species. Northern Rosella has a black head and bright white cheek patch, pale yellow underparts with a scarlet vent, yellow upperparts scaled black on the mantle, a blue tail, and a striking bright yellow rump, visible in flight. It also has a broad blue flash in the wing, visible even when perched. It is found in tropical open woodlands, scrubby areas, and grassy clearings within its small range, where it is regular but not common.

5 PALE-HEADED ROSELLA *Platycercus adscitus* 28–30cm/11–12in

A rosella with a pale yellow head and bluish underparts and wings. The vent is red and the tail dark blue. The se. QLD subspecies has white cheeks, the back and mantle bright yellow scalloped with black, and a blue rump obvious in flight. The ne. QLD subspecies has a bluish wash to the cheeks, the mantle and back are black edged in cream, and the rump is a pale cream. The species feeds on the ground and sometimes in trees. It is common in open eucalypt woodlands and tropical savannahs from Cape York Peninsula to sub-coastal n. NSW and is a regular bird in the drier parts of the Atherton Tableland.

6 WESTERN ROSELLA *Platycercus icterotis* 27–28/10.5–11in

A distinctive small tomato-red rosella unlike any other bird in sw. Australia. The male has a red head and body, a yellow cheek, green back and tail, and blue wings, while the female is duller. Unlike the other rosellas, which always look smart, this bird often appears a little scruffy. Western Rosella is a common bird of the wet and dry sclerophyll forests of sw. WA, ranging out into open eucalypt woodlands and croplands.

1
sw. NSW 1
Adelaide imm. 1
2
3
4
5
6

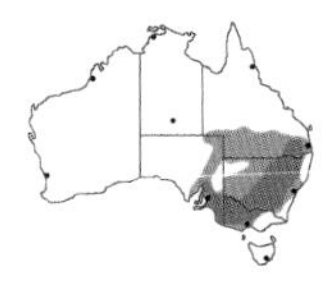

1 RED-RUMPED PARROT *Psephotus haematonotus* 24–29cm/9.5–11.5in

The male is a bright green parrot with a red rump, lemon-yellow belly and vent, and a subtle blue wash on the shoulder and forehead. It is told from male Mulga Parrot by its lack of red on the nape and vent and absence of a yellow shoulder patch. Females are dull brownish-green with little colour except some green on the rump. The lack of any strong shoulder mark or reddish nape patch separates this species from female Mulga Parrot. Red-rumped parrot is most likely to be found in pairs or small flocks. It readily perches in the open, is often conspicuous and approachable, and is more regularly found around country towns than Mulga Parrot. Red-rumped is a common species of the south-east, where it occurs in farmlands with scattered trees and grassy and other open woodlands, often around watercourses.

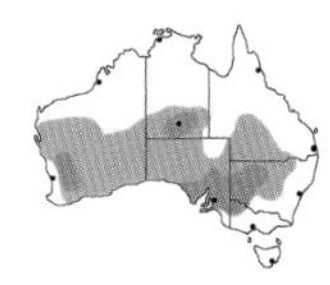

2 MULGA PARROT *Psephotus varius* 27–32cm/10.5–12.5in

Male Mulga Parrot is a bright green bird with a bright yellow bridge across the top of the bill, a scarlet nape patch, a yellow shoulder patch, and a red and yellow vent. The head markings and shoulder patch help to distinguish it from the male Red-rumped Parrot, the most likely confusion species. The female Mulga is dull greenish-brown and separated from the similar female Red-rumped by its red shoulder patch and pale pinkish-red nape patch. Females of both species show green rumps. Mulga Parrot inhabits dry scrublands and open woodlands, such as mallee and mulga, as well as timbered watercourses. It is fairly common, though shy and conspicuous, within dry arid country in the s. half of the mainland, where it is most likely to be found in pairs or family parties. It is quite common at Bowra Station in sw. QLD and also in woodlands around Griffith (NSW), such as Binya SF.

3 AUSTRALIAN RINGNECK *Barnardius zonarius* 30–42cm/12–16.5in

A large green parrot with a long, square-ended tail and a conspicuous yellow neck ring or collar. There are several subspecies, sometimes regarded as distinct species; the two western forms display darker hoods above the yellow collar, which is common to all forms.

3a Mallee Ringneck, subsp. *barnardi*

The most colourful of the subspecies, Mallee Ringneck has an aqua head, gold nape, prominent blue upperparts, and a red bridge across the top of the bill. It forages for seeds on the ground, and so is best detected when a group, inadvertently disturbed, flies up with distinctive deep wing beats, often while emitting raucous calls. It is a common, though rather inconspicuous parrot, most often found along vegetated watercourses, as well as in open woodlands, mallee, and farmlands with scattered trees. It occurs in an arc from nw. QLD through inland NSW and west to e. SA.

3b Concurry Ringneck, subsp. *macgillivrayi*

The most restricted of the ringnecks and also the dullest. Concurry looks like a small Mallee Ringneck but lacks the red areas on the face and the blue upperparts, though it shows a bright yellow underside. It occurs in riverine eucalypts in the stony uplands in the Mt Isa area of nw. QLD.

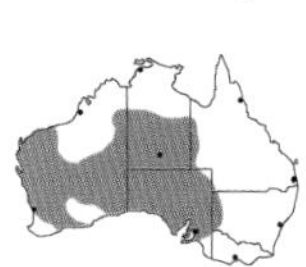

3c Port Lincoln Parrot, subsp. *zonarius*

Smaller than the other dark-headed form (Twenty-eight Parrot), this ringneck has a dark green throat and upper breast contrasting with the yellow breast and belly. It is a common species, most often found along vegetated watercourses, as well as in open woodlands, mallee, and farmlands with scattered trees. The most widespread of the ringnecks, it occurs from the w. coast east to e. SA, where it merges with the Mallee Ringneck.

3d Twenty-eight Parrot, subsp. *semitorquatus*

The larger of two western darker-headed forms, this subspecies is told from the smaller Port Lincoln Parrot by its underparts merging from a strong green at the throat to a slightly lighter green at the belly, and by showing more red on the bridge above the bill. It is found in the extreme sw. of the continent in the wetter, taller eucalypt forests.

♂ 1
♀ ♂ 2
3a
3b
3c
3d

1 HOODED PARROT *Psephotus dissimilis* 25–28cm/10–11in

A beautiful ultramarine parrot confined to the Top End of the NT. Males are unmistakeable: bright turquoise with brown wings, a striking gold wing patch, and a black cap. Females are duller, green birds lacking the bold wing patch and solid cap of the male but with an indistinct powder-blue wash on the face. Hooded Parrot occurs primarily in open savannah woodlands and areas of eucalypts on rocky ridges with a grassy understorey and an abundance of termite mounds, in which they nest. It is an uncommon bird, most often found in pairs or small groups, except when roosting communally. The stronghold of the species is around Pine Creek, where the birds can often be found foraging on lawns around the town.

1 ♂

2 GOLDEN-SHOULDERED PARROT *Psephotus chrysopterygius* 27cm/10.5in

A slender turquoise-blue parrot from Cape York Peninsula (QLD). Its sandy-brown back and golden wing patch are similar to those of Hooded Parrot, from which it differs by having a very dark grey (not black) cap that is separated from the bill by a yellow bridge. The female is a much duller, green parrot with an all-green head and a pale blue bill, belly, and vent. The lack of powder blue on the face differentiates it from the female Hooded Parrot of the NT. This rare and endangered species has a very limited distribution centred on c. Cape York Peninsula, where it occurs in open eucalypt woodlands and grassy savannahs. It needs large terrestrial termite mounds to nest and is associated with areas where they are present. It is most often seen around the Musgrave Station area.

3 PARADISE PARROT *Psephotus pulcherrimus* 27cm/10.5in

Unfortunately, with no confirmed records since the late 1920s this gorgeous parrot is considered extinct. It had a black crown, brown back and wings, a red wing patch, red vent, and turquoise-green breast. It was found in open woodlands and rangelands of s. QLD and n. NSW. We list it here because extensive potential habitat exists in regions that are rarely birded.

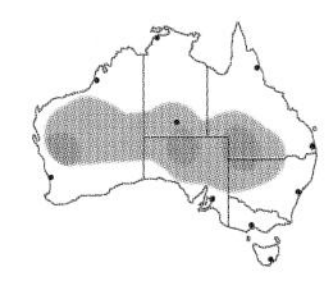

4 BOURKE'S PARROT *Neopsephotus bourkii* 23cm/9in

A small, understated powdery grey-brown parrot with a blue forehead. The feathers on the back and breast are fringed with soft grey, underparts are washed pink, and the lower belly and vent are powder blue. In flight the male shows blue patches on the upperwing. The female is much duller, with no blue on the forehead and much-reduced blue on the upperwing, but still has the pinkish wash. Both male and female have a white eye ring. This parrot feeds on the ground and is difficult to approach. It is a widespread species through c. Australia, with three distinct core areas. It favours arid woodlands, primarily mulga but sometimes other mixed woodlands with a shrubby or grassy understorey.

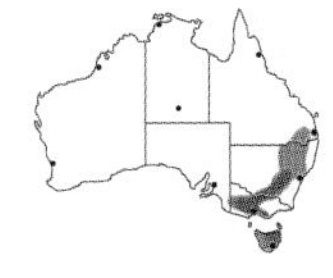

5 SWIFT PARROT *Lathamus discolor* 24–27cm/9.5–10.5in

A bright green arboreal parrot with red lores and chin bordered by a yellow moustachial streak, a red vent, and a small red patch on the wing. The red underwing and long black tail are very obvious in flight. A nectar-feeding species, it is usually seen fleetingly in small groups tearing throughout the canopy. This parrot breeds in the open eucalypt woodlands of TAS before migrating northward to winter in the se. mainland. As a nectar feeder it is a blossom nomad, following eucalypt blooms in the dry sclerophyll forests and so may be concentrated in different areas each year. On the mainland it occasionally makes it as far north as s. QLD but is often concentrated around the box-ironbark forests of n. VIC or the coastal woodlands of c. NSW.

♂ 1
♀
♂ 1
♂ 2
♀ 2
♂
♀ 4
5
5

PARROTS (PSITTACIDAE)

1 BUDGERIGAR *Melopsittacus undulatus* 17–19cm/6.75–7.5in

A familiar small long-tailed native parrot that is a popular cage bird. Although there are many colour morphs in captivity, wild birds are always bright green with a delicately barred yellow head and prominently scaled yellow wings and upperparts. In flight the long tail, bright green colouration, and conspicuous pale wing bars stand out. Budgerigar ranges widely in the more arid parts of Australia, occurring across much of the mainland except for the wetter coastal regions. However, it is highly nomadic and irruptive and therefore often absent from many suitable areas. It favours grasslands, farmlands, and open woodlands within inland Australia, where it regularly forages on the ground in groups, which can swell to massive flocks in the flush of new feeding areas following heavy rains.

2 EASTERN GROUND PARROT *Pezoporus wallicus* 29–33cm/11.5–13in

A long-tailed green parrot usually seen flying low over heath after being flushed. The back is green mottled with black and yellow, the throat and upper breast are grass green, and the lower breast to vent is yellow barred with dark green. It has a strong preference for dense, low-lying heaths and, in TAS, for button-grass plains. This dense habitat makes this bird notoriously difficult to find, as it spends most of the day well hidden, coming to feed on sides of paths only in very early morning or at dusk. The best means of locating birds is by listening at dusk or dawn for their distinctive buzzing calls. This parrot is patchily distributed from coastal regions of s. QLD, where rare, to TAS, where it is common in some areas. Well-known sites to search for it include Barren Grounds NR (NSW) and around Melaleuca (TAS).

3 WESTERN GROUND PARROT *Pezoporus flaviventris* 29–33cm/11.5–13in

Nearly identical to Eastern Ground Parrot, Western differs in having yellower underparts. This critically endangered, extremely rare, and secretive species now appears to be limited to areas around Cape Arid NP and Fitzgerald River NP in sw. WA. *(Refer to photo of preceding species)*

4 NIGHT PARROT *Pezoporus occidentalis* 24cm/9.5in

A chunky bull-headed green parrot with a short wedge-shaped tail. It has grass-green upperparts with rows of black arrow marks on the crown and mantle and black blotches on the back. The underparts are lime green with black arrows merging to unmarked vent. This is a highly secretive parrot that hides during the day and feeds at night. It is likely to flush only when approached very closely. The Scarlet Pimpernel of world birds, it is everywhere but nowhere and is recorded only every few decades or so. This bird has a massive possible distribution from nw. NSW to Mt Isa (nw. QLD) to the tropical WA coast. Little is known about its habitat preferences; it has been found in spinifex grasslands and chenopod (saltbush) shrublands. Its nocturnal habits make it extremely difficult to locate, even if you are lucky enough to be where it is.

5 RED-CAPPED PARROT *Purpureicephalus spurius* 37–38cm/14.5–15in

Stunning in its gaudiness. The scarlet cap, massive yellow cheek, and bright blue underparts (which appear electric blue in sunlight) of the male are hard to miss when it is perched. When in flight, the long tail, yellow rump, olive-green back, and red vent are also obvious. The females and immatures have the same general pattern and look like subdued, softer versions of the over-the-top male. Red-capped Parrot has a fairly wide distribution in the sw. corner of the continent, and is a common bird in woodlands, parks, and gardens.

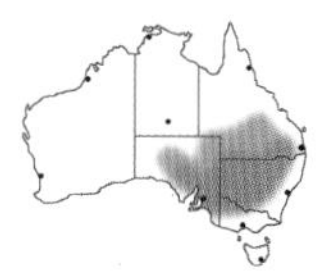

6 BLUEBONNET *Northiella haematogaster* 25–34cm/10–13.5in

This predominantly olive-brown parrot has a prominent blue face and bright yellow belly and vent with variable amounts of bright scarlet scattered across this vivid yellow area. The wings are deep blue in flight, and a dark red shoulder patch often shows when the bird is perched. The dull brownish colouration, blue face and wings, and red underbelly help to distinguish Bluebonnet from any other parrot. It is a south-eastern species found in semi-arid, dry country across e. SA, n. VIC, w. NSW, and sw. QLD. It is found in areas of open wooded country, mulga, farmlands with scattered trees, and tree-lined watercourses.

7 NARETHA PARROT *Northiella narethae* 29cm/11.5in

This rare olive-brown parrot is very similar to Bluebonnet and was previously considered a Bluebonnet subspecies. The upperparts are grey-brown, and it has a light blue forehead, prominent blue face, yellow belly, and red vent. It differs from Bluebonnet in having no red on the lower breast or belly and in its lighter blue forehead. This bird has very strict habitat requirements and is found only in samphire and bluebush sedgelands of the Nullarbor Plain. It has a very limited distribution in far se. WA and around the WA–SA border. *(Image computer-generated)*

1
2
♂ 5
♀ 5
6
7

PARROTS (PSITTACIDAE)

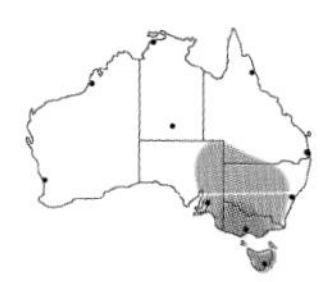

1 BLUE-WINGED PARROT *Neophema chrysostoma* 23cm/9in

This small green parrot is often flushed up suddenly from the ground where it feeds. It is dull yellowish-green with a yellow belly and blue shoulder patch. It has a faint yellowish mask bordered above by a blue line on the forehead. Females have less blue on the forehead and wings than males. Although the larger blue shoulder patch can help separate it from Elegant and Orange-bellied Parrots, be careful because this difference is often less obvious than depicted in field guides. Generally the Orange-bellied Parrot is brighter grass green and lacks the yellow mask, while the Elegant Parrot is brighter yellow-green and has two-toned blue edging to the wing. Blue-winged Parrot is an uncommon breeding bird of TAS and s. VIC, with summer dispersal into NSW and SA. It is most common in well-developed heath thickets merging into open woodlands with a heath understorey.

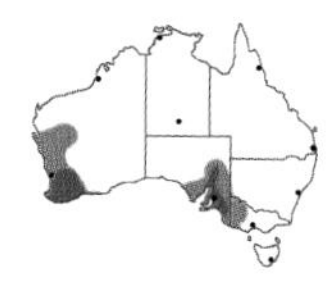

2 ELEGANT PARROT *Neophema elegans* 23cm/9in

A small bright yellow-green parrot with a yellow belly and thin, two-toned blue edging to the wing, dark blue on the edge and lighter blue above this. Look closely at the forehead: This parrot is bright yellow between the eye and bill and has a dark blue line across the forehead and a lighter blue line above this one that extends above the eye. On Blue-winged Parrot this line doesn't extend above the eye, and its wing is less obviously two-toned blue. Elegant Parrot overlaps with Blue-winged Parrot only around se. SA, so be careful in this area. Elegant Parrot is an uncommon bird of dry, open eucalypt woodlands with a population centre around se. SA and another in sw. WA. It is quite common in sw. WA and easily found in the Dryandra Woodland and Stirling Ranges.

3 ORANGE-BELLIED PARROT *Neophema chrysogaster* 23cm/9in

This is one of the rarest birds in the world, with a wild population of perhaps only 30 birds. It has a curious ecology, breeding in remote sw. TAS and migrating to the s. coast of VIC and far e. SA during the winter. It is brighter grass green than other *Neophema* parrot, rather than yellow-green or brown-olive. Be careful, as the orange belly is really obvious only on adult males. This critically endangered parrot nests in short heathlands of w. TAS and winters on the salt marshes of the w. VIC coast. The only reliable way to see it is by flying into Melaleuca in remote sw. TAS from about November to March.

4 ROCK PARROT *Neophema petrophila* 23cm/9in

Habitat is the key to locating this species; it is invariably found close to the coast, where it feeds among sand dunes or in grassy areas around rocky headlands. Rock Parrot is very similar to Elegant Parrot but its colour is duller, more of a brown-olive hue. Unlike Elegant and Blue-winged Parrots, it has no yellow in front of the eye, just pale blue with a dark blue band across the forehead. Rock Parrot is found along the s. coastline from just north of Perth (WA) to east of Adelaide (SA), though is missing from much of the Great Australian Bight shoreline. This bird is strictly coastal, preferring offshore islets, rocky shores, and heath associations on primary (closest to the sea) dunes. It can be elusive but is sometimes found on the grounds of the Cape Leeuwin Lighthouse in WA.

5 TURQUOISE PARROT *Neophema pulchella* 20cm/8in

This stunning little parrot is quite rare and local. It is usually found in open grassy woodlands on the slopes of the Great Dividing Range in se. Australia, from se. QLD to n. VIC. The male is unmistakeable, with a bright blue face, bright yellow breast, and green back. Its wings have blue edges and a red shoulder patch, which is not always very obvious. The female is duller and has no red shoulder patch. A very uncommon and patchily distributed parrot of the se. semi-coastal regions, it shows a strong preference for open eucalypt woodlands with an association of cypress and thick grassy ground cover. Regular sites include Munghorn Gap NR and Back Yamma SF in NSW and Chiltern–Mt Pilot NP in n. VIC.

6 SCARLET-CHESTED PARROT *Neophema splendida* 19–20cm/7.5–8in

Although occurring across a large area of arid s. Australia, this gem is one of the country's more elusive birds. It is only thinly spread throughout its range, most of which is remote desert and quite inaccessible. The male is unmistakeable, with a bright blue face, yellow belly, scarlet breast, and green wings with a blue edge. The female is duller and lacks the scarlet breast. This rare and nomadic, mallee- and she-oak–loving parrot has a large irruptive distribution but most of the time is concentrated in the Great Victoria Desert on the n. boundary of the Nullarbor Plain. It has been recorded recently in the mallee association at Gluepot Reserve (SA), though this is not a guaranteed location for the bird by any means.

♂ 1
imm. 1
2
3
♂ 4
♂ 4
♂ 5
♀ 5
♂ 6
♀ 6

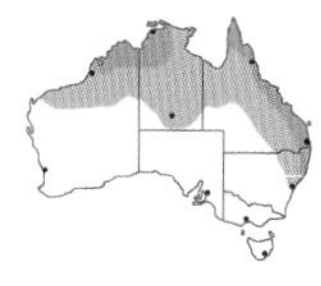

1 PHEASANT COUCAL *Centropus phasianinus* 55–69cm/21.5–27in

In breeding plumage this cuckoo is all black below and has heavily streaked brown upperparts and a heavily barred, very long tail. When not breeding it is a brown bird, heavily streaked all over, lacking the solid black underparts. Generally a skulking species, it can be observed for long periods from a distance, but when approached it often drops to the ground from a perch and runs away much in the manner of the Greater Roadrunner (*Geococcyx californianus*) of North America. Pheasant Coucal is found commonly through the tropical north and down the e. coastal region into n. NSW in such habitats as open woods, scrublands, heaths, swamps, and thickets.

1 nbr.

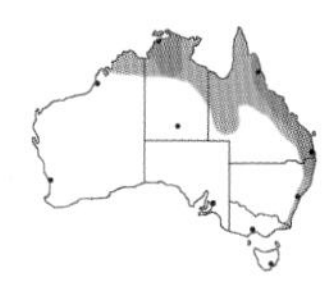

2 AUSTRALIAN KOEL *Eudynamys cyanocephalus* 41cm/16in

Australian Koel is a long-tailed cuckoo. The male is a large, fruit-eating, glossy black cuckoo with a blood-red eye, long tail, and pale horn-coloured bill. Satin Bowerbird males may appear black in poor light, but have short tails, blue eyes, and tend to forage on the ground. The female is a brown bird, spotted white all over, with a blackish cap, red eye, and black moustache. The *koel* call is loud and familiar in its range, and the bird is far more easily heard than seen in forests and other areas with tall trees, as it tends to sit well hidden while loudly and incessantly calling from the treetops. It is parasitic, laying its eggs in the nests of orioles, honeyeaters, Magpie-Larks, and riflebirds. Australian Koel sometimes joins groups of Channel-billed Cuckoos at fruiting fig trees. It is a breeding migrant to Australia (from New Guinea), found through the tropical north and down the continent's e. side to s. NSW and n. VIC.

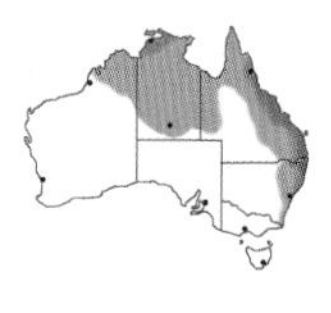

3 CHANNEL-BILLED CUCKOO *Scythrops novaehollandiae* 56–70cm/22.5–27.5in

A massive, loud, and distinctive species that is both the largest cuckoo and the largest brood parasite on earth. A real brute, it looks like a huge flying crucifix. It is a grey bird with a bulbous, straw-coloured bill (somewhat reminiscent of that of a hornbill, family Bucerotidae) and red skin around the eye near the bill base. The combination of massive size and odd bill shape make this a truly unmistakeable cuckoo. It is parasitic, like many other Old World cuckoos, laying its eggs in the nests of large birds such as magpies, currawongs, and crows, and taking no part in rearing the young. Channel-billed Cuckoo usually sits high in trees and often draws attention with its loud, raucous, and highly distinctive *awk* calls. It inhabits any habitat with large trees, such as rainforests, eucalypt forests, open woodlands, and wooded farmlands; congregations can sometimes be seen around fruiting figs and other trees. These cuckoos are migrants to the coastal areas of the tropical north and down the e. side of Australia through NSW to far e. VIC.

br. 1
nbr. 1
♂ 2
♀ 2
3
3

1 HORSFIELD'S BRONZE CUCKOO *Chrysococcyx basalis* 17cm/6.75in

Horsfield's Bronze Cuckoo has a conspicuous dark ear patch, broad white supercilium, and dull brown eyes, and it lacks an obvious eye ring. It has incomplete bars on the underparts that usually do not meet in the centre. Immatures have grey underparts with barring limited to the sides of breast, and a small white eyebrow. Shining and Little Bronze Cuckoos are both smart-looking birds, whereas Horsfield's often looks a little drab. It is one of the most widespread of all Australian birds, found in most habitats from arid scrublands to coastal heaths. Its high-pitched descending whistle is a good clue to its presence.

2 SHINING BRONZE CUCKOO *Chrysococcyx lucidus* 16.5–18cm/6.5–7in

A small, barred cuckoo with dull brown eyes and no obvious eye ring. This smart-looking bird is bronze green above and has dark bronze-green bars below that cross the entire breast right up to the throat. The undertail is buff-spotted rather than barred. Little Bronze Cuckoo is similar but has black bars on the underparts, a red eye ring, and heavier marking around the eye. Horsfield's has incomplete barring below and a prominent eyebrow and ear patch. Shining Bronze Cuckoo is a migrant species that occurs in e. coastal and sub-coastal habitats from Cape York Peninsula (QLD) around to Adelaide (SA), and is also found in WA from the Tropic of Capricorn around to Esperance. The birds that summer in se. and sw. Australia winter in n. QLD, Papua New Guinea, and Indonesia.

3 LITTLE BRONZE CUCKOO *Chrysococcyx minutillus* 15cm/6in

This small cuckoo is metallic green and possesses a red iris, a noticeable red eye ring, and heavy streaking around the eye. The cap is black with white smudging. Upperparts are bronze green, underparts are white scalloped with black all the way from throat to vent. Seen well, the outer undertail feathers are barred black and white, not buff-spotted as in Shining Bronze Cuckoo. Immature birds are much paler, with pale barring limited to sides of breast and belly. Subspecies *poecilurus*, sometimes called Gould's Bronze Cuckoo, has a pale russet wash to the back, sides of the breast, and cap. The species has a confusing distribution and movement pattern. The birds seen in se. QLD during summer are migrants, while subspecies *poecilurus* is resident in ne. QLD, and another subspecies is resident across the savannahs of the tropical north. Little Bronze Cuckoo can be seen from the Kimberley (WA) across the Top End (NT), in ne. QLD, and in summer right down the e. coast of QLD to n. NSW. It can be found in a wide variety of habitats including forest edges, monsoon forests, mangroves, and tropical savannahs.

4 BLACK-EARED CUCKOO *Chrysococcyx osculans* 20cm/8in

A small cuckoo without barring on underparts. It is grey-brown above, pale grey below, and has a very obvious white eyebrow and black mask. Immature birds are similar to adults, showing only slightly less contrast in markings. Black-eared is distinctive among the cuckoos, though it might be confused with Horsfield's Bronze Cuckoo, which has barring on the breast. This uncommon cuckoo is widespread in the interior, where it is found in dry woodlands, including mallee, mulga, and tropical savannahs. It is migratory, summering in s. Australia and wintering in n. Australia.

1
imm. 1
2
3
3
4
4

1 PALLID CUCKOO *Cacomantis pallidus* 30cm/12in

A largely uniform pale grey cuckoo with a dark line through the eye and a long, boldly barred tail. There is some plumage variation: Males may be variously darker or lighter, and a rufous morph occurs in females, with rufous on the back and scattered pale mottling above. Pallid Cuckoo is a brood parasite of smaller songbirds, such as honeyeaters, whistlers, and flycatchers. A very widespread cuckoo, it favours open habitats, in which it is commonly encountered perched prominently on roadside wires in both dry and wet areas of the entire continent. In s. Australia it is a summer breeding migrant.

1 imm.

2 FAN-TAILED CUCKOO *Cacomantis flabelliformis* 24–27cm/9.5–10.5in

A grey-backed cuckoo with a cinnamon-washed throat, neck, and breast. The breast colouration gradually merges into buff on the belly and vent. The face is grey with a contrasting yellow eye ring. The tail is dark grey heavily notched with white. Immatures have plain brown backs and grey underparts with dark brown bars. The species is found primarily in closed forests and wetter habitats from Cape York Peninsula (QLD) down through NSW to VIC, TAS, and s. SA, as well as sw. WA, and its mournful downward trill is a common forest sound.

3 CHESTNUT-BREASTED CUCKOO *Cacomantis castaneiventris* 24cm/9.5in

An oversaturated version of the more common Fan-tailed Cuckoo. The back is slate grey, and the underparts from throat to vent are rich chestnut. The eye ring is golden yellow. The slate-grey tail is notched with white. Immatures have uniform brown upperparts and are pale brown below without barring or mottling. The species is fairly common in the rainforests of Iron Range NP and ne. Cape York Peninsula (QLD). There also appear to be isolated populations farther south, in n. Atherton Tableland.

4 BRUSH CUCKOO *Cacomantis variolosus* 23cm/9in

A sombre-coloured cuckoo with grey-brown upperparts, a grey throat, buff-washed underparts, and a squarish tail without notching on its sides. The eye ring is much less distinctive than on Fan-tailed or Chestnut-breasted Cuckoos. Immatures have mottled brown backs and dark grey-buff underparts with chocolate vermiculations on the throat and dark grey barring on the breast. Brush Cuckoos are brood parasites, laying their eggs in the nests of smaller birds such as fairywrens, thornbills, and scrubwrens. The species is very vocal, and its feverish-sounding, downwardly inflected *peeo-peeo-peeo . . .*, often given incessantly, is a common spring and summer sound of n. and e. woodlands. Brush Cuckoo occurs in a wide variety of sub-coastal habitats, including tropical savannahs, open woodlands, rainforests, and monsoon forests. A common cuckoo, it is found from the Kimberley (WA) east and south through the e. coastal regions. Although a summer migrant to VIC, it is notably absent from TAS.

5 ORIENTAL CUCKOO *Cuculus optatus* 28–33cm/11–13in

A large cuckoo that can look like a small falcon. It is a long-tailed, pointy-winged bird with mid-grey upperparts, a pale grey throat and upper breast, and white from lower belly to vent; it is strongly barred black below. In an uncommon hepatic form, the females have rust upperparts irregularly barred black and pale buff underparts with fine black barring. Oriental Cuckoo is an uncommon summer migrant in the Kimberley (WA) and Top End (NT) and uncommon down the e. coast from Cape York to n. NSW. It is most often found in rainforests, monsoon forests, and mangroves of the north of the Top End and the QLD coast.

1
imm. 1
2
3
4
5

OWLS (STRIGIDAE)

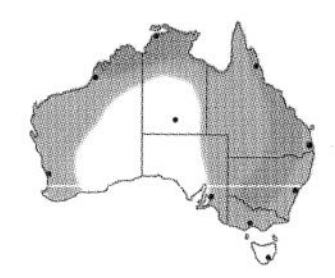

1 BARKING OWL *Ninox connivens* 39–44cm/15.5–17.5in

This large owl often appears small-headed. It is brownish above with some white spotting, and shows heavy streaks on the pale breast. It has large, staring eyes that are brilliant yellow, not greenish-toned as in Southern Boobook, which usually displays obvious dark markings (rather like black-eye bruises) around the eyes that are absent in the larger Barking Owl. Its call sounds exactly like the double-bark of a dog and is often given in duet: *woof-woof . . . woof-woof*. Less common than the more widespread Southern Boobook, Barking Owl is usually found in tropical woodlands and along wooded watercourses, mainly in the north and east of the mainland. It is significantly more abundant in the north of its range and is common in suburban Darwin (NT).

2 MOREPORK *Ninox novaeseelandiae* 29cm/11.5in

A recent split from Southern Boobook, with which it shares the prominent dark goggles, but its eyes are bright yellow instead of greenish yellow. It is darker than the widespread mainland subspecies of Southern Boobook and has a heavily white-spotted breast and upperparts. Its vocalisation is much more drawn out than that of Southern Boobook. Morepork is a fairly common bird in the forests of TAS and makes some limited winter movements to the very s. mainland, which are not well understood. *(No photo)*

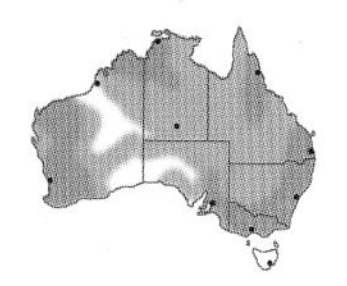

3 SOUTHERN BOOBOOK *Ninox boobook* 30–36cm/12–14in

A small, relatively small-headed, and highly variable brownish owl with greenish-yellow eyes. It is the smallest of the Australian owls, similar to Barking Owl, which has piercing bright yellow eyes. Southern Boobook displays dark markings around the eyes, reminiscent of black-eye bruises, which are usually surrounded by white markings. The *lurida* subspecies, from the rainforests of ne. QLD, is dark reddish-brown without spotting on the back and with faint white spotting on the breast. The call of Southern Boobook is a common night sound, a high-pitched *boobook-boobook*, repeated every few seconds. It is Australia's most common owl, found in almost any wooded habitat throughout the mainland.

4 POWERFUL OWL *Ninox strenua* 58–67cm/23–26.5in

A massive owl, often found roosting during the day with the half-eaten remains of a possum or flying fox in its talons. The bird has a unique pattern of brown chevrons on pale underparts. The upperparts are mainly dark brown with pale mottling and barring. The tail is brown with pale barring. It has large golden-yellow eyes. Immature birds have pale underparts with subtle brown streaking and large brown eyepatches. Powerful Owl is an uncommon bird of coastal and sub-coastal se. Australia, where it is found in wet and dry sclerophyll forests. It has adapted well to urban areas and can be found in suburban Sydney (NSW) and Melbourne (VIC). There is often a pair in the Royal Botanic Gardens in Sydney's city centre.

5 RUFOUS OWL *Ninox rufa* 46–56cm/18–22in

A very large, red-brown northern owl with fine cream barring on the throat and upper breast and broader cream barring on the lower breast and belly. It has uniform chestnut goggles and large yellow eyes. The Wet Tropics (QLD) subspecies (*queenslandica*) is a richer chestnut with less mottling on the back and less obvious goggles. Rufous Owl is uncommon in the Kimberley (WA), Top End (NT), and along the tropical QLD coast. It prefers dense forests, including rainforests, monsoon vine forests, and gallery forests. It appears to prey mainly on flying foxes, and can be found roosting near their colonies, often with a half-eaten flying fox in its talons.

1
c. 3
ne. QLD 3
se. 3
4
5

1 SOOTY OWL *Tyto tenebricosa* 33–48cm/13–19in

A large sooty-grey owl with a massive mask bordered in black. The upperparts are very dark, spotted with pale grey. In the northern Lesser Sooty Owl subspecies the facial disc is pale grey and the underparts are pale grey finely barred with dark grey, giving almost a pied look to the bird when it is seen in a spotlight. The southern Greater Sooty Owl subspecies has a breast the same colour as the upperparts with fine pale grey barring. The Lesser occurs only in rainforests of the Wet Tropics of ne. QLD and can be found in several places on the Atherton Tableland, including Mt Lewis and Mt Hypipamee. The Greater is quite rare and occurs in rainforests and wet sclerophyll forests on the coast and ranges from se. QLD south to Melbourne (VIC).

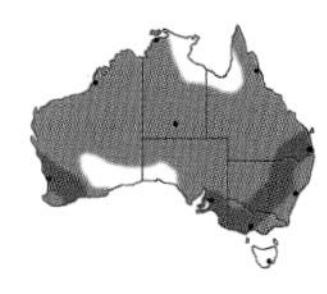

2 EASTERN BARN OWL *Tyto delicatula* 29–38cm/11.5–15in

This owl is ghostly white in colouration with indistinctly marked sandy upperparts and a characteristic heart-shaped face with a fine, faint border. Eastern Barn Owl occurs in open country, such as grasslands and open woodlands, throughout most of the Australian continent. It roosts and nests in hollow tree cavities and old buildings, including barns, as the name suggests. These owls are most likely to be seen either flying around country roads at night or perched on a roadside post, in which case their white colouration can be dazzling in the headlights of a passing vehicle.

3 AUSTRALIAN MASKED OWL *Tyto novaehollandiae* 38–47cm/15–18.5in

A large, variable *Tyto* owl with massive, powerful feet. There is a range of pale to dark morphs; the typical morph has a dark rufous back and pale underparts. Tasmanian birds are a separate subspecies (*castanops*) and a possible future species; they are very large, with dark upperparts and a rufous face and breast. The most likely confusion species, especially for the palest morphs, is Eastern Barn Owl, but Australian Masked Owl is much larger and more powerfully built and has a complete dark border to the mask. Eastern Grass Owl, although unlikely to be confused due to different habitat preferences, has much longer legs and a more elongated facial disk. Australian Masked Owl has a wide distribution but is nowhere common, and its habitat requirements are very poorly understood. Its strongholds appear to be the forests of the coasts and ranges in e. Australia and TAS, but it has been recorded locally across n., e., se., and sw. Australia.

3

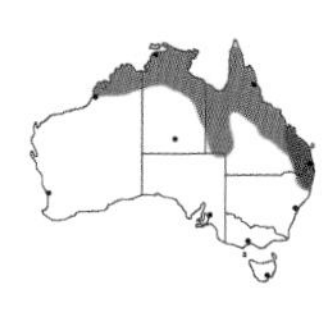

4 EASTERN GRASS OWL *Tyto longimembris* 34–38cm/13.5–15in

This bird is similar to Eastern Barn Owl but has much longer wings, generally darker plumage, and a different-shaped facial disk that gives it a 'long-faced' look. The male has dark rufous upperparts and white underparts, spotted brown. The female has rufous upperparts and orange-buff underparts, spotted brown. Both sexes are similar to Eastern Barn Owls but differ in having much longer, slimmer legs and generally darker, more heavily patterned upperparts. This owl has a very lazy flight and is usually seen flying low over grasslands with its legs hanging well below the body. It is a very uncommon owl in tropical and subtropical grasslands from the Kimberley (WA) around to n. NSW, occasionally found farther inland when conditions are suitable.

Greater 1
Lesser 1
2
2
3
4

KINGFISHERS (ALCEDINIDAE)

1 AZURE KINGFISHER *Ceyx azureus* 16–19cm/6.25–7.5in

A small, dumpy kingfisher with a very short, stumpy tail and a relatively massive bill. It has deep azure-blue upperparts and differs from Little Kingfisher, with which it sometimes occurs, in its bright apricot underside and prominent red feet. Azure Kingfisher is fairly common along rivers, creeks, small shady pools, and mangroves in n. and e. Australia. It usually hunts for fish from a low creek-side perch in the shade of an overhanging tree; it bobs its head regularly in order to pinpoint its prey and then suddenly plunges into the water, afterwards darting off at high speed to another concealed perch. It is found around coastal and sub-coastal Australia, from the Kimberley (WA) to w. VIC and TAS.

2 LITTLE KINGFISHER *Ceyx pusillus* 11.5–13cm/4.5–5in

A tiny bird and the smallest Australian kingfisher. It is deep royal blue to violet blue above and clean, shining white below, and shows a contrasting blaze of white on the sides of the neck and a bright white spot before the eye. Found along forest-lined creeks, at the edges of well-vegetated swamps, and in mangroves, Little Kingfisher can be frustratingly difficult to find as it perches low down near the water, often well hidden in the shadows. It occurs in coastal and sub-coastal areas of the Top End (NT), and again from Karumba (nw. QLD) up and around Cape York Peninsula to the c. QLD coast. It can often be seen on the Yellow Water cruise in Kakadu NP (NT) and from January to October on the Daintree River in n. QLD, with the help of local guides.

3 YELLOW-BILLED KINGFISHER *Syma torotoro* 18–22cm/7–8.5in

A distinctive small orange-headed kingfisher with a very restricted distribution. Both sexes have a green back and black nape. The male has a rufous wash to the underparts and an orange crown and orange-yellow bill. The female is white underneath and has a rufous half-collar and a prominent black cap. This kingfisher is found only in far n. Cape York Peninsula (QLD), where it is common in mangroves, rainforests, monsoon forests, and adjacent woodlands, though it is rather furtive and difficult to get a clear look at.

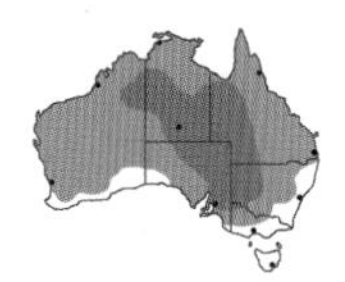

4 RED-BACKED KINGFISHER *Todiramphus pyrrhopygius* 23cm/9in

Red-backed Kingfisher has a blue back, white underside, and a white collar, much like several other kingfishers, but uniquely has a rufous rump, white supercilium, and a streaked crown. Red-backed is often found far from water in open woodlands and occurs sparsely over much of mainland Australia; it is most abundant in drier parts of the interior. It is most likely to be found sitting quietly and high on an open dead branch, from which it surveys the ground below for prey. It is found over most open habitats of the continent, and is notably absent from the Wet Tropics around Cairns (QLD), the south-west, south-east, and TAS.

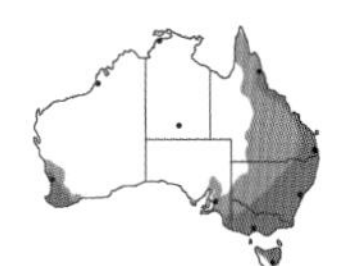

5 LAUGHING KOOKABURRA *Dacelo novaeguineae* 39–48cm/15.5–19in

Laughing Kookaburra is one of Australia's most familiar and famous birds. It differs from the similar Blue-winged Kookaburra in its dark eye, dark brownish mask around the eye, brown, barred tail, and lack of any blue on the wings or tail. It is famed for its extraordinary call, which is in fact a sequence of loud calls beginning with what has been described accurately as a 'merry chuckle' and rising to a crescendo of 'raucous laughter'. This rollicking call is often given by a pair, which then frequently triggers loud reactions from other neighbouring kookaburras, making for quite an explosive chorus. It is an abundant and familiar endemic that is widespread in the east and is common in every e. town and city. It has been introduced to sw. WA and TAS and has become well established and common there.

6 BLUE-WINGED KOOKABURRA *Dacelo leachii* 39cm/15.5in

A large kingfisher that gives very loud, raucous, obnoxious, and maniacal calls that draw substantial attention. It is told from the larger Laughing Kookaburra by its bright blue wings, its piercing white eye, and its plain, buff-coloured head, which has a capped appearance but lacks a dark eyepatch. The male Blue-winged also has a blue tail, unlike the brown-tailed Laughing, and Blue-winged is shyer and distinctly less approachable than the confiding Laughing Kookaburra. Where Laughing and Blue-winged co-exist, Laughing tends to prefer wetter areas. Blue-winged occurs in dry country and often far from water, where it hunts for a huge variety of prey items, including insects, snakes, small birds, small mammals, and even birds' eggs. Blue-winged Kookaburra is found only in the coastal and sub-coastal tropical north of Australia from the Pilbara (WA) around to near Brisbane (QLD), usually in open forests, wooded areas, and tree-lined watercourses, and is common in the Top End (NT) and the dry woodlands of the Atherton Tableland (ne. QLD).

1
2
♂ 3
4
5
6

1 BUFF-BREASTED PARADISE KINGFISHER *Tanysiptera sylvia* 30–36cm/12–14in

A sensational-looking bird decorated with long white tail streamers. It has a deep blue upperside, a long white tail, burnt-orange underparts, and a gaudy carrot-coloured bill that simply glows in the gloomy understorey. This odd kingfisher has the notable habit of nesting in active terrestrial termite mounds. Not tied to water, and usually found sitting inconspicuously in the forest understorey, it is often spotted when it betrays its position by its regular tail-flicking motion. A migrant from New Guinea, it comes to the rainforests of ne. QLD south to Rockhampton to breed from November to March.

1

2 FOREST KINGFISHER *Todiramphus macleayii* 19–22cm/7.5–8.5in

Forest Kingfisher is not tied to water. It is best separated from other kingfishers by one or more of these features: forehead markings, tone of the upperparts, rump colour, underparts colouration, and wing pattern. Forest is a clean blue-and-white kingfisher, deep blue above with a darker head and a very bold white spot on the forehead. Males have a complete white collar on the nape, like Collared and Sacred Kingfishers, although females and immatures have no collar and have a blue nape. Forest has pure, gleaming-white underparts, unlike the buff-tinged Sacred. Forest also shows a distinctive white patch on the upperwing in flight. This kingfisher can be found in open forests, open woodlands, and woodland edges; it is most likely to be encountered perched prominently on roadside wires. It is common across tropical n. Australia, from Kununurra (WA) around to n. NSW.

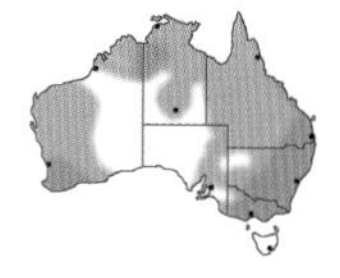

3 SACRED KINGFISHER *Todiramphus sanctus* 22cm/8.5in

A widespread kingfisher with pale blue-green upperparts, this species has buff-washed underparts and a buff collar that differentiate it from all others. The pale spot before the eye is buff-toned and indistinct in Sacred, in contrast to the bold white spot in Forest and Collared Kingfishers. Sacred is the most abundant, widespread kingfisher, found in a broad variety of habitats all over the Australian mainland and TAS, absent only from the Great Sandy Desert, Great Victorian Desert, and the Nullarbor Plain.

4 COLLARED KINGFISHER *Todiramphus chloris* 23–27cm/9–10.5in

A specialist of the coastal zone, this bird is found mainly around mangroves, which gives it its other name, Mangrove Kingfisher. The combination of deep green-aqua upperparts, a small white forehead mark, and a clean white collar and underparts firmly distinguishes this species from all others. It is a highly vocal kingfisher that often draws attention to itself by its strident calls emanating from the mangroves. Collared Kingfisher is found from the c. WA coast around the top of Australia to s. QLD. It can often be found in mangroves on the Cairns Esplanade (QLD) and at Nudgee in Brisbane (QLD).

1

2

3

4

DOLLARBIRD; BEE-EATER

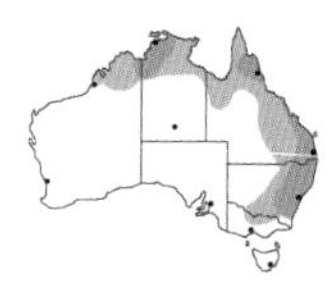

1 ORIENTAL DOLLARBIRD *Eurystomus orientalis* 25–29cm/10–11.5in

The species is named for the large, silver coin-shaped markings on its dark wings, which can be seen only in flight. It is the only representative of the roller family (Coraciidae) in Australia and therefore is a distinctive bird quite unlike any other. It is a stocky, well-built bird that is all deep blue-green glossed on the body, and has a bright, carrot-coloured bill and feet. It has a strange, rolling flight action that, along with the harsh cackling call, observers will find distinctive with experience. It occurs in all habitats within its range and uses exposed perches above the canopy. It is a common breeding migrant from New Guinea to the tropical north of Australia and the east of the country to VIC, becoming rarer in the south-east.

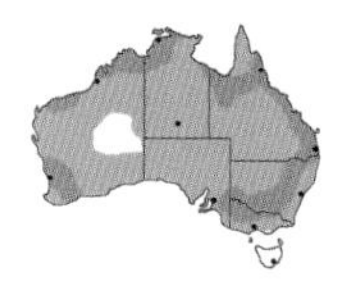

2 RAINBOW BEE-EATER *Merops ornatus* 22–25cm/8.5–10in

This representative of the bee-eater family (Meropidae) is well named, as it shows many bright and appealing colours: The head alone boasts a prominent black mask edged in turquoise blue, a rich rufous crown merging into a subtle jade-green forehead, and a bold yellow throat bordered with black and rufous. Most of the body is a greenish gem tone, turning bright azure blue on the rump and turquoise on the lower belly. When the bird takes to flight to snatch airborne prey, it reveals a distinctive body shape, two fine needle-like plumes protruding from its tail, which are shorter on the female and can be missing, as well as vivid coral-coloured underwings that contrast with the overall greenish colouration. Rainbow Bee-eater is common over nearly all of continental Australia in summer, wintering in n. Australia. Highly visible and vocal, it is found in many open areas, including around towns. The call, which sounds like a ringing telephone, draws attention to groups of these enchanting birds.

2

1
1
2
2

PITTAS (PITTIDAE)

1 NOISY PITTA *Pitta versicolor* 18–25cm/7–10in

A spectacularly colourful stub-tailed, long-legged ground-feeding bird. It is forest green above with a brilliant blue shoulder patch, which can glisten spectacularly when the sun hits it; pale buff underneath with a dark smudge on the lower belly and a deep scarlet vent; and black-headed with a conspicuous, broad chocolate-brown brow. It is especially striking in flight, when the bright iridescent blue forewings, bold white 'commas' in the primaries, and overall shape give it the impression of a tail-less kingfisher. The shape combined with the vivid colouration make it unlikely to be confused with any other bird, other than another pitta. Noisy Pitta is found in rainforests down the e. coast, and throughout its range is a partial migrant, usually breeding at higher elevations in summer and dispersing to coastal lowlands in the winter.

2 RAINBOW PITTA *Pitta iris* 16–18cm/6.25–7in

A gorgeous multi-coloured ground-feeding bird with long legs and a sturdy bill. Similar in some respects to Noisy Pitta, it is also green above with a glistening blue shoulder patch, scarlet vent, and chocolate-browed black head. However, the underside of this bird is solid black (buff in Noisy), and these species do not overlap in range; Rainbow is the only pitta species in its range. It is a locally common resident confined to the Kimberley (n. WA) and the Top End (NT), where it occurs in monsoon vine forests (where most abundant), mangroves, and bamboo stands. It is readily found around East Point in Darwin, Fogg Dam, and Howard Springs (all in the NT); it is most often seen foraging within the dingy leaf litter.

3 RED-BELLIED PITTA *Erythropitta erythrogaster* 17–19cm/6.75–7.5in

The most striking feature of this dazzlingly coloured bird is the bright crimson on the belly bordered above with a broad, soft blue band. It has deep forest-green upperparts with an all-dark head and a ginger nape. In flight it exposes blue inner wings and small white commas on the outer wing. It is unlikely to be confused with any other bird if the belly is seen. Red-bellied Pitta feeds on the ground but often calls from a raised perch, and although most easily detected when calling, it is always a difficult bird to find, much more reclusive than both Noisy and Rainbow Pittas. It is an uncommon summer migrant to n. and ne. Cape York Peninsula (QLD). It is usually found in the tropical lowland rainforests of Iron Range NP but also occurs in lower numbers in monsoon vine forests.

3

LYREBIRDS (MENURIDAE)

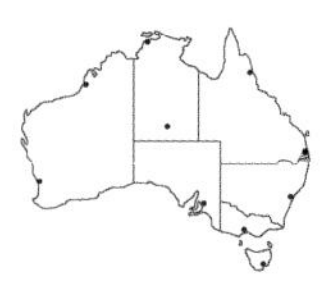

1 ALBERT'S LYREBIRD *Menura alberti* 65cm/25.5in (female); 90cm/35.5in (male)

Albert's is the smaller of the two lyrebirds. It is a large chestnut bird with a reddish throat and a long tail found on the forest floor, where it scratches in the leaf litter with its powerful feet and claws for invertebrates and seeds. It is a poor flyer, and more often runs swiftly than flies from danger. Albert's does not have quite the extensive arsenal of vocal mimicry that Superb Lyrebird is famous for, but it is known to regularly imitate Green Catbird and Satin Bowerbird among others, mixing these into a long sequence of calls with its own unique and varied vocabulary. Albert's Lyrebird is found in extreme se. QLD and extreme n. NSW only, where it is confined to wet forests on mountain ranges. With patience it can often be found on trails in Lamington NP (QLD).

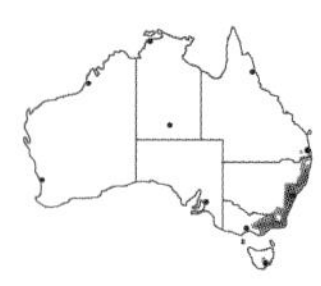

2 SUPERB LYREBIRD *Menura novaehollandiae*
76cm/30in (female); 103cm/40.5in (male)

The largest songbird on earth, Superb Lyrebird is a brown, long-tailed ground-dwelling bird, which prefers to run than fly when alarmed. While displaying (difficult to observe), it spreads its tail, arcs it over its head, and holds it like a feathered veil. Of the two lyrebirds it is the more accomplished mimic, and some remarkable individuals have been recorded accurately imitating chainsaws, camera shutters, ambulance sirens, and crying babies, along with a whole repertoire of local bird songs. Superb is not known to overlap anywhere with Albert's Lyrebird, which is very locally distributed. In areas where they occur close by, Superb prefers lower elevations and drier habitats. In the south of its range, however, it can occur in almost any forested habitat, from dry sclerophyll to temperate rainforests. Its range is to the south of Albert's, from se. QLD down into the Dandenong Ranges of VIC; there is an introduced population on TAS. Superb is more numerous and less shy than Albert's and in some areas has become quite habituated to people (e.g., Lady Carrington Drive, Royal NP, NSW).

1

2

AUSTRALASIAN TREECREEPERS (CLIMACTERIDAE)

1 WHITE-THROATED TREECREEPER *Cormobates leucophaea* 16–17cm/6.25–6.75in.

A dark-backed and dark-capped tree-hugging bird with a white throat. It has a rufous breast, belly, and vent streaked black and white on the sides. Unless seen well, it can look remarkably like Red-browed Treecreeper but lacks any form of brow and is generally far more common. The upward-spiralling feeding motion distinguishes the treecreeper from birds of all other families. This species tends to be alone or in pairs, rarely forming parties. It has two distinct populations: An uncommon smaller subspecies occurs in the rainforests and wet sclerophyll forests of the Wet Tropics in ne. QLD. The other, much more common subspecies is found in coastal and sub-coastal regions from the c. QLD coast to Adelaide (SA), in rainforests, wet and dry sclerophyll forests, and open woodlands.

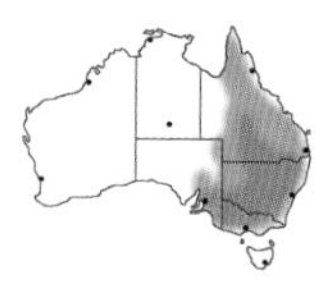

2 BROWN TREECREEPER *Climacteris picumnus* 14–18cm/5.5–7in

Largely brown in appearance, this treecreeper ranges from dark brown to grey-brown on the back and has paler brown underparts that are streaked with buff on the belly and chequered with buff on the undertail. The darker subspecies of n. QLD has a chocolate-brown back and rich buff underparts with a partial but very distinct necklace of black speckling between the buff breast and white throat. All populations display a fairly prominent pale eyebrow and indistinct buff streaking on the ears. Brown Treecreeper favours areas of open woodland where there is a lot of fallen timber and dead wood available for it to forage on. It often comes to the ground to feed. A widespread treecreeper, it is found all across the e. side of the mainland from w. QLD to e. SA.

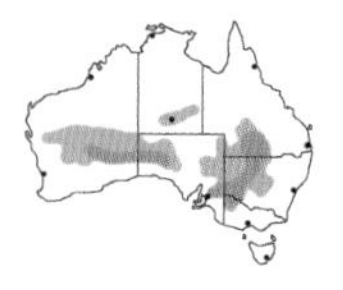

3 WHITE-BROWED TREECREEPER *Climacteris affinis* 15cm/6in

White-browed looks like a more intensely coloured and patterned Brown Treecreeper. It has grey-brown upperparts and pale grey-buff underparts that are much more heavily streaked than in Brown. Its eyebrow is white, as opposed to the cream or buff brow of the Brown Treecreeper, and it has noticeable streaking on the ear. This species has two, possibly three, very distinct population centres associated with the distribution of its favoured habitat: arid open woodlands or shrublands dominated by acacias such as Mulga (*Acacia aneura*). It can be found in stands of acacia throughout the mallee and mulga belts of nw. VIC, w. NSW, and sw. QLD; Bowra Station (QLD) is a good place to find it. It is also found in similar habitat in s. NT and from c. SA west to wc. WA.

4 RUFOUS TREECREEPER *Climacteris rufus* 16–18cm/6.25–7in

A distinctive, rich rufous treecreeper with a darker face, nape, and back, and subtle but beautiful black-and-cream streaking on the breast. Over most of its range it is the only treecreeper present, and the distinctive upward-spiralling feeding style separates it from birds of other families. It is most likely to be found in small groups, feeding either low down on tree trunks or on fallen wood in areas with plenty of dead timber, where it frequently forages on the ground. It occurs in a range of wooded habitats, including mallee, open eucalypt forests, and eucalypt woodlands dominated by Wandoo, as well as timbered watercourses, parks, and golf courses. Locally common in areas of sub-coastal mulga of WA and SA, it is especially abundant in Dryandra Woodland and Porongurup NP (WA) and rarer in SA, where it is restricted to remnant mallee areas on the Eyre Peninsula.

5 RED-BROWED TREECREEPER *Climacteris erythrops* 15cm/6in

A very similar bird to White-throated Treecreeper, unless seen in good light, in which the red brow can be discerned. This is often difficult because of this species' preference for feeding high in the canopy. Red-browed has darker and much more streaked underparts than White-throated. A communal species, generally found feeding in pairs or small groups, this treecreeper occurs primarily in mountainous country on the edge of rainforests and wet and dry sclerophyll forests. Patchily distributed and generally uncommon, it is found between Brisbane (QLD) and Melbourne (VIC). It can often be seen around Cunninghams Gap in Main Range NP west of Brisbane, or on Duck Creek Road in Lamington NP (QLD), or along Lady Carrington Drive in Royal NP (NSW).

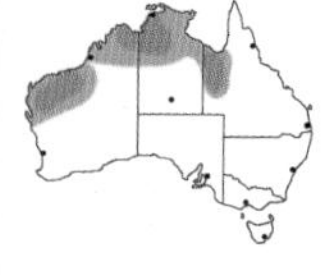

6 BLACK-TAILED TREECREEPER *Climacteris melanurus* 18–20cm/7–8in

The prettiest of the treecreepers; its upperparts are chocolate brown, underparts are chestnut. The male has a black bib delicately streaked with white, and the female has a white bib bordered on the bottom with white streaks above the chestnut breast. A smaller separate subspecies occurs in the Pilbara (WA). Pilbara males have a white bib bordered by a necklace of black-and-white streaking. Both subspecies are dark and flash distinctive large white patches on the wing when flying. This is a widespread bird in tropical savannahs across n. Australia to w. QLD. In the Pilbara it occurs in the lightly wooded gullies and watercourses in otherwise sparse countryside. It can often be seen in woodlands around Kakadu NP and south to Katherine (NT).

1
2
3
4
5
6

BOWERBIRDS (PTILONORHYNCHIDAE)

1 GREEN CATBIRD *Ailuroedus crassirostris* 28–32cm/11–12.5in

A large, fat-bodied bright green bowerbird with a stout, ivory-coloured bill, lightly white-streaked brownish-green underparts, and a prominent deep red eye. It is very similar in appearance to Spotted Catbird, from which it differs in being plainer-headed, lacking the boldly marked top to the head, and having streaked or spotted, not scalloped, underparts. Green Catbird also displays conspicuous white spotting on the wings compared to Spotted, which has more subdued wing markings. Green is found to the south of Spotted Catbird's range, in the coastal temperate rainforests of s. QLD and NSW, and is commonly seen around O'Reilly's, in Lamington NP, near Brisbane.

2 TOOTH-BILLED BOWERBIRD *Scenopoeetes dentirostris* 25–27cm/10–10.5in

Also known as Tooth-billed Catbird. A fat-bodied, nondescript brown bird with few bold features except for its heavily streaked off-white underparts and large, powerful bill. It is best identified by its distinctive, large-bodied, big-billed shape, rather than colouration. It is an uncommon bird, endemic to the Wet Tropics of ne. QLD but easily located in the mountain rainforests because of its near-incessant calling when near its bower, which is comprised of nothing more than a stage of large leaves laid down on the forest floor. It is regularly seen on Mt Lewis and in Mt Hypipamee NP.

3 SPOTTED CATBIRD *Ailuroedus melanotis* 25–30cm/10–12in

A large green bowerbird with an arrow-marked breast, a heavily streaked head with blackish sides, and a conspicuous red eye. Spotted is very similar to the closely related Green Catbird but differs in appearance in having a boldly spotted white forehead, an arrow-marked, not streaked, underside, and less conspicuous wing markings; the two are also separated in geographical range. Spotted Catbird is best detected by its strange cat-like squeals, which can even sound rather like a crying baby at times. Spotted is confined to the tropical rainforests of ne. QLD, in two populations. The northern is centred on the rainforests of Iron Range NP on Cape York Peninsula; the other is centred on the Wet Tropics from Cooktown to Paluma Range NP, where it is a widespread, locally common species.

4 GOLDEN BOWERBIRD *Prionodura newtoniana* 23–25cm/9–10in

The male Golden Bowerbird is striking, with a vivid golden-yellow belly, collar, and crown patch. The rest of its plumage is dull olive green. Females are dull-plumaged, confusing, nondescript birds best identified by a combination of structure, underparts colouration, and yellow eye. They can be distinguished from both males and females of the sympatric Tooth-billed Bowerbird by their much smaller size and their plain, not streaked, underparts. Golden Bowerbird builds a bower with a central-support 'maypole' design, re-using the same bower for years, and favours pale objects in shades of white, green, or yellow in decorating it. An uncommon and very local bowerbird endemic to the Wet Tropics of ne. QLD, it is found in mountain rainforests and is hard to find away from known bowers. Local guides usually know of bowers around Mt Lewis and Mt Hypipamee NP on the Atherton Tableland and around Paluma near Townsville.

5 REGENT BOWERBIRD *Sericulus chrysocephalus* 24–28cm/9.5–11in

One of the most spectacular birds in Australia, Regent Bowerbird is all glossy black except for a gaudy golden hood and a broad golden flash in the wing that is dazzling in flight. Females are all greyish brown, mottled and spotted with pale markings on the back, and possess a bold black cap that constitutes their most prominent feature. Unlike some bowerbirds, Regents do not maintain their bowers for a long period (usually 10 days or less) and regularly destroy their own bowers when rivals detect them. This is a scarce species restricted to coastal temperate rainforests in s. QLD and n. NSW, where it is often inconspicuous despite the male's striking appearance. It is easy to see during summer at O'Reilly's in Lamington NP, near Brisbane.

1
2
3
3
♂ 4
♀ 4
♂ 5
♀ right 5

BOWERBIRDS (PTILONORHYNCHIDAE)

1 SATIN BOWERBIRD *Ptilonorhynchus violaceus* 27–33cm/10.5–13in

Satin Bowerbird male is a stocky, glossy deep-blue bird with a stout pale horn-coloured bill, similarly coloured legs, and a beautiful, soft violet-coloured eye. The female is equally beautiful: olive on the upperparts, with a darker bill, rusty-coloured wings and tail, boldly scaled underparts, and a similarly seductive violet eye. Satin Bowerbird favours blue objects, often decorating its bower with blue bottle tops, ring pulls, and straws, and the bowers are easy to locate. The birds are most likely to be seen bounding around on the ground in the early morning, foraging for prey such as fruit or insects, or searching for items to ornament their bowers. The species is common in the mountain forests of ne. QLD and temperate rainforests and wet eucalypt forests of se. QLD, e. NSW, and VIC. It is a large and distinctive bowerbird that is familiar to many people in its range, as it can be remarkably approachable and conspicuous in some areas (e.g., Lamington NP).

2 GREAT BOWERBIRD *Chlamydera nuchalis* 32–37cm/12.5–14.5in

The largest bowerbird. It is a dowdy species, grey-brown in colouration, with darker mottling on the upperparts and a well-concealed violet-pink crest on the nape that is rarely seen outside of courtship displays. It builds a U-shaped 'avenue' bower that it re-uses for years. It inhabits dry woodlands, wooded watercourses, and tropical savannahs. It frequently comes to picnic grounds, and the collection of white objects around its bowers can often be found near parks and schools. Great Bowerbird is common in its range, which covers the tropical north of Australia from n. WA eastward through n. QLD.

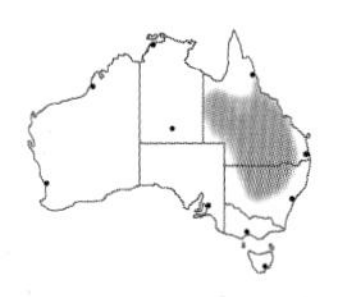

3 SPOTTED BOWERBIRD *Chlamydera maculata* 28–30cm/11–12in

An indistinctive inland bowerbird that is brown above with large fawn spots down its nape and back. The throat and breast are rich cream vermiculated with brown, which merges into a buff belly with faint brown vermiculations. The male has a pink nape that is generally indistinct. Spotted Bowerbird occurs in a variety of open woodlands, often in areas with mulga or she-oak groves. It likes to decorate its bower with white and shiny objects. It is a generally uncommon bird found in inland e. Australia from c. NSW north to the tropics. It seems most common around the NSW–QLD border and is regularly seen at Bowra Station in sw. QLD.

4 WESTERN BOWERBIRD *Chlamydera guttata* 27–28cm/10.5–11in

Western Bowerbird seems a more interesting, better-looking version of Spotted Bowerbird. Its back is brown, heavily spotted with cream. The throat and breast are brown, heavily marked with almost arrow-shaped cream spots. The nape is violet, much brighter than Spotted's nape. A common bowerbird of c. Australia, Western Bowerbird is regularly found in the mulga groves around the NT's West MacDonnell Ranges south to Uluru (Ayers Rock). It likes to decorate its bower with white and green objects. A disjunct and less-common population exists in the mulga district of c. WA.

5 FAWN-BREASTED BOWERBIRD *Chlamydera cerviniventris* 25–30cm/10–12in

This bowerbird has a grey-brown back with some faint buff scaling. From chin to vent it is cream; fine brown-and-buff streaking on the throat and breast merge with streaking on the face to give the bird a pale-cheeked appearance. The smaller size and richer colours separate it from Great Bowerbird, which is the only likely confusion species in Fawn-breasted's range. It has a very restricted distribution within Australia, only on far n. Cape York Peninsula (QLD), where it is uncommon in monsoon thickets, mangroves, dry heaths, and adjacent savannah woodlands.

♂ 1
♀ 1
2
3
4
5

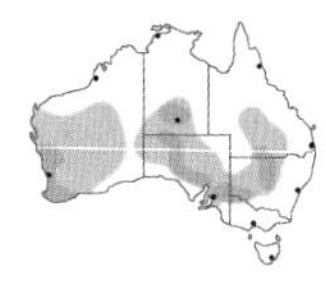

1 SPLENDID FAIRYWREN *Malurus splendens* 11–14cm/4.25–5.5in

Some birders consider this the most beautiful bird of them all. The male is an almost entirely cerulean-blue wren with striking powder-blue crown and cheeks and just a few narrow black markings—through the eye, across the nape, and across the breast—breaking up this bright blue colouration. The intensity of the blue varies with subspecies. The eastern bird, known as Black-backed Fairywren, is powdery blue on the crown, its cheeks and belly are paler, and it has a black rump and lower back. All females are plain pale brown on the back, whitish underneath, and have a blue tail and a dull reddish bill and patch around the eye. Although it occurs in scrubby habitats, Splendid Fairywren tends to be more abundant in wooded habitats, such as mulga and mallee, and is especially common in open eucalypt woodlands in sw. WA. It is common and quite vocal and often draws attention when in groups with its high-pitched twittering calls.

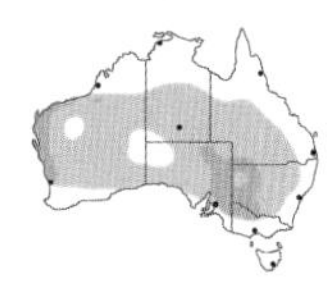

2 WHITE-WINGED FAIRYWREN *Malurus leucopterus* 11–13cm/4.25–5in

From a distance these birds look like blue Ping-Pong balls in the fields. The male is such a conspicuous and striking bird—bright blue with white wings—that it is unlikely to be confused with any other. The subspecies (regarded as a separate species by some people) found on Dirk Hartog and Barrow Islands off WA is black with white wings. The female is less obvious, pale brown on the back and plain whitish below, with a bluish wash to the tail and no markings around the eyes or lores. These are common birds of semi-arid, treeless scrubby or grassy country, especially areas of saltbush, spinifex, cane grass, or bluebush. They are found from the WA coast (though absent in the extreme south-west) across the continent to about Hay (NSW) and St. George (QLD).

3 RED-BACKED FAIRYWREN *Malurus melanocephalus* 10–13cm/4–5in

The male Red-backed Fairywren is distinctively marked: It is a blackish bird with a striking crimson saddle across the back. The female is pale with a strong fawn wash, a reddish-pink bill, and no markings around the eyes. The species responds well to pishing. It occurs in tall grasses within open woodlands, tropical savannahs, rainforest edges, scrublands, lantana thickets, and thickets along the edges of eucalypt forests, as well as town parks and gardens in some areas. In the s. part of its range it occurs in wet heathlands and around cane fields. Red-backed Fairywren is common across the tropical north from n. WA east to n. QLD, and in the east from n. QLD south to n. coastal NSW.

3 imm.

♂ w. 1
♂ e. 1
♀ 1
♂ 2
♀ 2
♂ 3
♀ 3

1 PURPLE-CROWNED FAIRYWREN *Malurus coronatus* 14–15cm/5.5–6in

Purple-crowned Fairywren is very different from all other fairywrens. The male is strikingly marked with a black mask and a broad lilac ring surrounding an indistinct, narrow black crown stripe. Otherwise it is brownish with a blue tail and pale whitish underparts. Females are also gorgeous, with their rich chestnut ear patches contrasting with the pure white throat and a thin white brow running from the bill to above the eye. Less gregarious than other fairywrens, Purple-crowned is more likely to be encountered in pairs than in large family groups. It is a habitat specialist, inhabiting tall stands of cane grass within thickets of *Pandanus* and paperbarks along watercourses. It has two tiny, discrete populations: one along the Gulf of Carpentaria (ne. NT to nw. QLD), and one in the Kimberley region (nw. WA) and adjacent NT. It is unlikely to be seen unless specifically targeted at one of the few known sites, such as the Victoria River Crossing in the sw. Top End (NT).

2 SUPERB FAIRYWREN *Malurus cyaneus* 11.5–14cm/4.5–5.5in

Superb is a common eastern fairywren. Males differ from all other fairywrens in having two well-separated blue areas on each side of the head delineated by a broad black eye line. It also lacks the bright chestnut shoulder present in the males of many other fairywrens. Females are more challenging if encountered with no males (which rarely occurs); they are all brownish with a dark reddish bill and eye ring, and female Superb lacks the white tail tips of female Variegated. Superb is the only fairywren represented on TAS. This common blue bird inhabits thickets in a wide variety of habitats that include eucalypt forests, salt marsh, heathlands, scrubby areas, woodland edges, and even fringes of mallee, where confiding and vocal family groups are usually encountered feeding low down in vegetation and on the ground. This common and tame fairywren is found in se. QLD, e. NSW, VIC, se. SA, and TAS.

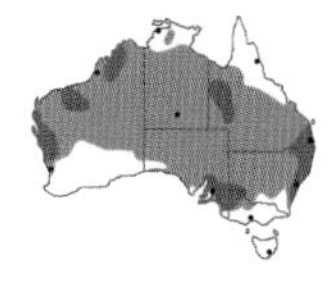

3 VARIEGATED FAIRYWREN *Malurus lamberti* 11.5–14cm/4.5–5.5in

Australia's most widespread fairywren. Like males of several other fairywrens, the male Variegated has a blue crown contiguous with a pointed blue ear patch, a black throat and upper breast, a white lower breast, and a large chestnut shoulder patch. The blue markings are not uniform and have a bright, shimmering iridescent quality. There are distinct subspecies, which differ in the intensity of the blue crown colour. The males can be separated from those of Lovely, Blue-breasted, and Red-winged Fairywrens by the light blue edges at the sides of the breast, lacking in those species. Females differ much more among subspecies; over most of Australia they are sandy brown above and buff below, with a white-tipped blue tail and chestnut lores and eye rings that create a masked appearance. Superb Fairywren females differ in lacking a fine white tip to the tail. Variegated females from Arnhem Land (NT) and the Kimberley region (WA) are white below and have black wings and beautiful powder-blue upperparts. Small groups of this common bird generally favour undergrowth and clearings within habitats such as open woodlands, eucalypt forests, mallee, mulga, and even rainforest edges in e. Australia. Variegated is found all over mainland Australia except the s. and n. extremities.

♂ 1
♀ 1
♂
♀ 2
♂ e. 3
♂ Kimberley 3
♂ c. 3
♀ 3

1 LOVELY FAIRYWREN *Malurus amabilis* 13cm/5in

A tropical rainforest fairywren. The male is similar to the much more common Variegated Fairywren but has a more pied appearance on the underparts, a shorter tail, and a uniform electric-blue crown and ear colour. The female is much more distinctive, having white underparts and a blue head, nape, back, and tail. Her wings are brown, and she has very a prominent white eye ring and lores that create a white-masked appearance. Lovely Fairywren is common on n. Cape York Peninsula and regular south along the coast to near Townsville (QLD). It prefers margins and clearings in rainforests, mangrove edges, and monsoonal thickets with adjoining open forests.

2 RED-WINGED FAIRYWREN *Malurus elegans* 14–16cm/5.5–6.25cm

Very similar to the smaller Blue-breasted Fairywren. Red-winged male has more of an aqua-blue hue to the crown and contiguous ear patch. The shape of the ear patch is more pointed (less splayed) than in Blue-breasted, and the belly and vent are buff rather than white. Females have a black bill and a grey-brown tail, distinguishing them from the very similar Blue-breasted female, which has a bluish tail. Red-winged Fairywren prefers thickets in wet and dry sclerophyll forests, and wet heaths surrounding swamps. It is harder to locate and somewhat shyer than Blue-breasted Fairywren. Red-winged is fairly common in extreme sw. WA.

3 BLUE-BREASTED FAIRYWREN *Malurus pulcherrimus* 15cm/6in

The common blue-crowned fairywren of the south-west. The male looks very similar to the male of the more widespread Variegated, with a blue crown and contiguous splayed ear patch, and chestnut wing patches. However, Blue-breasted's throat and breast are blue-black, with a sheen when seen in good light, and the bird shows a distinct demarcation between the breast and the white underparts, and has no blue to the sides of breast. It is told from the larger Red-winged Fairywren by having less of an aqua tint to the blue of the crown and back, and by the blue sheen to the breast. The female looks practically identical to the overlapping subspecies of Variegated Fairywren, with chestnut lores and bluish tail, though if seen extremely well together for comparison, Blue-breasted female appears colder-toned on the underparts. The species is a common bird of dry sclerophyll forests, open eucalypt woodlands, mallee, and coastal heathlands of sw. Australia and mallee of the Eyre Peninsula (SA). It is easily drawn out by pishing and readily seen.

3 imm.

♂ 1
♀ 1
♂ 2
♀ 2
♀
♂ 3

1 SOUTHERN EMU-WREN *Stipiturus malachurus* 16–19cm/6.25–7.5in

A tiny bird with a ridiculously long tail for its size; the tail's filamentous feathers resemble the quills of an Emu, hence the name 'emu-wren'. The male is chestnut below with a white belly and an extensive, well-defined powder-blue bib; the upperparts are sandy brown with intense black streaking. The crown is rufous with black streaking, and the face and ear patch are indistinct powder blue with fine black streaking. Females lack the powder-blue face and bib and have a more uniform rufous appearance to the face and underparts. Southern Emu-Wren is very unobtrusive and is usually more common than it appears. It is found around the coastal heathlands of s. Australia from Perth (WA) to n. NSW, and tends to prefer those with a cover between 30 and 90cm (1–3ft) tall.

1 ♂

2 MALLEE EMU-WREN *Stipiturus mallee* 13cm/5in

Mallee Emu-Wren seems a very washed-out version of Southern Emu-Wren. The male is pale sandy brown above and buff below with a powder-blue face and bib. The blue bib merges into the buff underparts, showing less contrast than in Southern. The female is buff below with a faint blue wash to the face and cheek patch. This rare and shy emu-wren has a very limited distribution based around mallee stands with healthy spinifex ground cover in nw. VIC and se. SA.

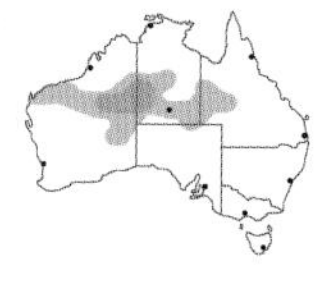

3 RUFOUS-CROWNED EMU-WREN *Stipiturus ruficeps* 13cm/5in

A gorgeous desert emu-wren with a shorter tail than Mallee or Southern Emu-Wrens. The male has an unstreaked bright rufous crown and upper nape; the sandy-brown nape and back are streaked black, though not as intensely as in Southern. The large blue face and bib area show limited black streaking around the face and ear, and fine streaking on the lower edge of the bib. Underparts are a soft rufous with a white central belly. The female is rufous above with brown streaking, buff below, and has a soft blue face and ear patch with indistinct black streaking. This is a locally irruptive species, with large population fluctuations within its range in arid c. and w. Australia. Its core range is to the west of Alice Springs (NT), and it is usually found in healthy spinifex clumps with sparse mulga or eucalypt cover.

♂ 1
♀ 1
♂ 2
♀ 2
♂ 3

1 WHITE-THROATED GRASSWREN *Amytornis woodwardi* 20–23cm/8–9in

A stunning species, this bird is found on the spinifex-clad sandstone Arnhem Land escarpment in the NT. Like Black Grasswren, it is mainly black and chestnut and covered in fine white streaks, but unlike that species it has a prominent white throat. There are no easily accessible sites to see this grasswren, and where it does occur it is difficult to see, scurrying around the escarpment like a rat and hiding behind clumps of spinifex.

2 BLACK GRASSWREN *Amytornis housei* 20cm/8in

This beautiful bird is found only in the remote and rugged sandstone escarpments of the Kimberley in nw. Australia. The male has a black head and underparts and a rich chestnut-coloured back. The female is similar but also has a chestnut breast. Both sexes are covered in fine white streaks. It is quite easy to see for a grasswren, often found in small vocal groups that actively move around their territories; the difficulty lies in getting to where it occurs. Mitchell Falls (WA) is a very reliable place to see it.

3 GREY GRASSWREN *Amytornis barbatus* 19cm/7.5in

This secretive bird has pale rufous upperparts and a white breast and face. The face has an intricate pattern of black-and-white markings, while the rest of the bird is covered in fine white streaks. It lives in lignum and cane grass, both spindly dense grasses that grow in and around remote desert swamps, in the Bulloo Overflow region of sw. QLD and nw. SA, where its distribution is very localised. The furtive and shy birds live in small groups deep in these thickets and are very difficult to see. The species is extremely unlikely to be encountered unless specifically targeted.

3

4 CARPENTARIAN GRASSWREN *Amytornis dorotheae* 18cm/7in

This small grasswren has rufous upperparts covered in fine white streaks and a slightly darker head. The throat and breast are white, bordered by a black whisker mark below the eye. The species is found on sandstone ridges and escarpments along the s. edge of the Gulf of Carpentaria in the NT and QLD, and spinifex clumps are prominent wherever it is found. It is extremely shy and difficult to see in this habitat, bouncing away behind the grasses at the slightest disturbance.

1
♀ 2
3
4

1 STRIATED GRASSWREN *Amytornis striatus* 15–20cm/6–8in

A small grasswren with several subspecies that range widely throughout arid c. Australia. The colour of the upperparts varies across its range, from rich red-brown in nw. Australia to duller greyish brown in the south-east. The throat is white, and the breast is brownish, while the head, back, and upper breast are finely streaked with white. Spinifex is an important component of its habitat, and like most grasswrens it is difficult to see as it skips between clumps. In the south-east it can be found in Hattah-Kulkyne NP in VIC and also at Gluepot Reserve in SA. It is also seen occasionally around Uluru (Ayers Rock) in c. Australia.

1

2 SHORT-TAILED GRASSWREN *Amytornis merrotsyi* 15cm/6in

In spite of the name, this bird still has a relatively long tail, though it is short compared to that of the closely related Striated Grasswren. The plumage is very similar to that of Striated: brownish upperparts and breast, and a white throat. The upperparts and upper breast are covered in fine white streaks. The species is found in the Flinders and Gawler Ranges of s. SA; Stokes Hill Lookout in the Flinders Ranges is the most reliable place to see it.

3 EYREAN GRASSWREN *Amytornis goyderi* 15cm/6in

This small grasswren has pale rufous upperparts covered in fine white streaks. The throat and breast are white. Unlike other grasswrens this species has quite a thick, stubby bill that is a pale grey colour. Occurring in the Simpson and Strzelecki Deserts of c. Australia, the bird is usually found in clumps of cane grass growing on sand dunes. There are several known locations on the Strzelecki and Birdsville Tracks where it can be found fairly reliably.

s. 1
QLD 1
2
3

1 KALKADOON GRASSWREN *Amytornis ballarae* 16cm/6.25in

This grasswren has a very restricted range confined to the rugged slopes in the Mt Isa area of nw. QLD. It is slightly richer in colouration than the similar Dusky Grasswren, but the two don't overlap in range. There are several locations close to Mt Isa where this bird can be seen, such as Mica Creek, although like most grasswrens it can be quite shy.

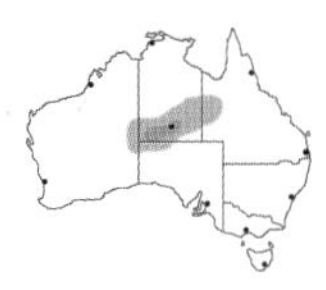

2 DUSKY GRASSWREN *Amytornis purnelli* 16–18cm/6.25–7in

This bird is found in the rocky ranges of c. Australia, living among clumps of spinifex and various small shrubs. It is rich rufous with greyer underparts and a slightly darker head. Like all grasswrens it is covered in fine white streaks. The female has rufous flanks. The species is fairly common within its range and can be seen rather easily in the West MacDonnell Ranges near Alice Springs (NT); Simpsons Gap and Ormiston Gorge are good places to look for it. It can also be found at Kings Canyon on the way to Uluru (Ayers Rock) from Alice Springs.

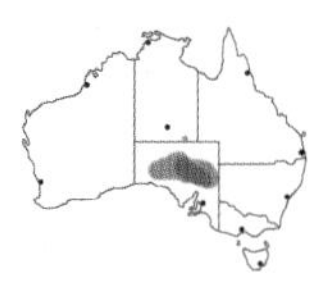

3 THICK-BILLED GRASSWREN *Amytornis modestus* 16–19cm/6.25–7.5cm

This grasswren is found mainly in shrubby areas in the stony deserts of c. SA, and although nearly identical to Western Grasswren, it is separated from the Monkey Mia (WA) population of that species by a vast distance. Thick-billed is typically very shy and furtive as it hides or bounces around between shrubs. It is a pale grey-brown, slightly paler below, and covered in fine white streaks. The female has rufous flanks. Although difficult to see, it can be found in a few spots on the s. Strzelecki Track, such as around Lyndhurst. *(See photo of following species)*

4 WESTERN GRASSWREN *Amytornis textilis* 16–19cm/6.25–7.5in

This species has been recently split from the very similar Thick-billed Grasswren, and the two look nearly identical. The affinity of the s. SA birds is still unknown, but the main Monkey Mia population of this species and the Gawler Ranges population of Thick-billed Grasswren are separated by thousands of kilometres. Western Grasswren is generally rufous above, slightly paler below, and covered in fine white streaks. The female has rich rufous flanks. It is very shy, skittering around and hiding in low shrubby vegetation on the sand plains of its habitat. The best place to see it is Monkey Mia in Shark Bay on the coast of WA.

♀
♂
1

2

4

SCRUBBIRDS; BRISTLEBIRDS

1 RUFOUS SCRUBBIRD *Atrichornis rufescens* 18cm/7in

Good luck! Even when you are close to this bird, it remains exceptionally (and frustratingly) difficult to observe. Rufous Scubbird (family Atrichornithidae) is dark rufous above and finely and indistinctly vermiculated with chocolate brown. It has a white moustachial streak and a black chin merging into the pale rufous breast, all of which are finely barred. The females are more diffusely coloured, dark rufous with a paler rufous throat and bib. To glimpse this bird you should expect to spend many stationary hours waiting for it to peek out from cover. Rare and highly restricted to mountain rainforests of n. NSW and s. QLD, it is usually seen in fern groves at the base of Antarctic Beech (*Nothofagus moorei*) stands in Border Ranges NP or Lamington NP.

2 NOISY SCRUBBIRD *Atrichornis clamosus* 22–23cm/8.5–9in

Much easier to see than Rufous but still notoriously elusive. Noisy Scrubbird has short wings and a long tail and is almost never seen flying. Birds are dark brown above and finely vermiculated with chocolate brown. The male has a prominent black bib and very obvious white moustachial streak; the chest is buff, merging into a dark brown belly and vent. The female has a buff throat and breast Highly restricted to coastal heathlands of extreme sw. WA, it prefers low heath with some emerging eucalypt species. Although still rare, it is most common near Cheynes Beach in Waychinicup NP. Although easy to hear here in spring, this bird is extremely difficult to locate, best found by waiting for it to cross the road at very regular locations. Note that although it is very vocal in the heathland, it tends to stop calling a good two minutes before it runs across the road. Stay focused!

3 RUFOUS BRISTLEBIRD *Dasyornis broadbenti* 25–28cm/10–11in

The largest, most distinctively coloured, and easiest to see of the bristlebirds (family Dasyornithidae). It has olive nape, shoulders, and back, rufous wings and tail, and a prominent rufous cap and cheek patch. The white eye line and eye ring give it a goggled appearance, even at a distance. Underparts are paler rufous, with heavy grey scalloping on the throat and upper breast, gradually becoming more diffuse and less distinct down to the belly. Although easier to see than the other two bristlebirds, Rufous is still a very difficult bird to get a good look at. It is the most widespread bristlebird, though that is not saying much, as it is found only from e. coastal SA to near Melbourne (VIC).

4 WESTERN BRISTLEBIRD *Dasyornis longirostris* 18–20cm/7–8in

Another skulker unlikely to be seen well. Although it has a fine bill and long tail, it gives the impression of an immature *Zoothera* thrush with its feeding action and its plumage, which shows grey arrow-marked scalloping on the rufous upperparts and buff vermiculated scalloping on the grey-brown underparts. This bristlebird is highly restricted to coastal heathlands of extreme sw. WA, where it prefers low heath with some emerging eucalypt trees. Although still rare, it is more common at Two Peoples Bay NR.

5 EASTERN BRISTLEBIRD *Dasyornis brachypterus* 22cm/8.5in

A skulking bird unlikely to be seen casually. Upperparts are plain brown with a rufous wash to the wings. Underparts are grey-buff with irregular and indistinct darker brown blotching. The tail is long but never cocked. It has a pale eyebrow, which is discernible only at close range. This rare and endangered species has an extremely patchy and localised distribution from s. QLD to Howe Flat in e. Vic. In the north it lives in tussock grass in dry sclerophyll forests on the edge of the rainforests. In the south it is limited to low upland heathlands where the low heath merges with the taller banksias.

♀ 1
♀ 2
3
4
5

AUSTRALASIAN WARBLERS (ACANTHIZIDAE)

1 ROCKWARBLER *Origma solitaria* 13–14cm/5–5.5in

Formerly known as Origma. This is a plain, uniform reddish bird with a whitish throat and a relatively long tail. It really is the featurelessness that makes this bird distinctive. An extremely localised, warbler-like bird, it is confined to areas of Hawkesbury sandstone outcroppings in se. NSW. The entire population occurs within an approximately 250km (155mi) radius of Sydney. It is locally common in its limited range, where it occurs on outcrops and is most often seen hopping over them. It is generally not active until after the sun hits the rocks in the morning.

2 REDTHROAT *Pyrrholaemus brunneus* 11.5cm/4.5in

A truly dull-looking bird. It has olive-grey upperparts, pale grey underparts, and white outer tips to the tail. Both male and female have obvious white lores and a partial eyebrow. Males have an orange throat; the females lack even this. This uncommon species has a wide distribution through c. Australia and has a preference for mulga associations in the w. part of its range. In the e. part it has a stronger association to mallee areas with spinifex and saltbush.

3 PILOTBIRD *Pycnoptilus floccosus* 17cm/6.75in

A stocky, terrestrial thrush-like species. Upperparts are uniform chocolate brown; the frons and throat to vent are rufous, and there is faint cream barring on the breast. It has a slightly paler rufous eye ring and lores surrounding the red eye. Pilotbird has an occasional habit of feeding with Superb Lyrebirds, though is often found well away from the nearest lyrebird. It occurs in wet sclerophyll forests, rainforests, and dense wet heath groves. It is patchily distributed in the coastal regions from Sydney (NSW) to Melbourne (VIC).

4 FERNWREN *Oreoscopus gutturalis* 14cm/5.5in

A hard-to-see but beautiful little skulker. Fernwren is usually seen in very deep shade, and at first impression may appear black, but it is chocolate brown. Short-tailed and long-billed, the bird has a front-heavy appearance. The male has a black bib, glowing white throat, and thin but very obvious white eyebrow. The female is extremely similar but has a reduced bib, buffier throat, and cream eyebrow. Fernwren walks along the ground under leaf litter on the floors of the mountain rainforests and is best located by its high-pitched whistle. A noisy feeder, it can also often be found by the rustling of leaves on the forest floor. It is readily seen on Mt Lewis and Mt Hypipamee NP (QLD).

5 YELLOW-THROATED SCRUBWREN *Sericornis citreogularis* 12–15cm/4.75–6in

A cute yellow ground bird that glows in the understorey. Yellow-throated Scrubwren is washed yellow below, most strongly on the throat, is olive-backed, and has a yellow eyebrow that gives the black face a prominent masked appearance. It prefers the denser interior forest areas well away from the forest edge. It has two distinct population centres. The northern population is centred in the mountain rainforests of the Atherton Tableland from Cooktown to Paluma (QLD). The southern population is in the subtropical rainforests from s. QLD to around Wollongong (NSW).

5

1
♂ 2
3
♂ 4
5

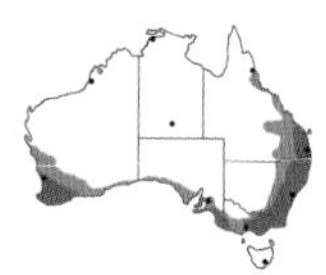

1 WHITE-BROWED SCRUBWREN *Sericornis frontalis* 12–13cm/4.75–5in

The common scrubwren of e. and s. Australia. It has a dull whitish throat and uniform brown upperparts, a pale beady eye, and a dark facial mask bordered above and below by a very prominent white eyebrow and obvious white moustachial streak. Many subspecies occur; birds from ne. QLD have a yellow wash to the breast and a less obvious mask; se. birds have a very obvious mask; and those of the western subspecies have black striations down a buff breast. The species is very liberal in choice of habitats, found in both wet and semi-arid environments, anywhere with a shrubby understorey.

2 TASMANIAN SCRUBWREN *Sericornis humilis* 13cm/5in

From a birder's perspective, this is the TAS equivalent of White-browed Scrubwren. It differs in its slightly larger size and its more uniformly pale brown colouration below. A common TAS endemic, it is found in most types of undergrowth around the island.

3 SCRUBTIT *Acanthornis magna* 11cm/4.25in

A very distinctive scrubwren of TAS with very obvious white underparts. It has a brown back with white chevrons on the shoulder, white edging to the primaries, and a black subterminal band on the chestnut tail. The face shows a prominent white eye and thick white lores extending to a white eyebrow. Shyer than other scrubwrens, it feeds inconspicuously near and on the forest floor. Endemic to TAS, it is common in the temperate rainforests and wet sclerophyll forests of the island's west. It occurs in lower numbers in the east, though is regularly found on Mt Wellington.

3

4 TROPICAL SCRUBWREN *Sericornis beccarii* 11.5cm/4.5in

A very small, relatively ornate scrubwren. It has olive-brown upperparts with white chevrons on the darker shoulders and nankeen underparts. The face shows black lores, a white line above the lores, a fine eyebrow, and a broken eye ring. The pale throat extends to just under the brown ear patch. It is most likely to be confused with an immature White-browed Scrubwren, though the ranges do not overlap. Tropical Scrubwren is limited to the rainforest understorey of Iron Range NP and the monsoon forests and thickets of n. Cape York Peninsula. It is common in most of its range.

5 ATHERTON SCRUBWREN *Sericornis keri* 14cm/5.5in

This bird has the most restricted range of all brown scrubwrens. Its upperparts are green-brown; the underparts are fawn-brown. It has a subtle grey wash to the lores, face, and ear patch. Atherton is very similar to the more widespread Large-billed Scrubwren but has a very different feeding style. This species feeds singly or in pairs low in rainforest bushes and on the forest floor, whereas Large-billed feeds in noisy groups higher in the trees. Atherton Scrubwren is endemic to the upland rainforests of the Atherton Tableland (QLD).

6 LARGE-BILLED SCRUBWREN *Sericornis magnirostra* 12cm/4.75in

A very sombre, uniformly coloured scrubwren. It has a long, pointed bill and a red eye, and is darker brown on the upperparts than the underparts. This scrubwren feeds sometimes very high in trees and almost never on the ground. Found from the Wet Tropics (QLD) down the e. coast to near Melbourne (VIC), it is a common species of the se. rainforests and the upland rainforests of the Atherton Tableland.

w. 1
se. 1
se. 1
2
3
4
5
6

AUSTRALASIAN WARBLERS (ACANTHIZIDAE)

1 CHESTNUT-RUMPED HEATHWREN *Calamanthus pyrrhopygius* 14cm/5.5in

A shy, inconspicuous ground bird with a cocked tail that has conspicuous white tips to the black outer tail feathers. The upperparts are dull brown with a chestnut wash to the shoulder and a bright chestnut rump and base to tail. The throat and breast are buff with blackish streaking, merging into a fawn belly. It has a buff eyebrow and indistinct scalloping on the base of the primaries. This is an uncommon and patchily distributed bird in coastal and sub-coastal NSW, VIC, and SA. It occurs in heath and in eucalypt woodlands with a heath understorey. A true skulker, it rarely exposes itself for any period. It is regularly sighted is the heathlands of Royal NP (NSW).

2 STRIATED FIELDWREN *Calamanthus fuliginosus* 13.5cm/5.25in

A gorgeous little thing that pops up on the top of the heath to call. This bird is heavily streaked all over, with black streaks on the deep olive upperparts and on the buffy-yellow underparts. The tail appears shorter than in similar species but is still cocked most of the time. It is an uncommon species found along the se. Australian coast and TAS. On the mainland it prefers coastal heath and swamp edges. In TAS, where it is more common, it also occurs in alpine heathlands and can be found at the top of Mt Wellington.

3 RUFOUS FIELDWREN *Calamanthus campestris* 13.5cm/5.25in

This species is less intensely coloured than Striated Fieldwren and looks like a washed-out version with much finer streaking both above and below. Because of the finer streaking, the white eyebrow looks far more obvious in this species than in Striated. Rufous Fieldwren is widely distributed within arid s. Australia and possibly has three disjunct populations. In the west it prefers coastal heaths and dry, open woodlands with a heath-dominated understorey; and in the east of its range it has a strong preference for rocky areas with sparse eucalypts and a spinifex ground cover.

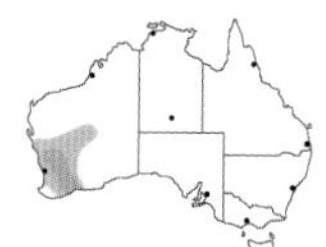

4 WESTERN FIELDWREN *Calamanthus montanellus* 13.5cm/5.25in

Formerly regarded as a subspecies of Rufous Fieldwren, it looks very similar, with the fine black streaking that differentiates both from the more thickly streaked Striated Fieldwren. Western differs from Rufous in being more intensely coloured with darker streaks. It is found in WA from south of Geraldton across to just east of Fitzgerald River NP.

5 SHY HEATHWREN *Calamanthus cautus* 12cm/4.75in

A very distinctive bird within its range. It has clean white underparts with stripes of black pearl-shaped spots, and is uniform brown above, with an extremely prominent clean white eyebrow, black-and-white scalloping on the bases of the black primaries, which also shows in flight, bright rufous rump and large white corners to the dark tail. Not as shy as the name suggests, it is easier to lure with pishing than Chestnut-rumped. It differs from Chestnut-rumped in the very strong white eyebrow and colder-toned underparts. Shy Heathwren is a widely distributed bird within the mallee belts from coastal WA to w. NSW and VIC. It prefers stands of healthy mallee, where it can be locally common, such as Round Hill (NSW) and Gluepot Reserve (SA).

6 SPECKLED WARBLER *Pyrrholaemus sagittatus* 12cm/4.75in

A very distinctive bird reminiscent of Louisiana Waterthrush (*Seiurus motacilla*) of the Americas. Its upperparts are rufous-brown streaked with chocolate; the cap is black streaked with cream; and the underparts are cream streaked with black. The buff lores, brow, and underbrow, which give the face a masked appearance, contrast with a blackish ear patch finely streaked with white. If seen even moderately well it is highly unlikely to be confused with any other species. Speckled Warbler occurs in the rangelands of se. Australia. It favours open forests and woodlands with a thick, dry leaf litter. Canopy type does not seem to be a limiting factor, as the bird is found in mulga and mallee of c. NSW, and *Callitris*-dominated open woodlands such as those in Binya SF (NSW).

1
2
3
4
5
6

1 BROWN GERYGONE *Gerygone mouki* 10cm/4in

Brown Gerygone is fairly common in dense forests and rainforests down the e. coast of Australia. It can be difficult to see well in the gloom of the rainforest, particularly as it rarely sits still. The often-repeated *wit-chit-chit, wit-chit-chit* is a common sound in rainforests of the south-east. This bird is dull brown above with a small, often indistinct white eyebrow and a grey face and breast. In the north the face and breast tend to be browner. The species has white spots at the tip of the tail feathers, a good field mark that stands out in the forest and separates it from Brown Thornbill and Large-billed Scrubwren, with which it often associates.

2 DUSKY GERYGONE *Gerygone tenebrosa* 12cm/4.75in

Another plain gerygone, this species is dull brown above and has white underparts. It is a mangrove specialist and is found only in nw. Australia. It has no markings in the tail and is the only gerygone to have a white eye, so can be easily separated from Large-billed and Mangrove Gerygones, with which it overlaps on the Kimberley coast (WA).

3 MANGROVE GERYGONE *Gerygone levigaster* 11cm/4.25in

This plain little bird is restricted to mangroves around n. and e. Australia as far south as Sydney (NSW). It is plain brown above and has a slightly paler breast and an indistinct white eyebrow. Its beautiful song is often the first indicator of its presence. A very inquisitive bird, it reacts well to pishing or squeaking, often coming in quite close to investigate. It is easy to find in the Nudgee area of Brisbane (QLD).

4 LARGE-BILLED GERYGONE *Gerygone magnirostris* 12cm/4.75in

This small gerygone is found only in far n. and far ne. Australia. It is perhaps the dullest of its genus, with plain brown upperparts, paler underparts, and a broken white eye ring. The bill is large compared to the bills of other gerygones, but this is not an obvious feature in the field. Usually associated with water, Large-billed Gerygone is found in mangroves and in gallery forests and rainforests beside streams. It can usually be seen along forest-lined creeks in the Daintree area (QLD) and often around mangroves in Darwin (NT).

1
2
3
4

1 FAIRY GERYGONE *Gerygone palpebrosa* 11cm/4.25in

Occurring along the n. and e. coast of QLD, this gerygone is found mostly in fairly lush vegetation, such as rainforests, monsoon forests, and tropical gardens. It is quite easy to find around Cairns, where Centenary Lakes is a good place to look for it. Across most of its range, Fairy Gerygone is fairly plain: dull green above and pale yellow below with a white throat and white spot in front of the red eye. The male of the subspecies in far ne. QLD has a dark throat and a small white moustachial streak, but these features becomes less prominent farther south.

2 GREEN-BACKED GERYGONE *Gerygone chloronota* 10cm/4in

Found only in far nw. Australia, this gerygone prefers tall, dense vegetation such as monsoon forests or lush gardens. It is difficult to see well, often staying in the canopy, where the overall impression is of a small bird with pale underparts and dark upperparts. If seen well, the wings are dull green, the head is grey, and the throat and breast are white. As with other gerygones, its call is the first sign it is around. The song is more rhythmic and repetitive than that of other species. It is quite common in the Darwin area, found easily at places such as East Point and Buffalo Creek (NT).

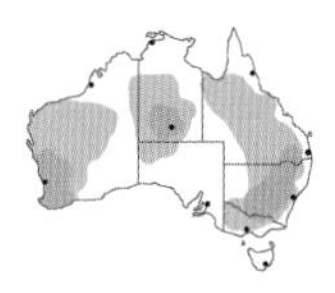

3 WESTERN GERYGONE *Gerygone fusca* 10–11cm/4–4.25cm

The only gerygone in much of its range, this species prefers drier habitats than the other species. It is usually detected through its beautiful song, a lilting, sweet melody of jumbled notes that eventually descends the scale. The bird itself is quite dull, greyish above and white below. The underside of the tail is dark with a pale base and a band of pale spots near the tip. The species is quite common throughout its range, particularly in s. NSW and sw. WA.

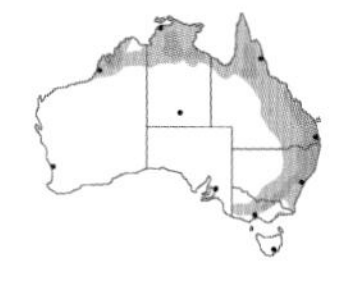

4 WHITE-THROATED GERYGONE *Gerygone olivacea* 11cm/4.25in

This is the common gerygone in coastal se. Australia during the summer, when its 'falling leaf' call (a descending warble) is often heard. It is more brightly coloured than other gerygones, with its yellow breast, olive-green wings and head, and white throat. It also has a small white spot in front of the eye. While the south-eastern birds migrate north for the winter, there are resident birds across n. Australia. Throughout its range the species is found in open woodlands.

ne. QLD 1
e. QLD 1
♀ 1
2
3
4

AUSTRALASIAN WARBLERS (ACANTHIZIDAE)

1 BROWN THORNBILL *Acanthiza pusilla* 10cm/4in

This species is a good one to learn, as it is a common bird in most wooded habitats in se. Australia, including suburban gardens. It is grey-brown above with a pattern of rufous speckling on the forehead and a red eye. It has a rufous rump and dark tail. The throat and breast are pale with fine streaking, while the belly is pale buff. Brown Thornbill has a distinctive song, which although quite varied has a quality that becomes familiar with experience.

2 MOUNTAIN THORNBILL *Acanthiza katherina* 10cm/4in

This range-restricted species is endemic to the Wet Tropics of ne. QLD where it is found only in rainforests at high elevations. It is the only thornbill likely to be seen there, but does occur with some scrubwrens and gerygones, none of which has the pale eye of Mountain Thornbill. Otherwise it is fairly nondescript, with brownish-green upperparts and pale underparts. The main parking lot at Mt Hypipamee NP is a good place to see this thornbill.

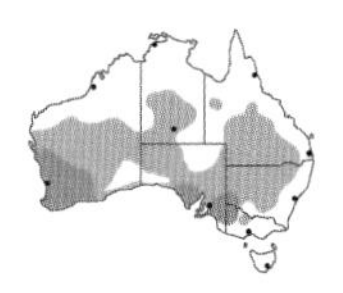

3 INLAND THORNBILL *Acanthiza apicalis* 11.5cm/4.5in

This species replaces Brown Thornbill in the drier areas of inland Australia across to the sw. coast. The two species are very similar in appearance; Inland Thornbill is greyer above, has white speckling on the forehead (rufous in Brown), and also has a whiter belly and flanks. These features are difficult to discern in the field, and beginning birders will find range and habitat the best way of separating these birds at least initially. In the e. part of their range Inland Thornbills are likely to be found in drier habitats, although in sw. WA, where Brown doesn't occur, they can be found in wetter forests.

4 TASMANIAN THORNBILL *Acanthiza ewingii* 11.5cm/4.5in

This species is found only in TAS, where it is very difficult to separate from Brown Thornbill, which also occurs there; even the most experienced observers have trouble with this one. Habitat is a good place to start: Tasmanian prefers dense, wet forests and scrubs. Though the two species are very similar in plumage, there are a couple of features to look for. Tasmanian Thornbill tends to have a whiter belly and undertail, and also has rufous edging to the primary wing feathers. There are also slight differences in bill size and intensity of streaking on the breast, but these are difficult to discern without direct comparison. A good place to find Tasmanian Thornbill is in the wet forests around Fern Tree at the base of Mt Wellington near Hobart.

5 BUFF-RUMPED THORNBILL *Acanthiza reguloides* 11.5cm/4.5in

A nondescript thornbill, plain greenish brown above and very pale yellow below. The rump is also pale yellow, and the tail is dark. With good views, the pale eye of the adult can be seen, but remember that young birds have dark eyes. This thornbill is usually found in small groups that are constantly twittering as they forage among debris on the ground or in low foliage. The distinctive call, a rapid *pitta-pitta-pitta-pit*, is a good way to locate and identify the species. Common in coastal and sub-coastal e. Australia, it is most often found in dry sclerophyll forests and open eucalypt woodlands.

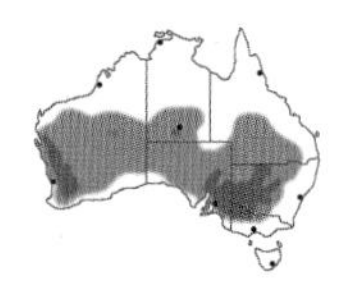

6 CHESTNUT-RUMPED THORNBILL *Acanthiza uropygialis* 11.5cm/4.5in

This species' chestnut rump is not a good field mark, as it is shared with other similar species, including Inland Thornbill, but among them only Chestnut-rumped Thornbill has a pale eye. It also has paler underparts with no streaking. Be careful though, as young Chestnut-rumped Thornbills can have dark eyes. The species is found in dry habitats such as mallee and mulga and is often the most common thornbill where it occurs. This thornbill is quite common throughout s. inland Australia, where it is often found in small active groups that forage in low vegetation or on the ground.

1
2
3
4
5
6

1 WESTERN THORNBILL *Acanthiza inornata* 10cm/4in

This is the plainest of the thornbills, but is quite easy to identify within its restricted range in far sw. WA. It is plain greyish brown above and pale grey below with a pale rump and dark tail. Adults have a pale eye. It is found in a wide variety of habitats of sw. WA, from tall wet forests to dry open woodlands and heaths, and is generally common. The Dryandra Woodland is a good place to find it.

2 SLENDER-BILLED THORNBILL *Acanthiza iredalei* 9.5–10cm/3.75–4in

This thornbill is greyish olive above with a pale eye and a pale greyish, faintly mottled throat. The belly and rump are very pale yellow, and the tail is dark. Birds in the w. part of the species' range are palest, while birds around Adelaide (SA) are darkest. The small population in se. SA is intermediate in colouration. Widespread, with disjunct populations throughout arid s. Australia, this thornbill has quite specific habitat preferences for low shrubby samphire, saltbush, and other similar vegetation associations. The more easterly populations prefer saltbush, samphire plains, and mallee heaths. The western population's more varied habitat selections also include mulga areas and coastal and semi-arid heaths.

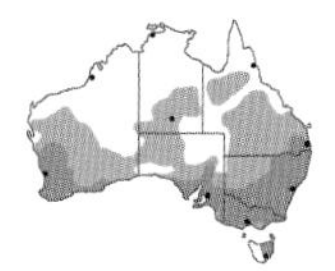

3 YELLOW-RUMPED THORNBILL *Acanthiza chrysorrhoa* 12cm/4.75in

This is one of the more brightly coloured thornbills and is quite common across s. Australia, being a familiar bird of parks and gardens. These birds are often seen hopping around on the ground in small parties, constantly twittering. They have a bright yellow rump that is particularly obvious as they fly up from the ground. Otherwise they have pale underparts and olive-brown upperparts, with a black crown covered in white speckles.

4 STRIATED THORNBILL *Acanthiza lineata* 10cm/4in

Although it is quite common, this species can be difficult to observe because it tends to forage high in the canopy. It is olive above and pale yellow below, with faint streaking on the throat, face, and crown. This streaking can be difficult to see when the bird is high up in the canopy and is most obvious just behind the eye. Like other thornbills this species is usually seen in small flocks constantly moving through the canopy. It is found in the coastal and sub-coastal areas from s. QLD to Adelaide (SA), mainly in sclerophyll forests though in eucalypt woodlands and mallee in the south.

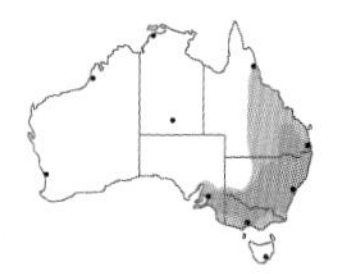

5 YELLOW THORNBILL *Acanthiza nana* 10cm/4in

Unlike many of the other thornbill species found in the same areas, Yellow Thornbill is rarely seen on the ground, instead working its way actively through the mid-storey. It prefers drier, open forests and is often associated with acacias. It is yellowish overall, with slightly greener upperparts, and at close range one can see streaks on the cheek just behind the eye. It is quite variable in colouration across its range, with inland birds paler than birds near the coast. It is most common in dry sclerophyll forests, open woodlands, and mulga, though it occurs in most wooded habitats within its range. This species is relatively common throughout its range, particularly on the w. slopes of the Great Dividing Range; it is found from coastal to inland areas of e. Australia, from Cairns (QLD) to Adelaide (SA).

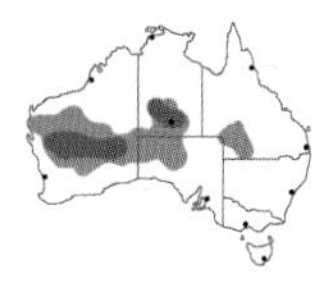

6 SLATY-BACKED THORNBILL *Acanthiza robustirostris* 9.5–10cm/3.75–4in

This is a very difficult species to identify, as is often confused with Chestnut-rumped Thornbill. Adult Chestnut-rumped has a pale eye, while Slaty-backed has a dark eye; however, young Chestnut-rumped has a dark eye also, which can cause confusion. Slaty-backed Thornbill is greyish above, with pale underparts, a chestnut rump, and a dark tail. Its crown is finely streaked, while Chestnut-rumped Thornbill has a more speckled crown. Slaty-backed prefers thick mulga, and is also found in shrublands in low-lying areas. It ranges across the c. inland mulga belt, from the WA coast to sw. QLD, though it is somewhat localised and always uncommon.

1
2
3
4
5
6

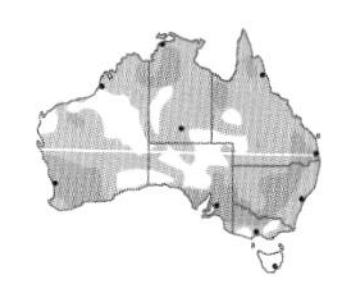

1 WEEBILL *Smicrornis brevirostris* 7.5–9cm/3–3.5in

The smallest Australian bird, this tiny yellow thornbill is similar to the closely related *Acanthiza* thornbills and the gerygones, from which it can be differentiated by its short, stubby, horn-coloured bill. The bill and small size give it a shape rather like the more intricately patterned pardalotes. Weebill also shows a creamy iris, a pale supercilium, and a white tip to the tail, very evident when it flutters around foliage, as it typically does when foraging for insects. A gregarious bird, most often encountered in pairs or small groups, it occurs in drier woodlands throughout the mainland but is absent from the driest desert areas. Its *weebill* or *mini-me* call is a common sound of inland Australia.

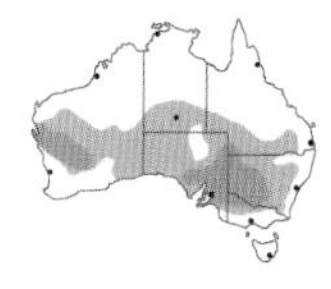

2 SOUTHERN WHITEFACE *Aphelocephala leucopsis* 10–12cm/4–4.75in

A small, dull finch-like bird that often feeds on the ground. It is grey-brown above and white below with fawn flanks and an obvious pale eye. The short tail is black tipped with white and is flared in flight. The large white lores bordered above by a black bridge across the crown give the bird a comical, owl-like appearance when it is viewed straight on. The much rarer Chestnut-breasted Whiteface looks very similar but has a wide pale rufous breast band and rufous flanks. Southern Whiteface is found in a variety of arid habitats, from cypress-pine (*Callitris*) associations to mallee, mulga, and shrublands. It often associates with ground-feeding thornbills and Speckled Warblers. It occurs from the w. Australian coast right across inland s. Australia to the w. edge of the Blue Mtns (NSW).

2

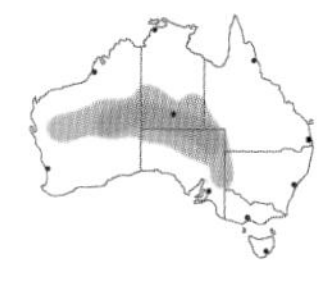

3 BANDED WHITEFACE *Aphelocephala nigricincta* 10cm/4in

A small, ground-feeding bird. Pale brown above, it has a rufous rump and short black tail tipped with white. It resembles Southern Whiteface with its pale eye, white lores, and black bridge across the crown but differs in having a very prominent black breast band that contrasts with the white underparts. When seen well, the fawn scalloping on the flanks is diagnostic. Banded Whiteface shies away from denser vegetation where the Southern Whiteface can be found, but they both occur in gibber plains, saltbush, and sand dunes where their distribution overlaps.

4 CHESTNUT-BREASTED WHITEFACE *Aphelocephala pectoralis* 10–13cm/4–5in

A scarce and local bird unlikely to be found without a specific search. At first appearance, this whiteface looks very similar to Southern and Banded Whitefaces, with pale eyes, white lores, and a black bridge across the crown, but it is more richly coloured than both those species and has a wide, pale rufous breast band and rufous flanks. It prefers and is restricted to much more barren habitats than the other two, but the others do occur within its restricted range. Chestnut-breasted Whiteface is usually found on gibber plains or rocky ridges with very sparse vegetation and appears to be more furtive and shyer than the other whitefaces. Very restricted to areas of c. and ne. SA, it is most often located on the Strzelecki and Birdsville Tracks.

1
2
3
4

PARDALOTES (PARDALOTIDAE)

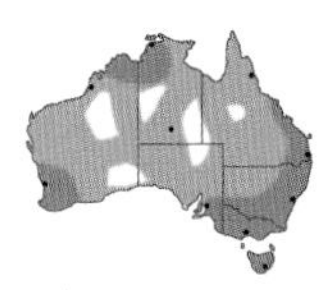

1 STRIATED PARDALOTE *Pardalotus striatus* 9–11.5cm/3.5–4.5in

Striated is a striking bird with many subspecies that vary in appearance. It always displays a plain brown mantle, and it lacks the white spotting on the crown and wings that Spotted Pardalote shows. The crown of Striated varies from solid black to lightly flecked with white. The wings are black, with long flashes of white and a small red mark near the shoulder (yellow in the nominate subspecies that breeds on TAS and Bass Strait islands). It always shows a broad white eyebrow that meets a warm yellow mark over the bill. It is usually encountered feeding in the crowns of eucalypt trees in groups that usually draw attention to themselves with their incessant chipping calls. Striated is the most widespread pardalote species and is found in forests and woodlands throughout the mainland and TAS.

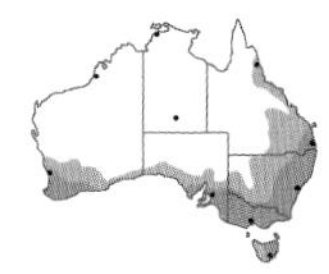

2 SPOTTED PARDALOTE *Pardalotus punctatus* 8–10cm/3.25–4in

A stunning tiny songbird with strikingly marked plumage. As are all pardalotes, it is blunt-billed, plump-bodied, and short-tailed. It is even more boldly patterned than Striated Pardalote, from which it differs in having a boldly spotted crown, a scalloped grey mantle, and conspicuous white spots on the sides of the wings. It also displays a bolder, clean-cut yellow throat (larger in males), and a rich red upperside to the tail (yellow in the inland subspecies). Spotted also possesses a white supercilium, but it never shows a bold yellow spot at the front end as in Striated. Like other pardalotes, Spotted is usually found in groups, feeding in the canopies of eucalyptus trees. It is a common species in the coastal zone from near Cairns (QLD) south through e. Australia to TAS, then across s. Australia to sw. WA. It is absent from much of n. Australia and from the dry interior. It is usually found in eucalypt forests, woodlands, and arid associations such as mallee. It may also be found in parks and gardens.

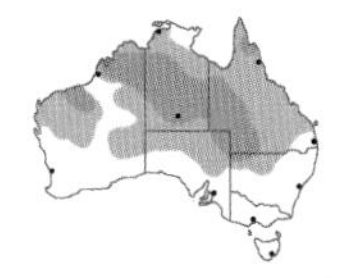

3 RED-BROWED PARDALOTE *Pardalotus rubricatus* 12cm/4.75in

This bird looks like a subdued cross between Striated and Spotted Pardalotes. It is a large pardalote with a uniform dull olive back, black wings edged with white and orange, a yellow wash to the breast, and a small red brow. Its most diagnostics features are the black crown heavily spotted with white, which differentiates it from Striated Pardalote, and the lack of spots on the wings, which differentiates it from Spotted Pardalote. It is found in drier woodlands across inland n. Australia, including mulga, open woodlands, and savannahs, though usually outnumbered by Striated Pardalotes. Although fairly common in places, it can be difficult to track down.

4 FORTY-SPOTTED PARDALOTE *Pardalotus quadragintus* 10cm/4in

The most subdued of the pardalotes. It is olive-brown above with a yellowish wash to the face and has pale grey-buff underparts. It has a black tail and wings, which are prominently tipped with white. It differs from Striated Pardalote in having spotted wings and from female Spotted Pardalote in the olive-brown back and lack of spots on the crown. Forty-spotted Pardalote is a rare and declining TAS (and Flinders Island) endemic mainly confined to the se. part of the island. It occurs in dry sclerophyll forests and open eucalypt woodlands, though is missing from what appears to be suitable habitat. The most accessible and reliable sites are the Peter Murrell Reserves and Bruny Island.

sw. 1
n. 1
TAS 1
se. 1
♂ inland 2
♀ 2
3
4

1 PURPLE-GAPED HONEYEATER *Lichenostomus cratitius* 17–19cm/6.75–7.5in

A honeyeater with olive-fawn underparts, grey-brown upperparts, and a grey head with a sooty-black mask and a yellow moustachial stripe bordered above by a purple line ending in a yellow ear plume. The purple gape line (yellowish in juveniles) distinguishes it from the very similar Grey-headed Honeyeater, and it differs from Grey-fronted and Yellow-plumed Honeyeaters in lacking streaking or striping below. An uncommon, somewhat nomadic bird of the s. mallee regions, it is generally a solitary species but congregates during periods of bloom. It can sometimes be found in the mallee reserves of nw. VIC and se. SA, such as Gluepot Reserve (SA), and around the Stirling Ranges in sw. WA.

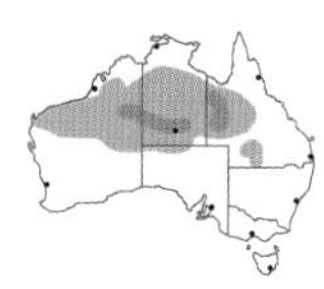

2 GREY-HEADED HONEYEATER *Lichenostomus keartlandi* 16cm/6.25in

A honeyeater with faintly streaked pale underparts with a slight lemon-yellow wash. Upperparts are grey-brown with yellow edging on the primaries. The grey head has a black mask and a yellow ear crescent. This species differs from Grey-fronted Honeyeater in having a complete ear crescent and dark mask and in the lemon-washed underparts. Like many arid-zone species, this bird is nomadic, and its numbers fluctuate according to conditions. It is found in a variety of arid, open woodland habitats often associated with hills and gorges. It can frequently be seen in the West MacDonnell Ranges near Alice Springs (NT) and around Mt Isa in nw. QLD.

3 GREY-FRONTED HONEYEATER *Lichenostomus plumulus* 14–15cm/5.5–6in

This honeyeater has olive upperparts and fawn underparts with heavy grey-brown streaking. The face shows black lores and a large yellow ear patch with a narrow black upper border. It differs from Grey-headed in lacking the black mask and in its larger ear patch and from Yellow-plumed in having much more faintly streaked underparts and a black border above the yellow ear patch. Though Grey-headed is widespread in interior mainland Australia, it is nomadic and has an unpredictable distribution. It is found in a variety of arid woodland habitats, including savannahs, mulga, and mallee, often with a grassy or spinifex ground cover.

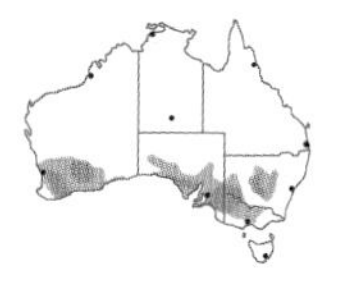

4 YELLOW-PLUMED HONEYEATER *Lichenostomus ornatus* 13–16.5cm/5–6.5in

This honeyeater is boisterous and very obvious when present. Heavy grey-brown streaking on the white breast and a contrasting green head with a very strong yellow neck plume give the bird a stunning, clean look. This cleaner look sets it apart from Grey-fronted, Grey-headed, and Purple-gaped Honeyeaters. Yellow-plumed is a common and somewhat nomadic species of mallee areas, ranging into heaths and sclerophyll forests in WA.

5 YELLOW-TUFTED HONEYEATER *Lichenostomus melanops* 18–22cm/7–8.5in

An extremely distinctive honeyeater. It has a prominent black mask with a contrasting yellow throat, crown, and ear patch. It is grey-brown above and has olive-fawn underparts. The helmeted subspecies (*cassidix*) has stronger yellow on the breast and a yellow tuft on the forehead. If seen well, Yellow-tufted Honeyeater should not be confused with any other species. Occurring along the coast and ranges of se. Australia, with a very patchy distribution, it is usually found in dry sclerophyll forests, often in gullies or on hillsides.

♂ 1
♀ 2
♂ 3
♀ 4
♂ 5
helmeted 5

1 GREEN-BACKED HONEYEATER *Glycichaera fallax* 11–12cm/4.25–4.75in

A dull, nondescript honeyeater, somewhat reminiscent of a North American vireo. Easily overlooked because of its drab appearance, it has olive-green upperparts with brown wings and tail, and fawn underparts with a dull yellow wash. It has an off-white iris and a prominent white eye ring. Green-backed Honeyeater has a very restricted range within extreme n. QLD, from the rainforests of Iron Range NP west to the monsoon forests of Wenlock River Crossing; the latter site is a good place to look for it.

2 TAWNY-CROWNED HONEYEATER *Gliciphila melanops* 16–18cm/6.25–7in

A striking honeyeater with a bright, bronzy-yellow crown and gleaming white underparts bordered by a thick blackish-brown border on the neck sides and along the flanks. Otherwise it is brown above, save for a small white crescent on the cheeks. It has a long, decurved bill, which approaches those of the spinebills in length. On the e. coast of NSW it occurs almost exclusively in stunted heathlands; in VIC to WA it also inhabits adjacent mallee areas. It is a very inquisitive species and often responds to pishing by sitting on an exposed perch above the heath.

3 EASTERN SPINEBILL *Acanthorhynchus tenuirostris* 14–16cm/5.5–6.25in

A gorgeous, strikingly patterned sunbird-shaped honeyeater; it has a long, down-curved bill used for probing flowers for nectar. Eastern Spinebill displays a bold white throat with a soft rosy patch in its centre that is bordered below by two broad black braces and solid cinnamon underparts below that. It has a deep blue-grey crown, black mask, red eyes, and an olive back bordered above by a rich rufous collar. It displays prominent white flashes in its tail, often exposed as it flits between flowers. It is unlikely to be confused with any other bird, as it does not overlap with the closely related Western Spinebill. Common within rainforests, woodlands, heaths, and gardens in e. Australia, it has two distinct populations, one centred in the highlands around Cairns (QLD) and another from se. QLD south to VIC and TAS and west into se. SA. It is most often encountered at flowering shrubs in pairs or singles.

4 WESTERN SPINEBILL *Acanthorhynchus superciliosus* 13–16cm/5–6.25in

A stunning, multi-coloured honeyeater with a characteristic long, down-curved bill. The male is arguably the most beautiful of all the honeyeaters, possessing a vibrant coppery-brown throat bordered with two strong crescents below, first white, then black, and a clean white belly below that. It has a deep blue head, which is broken up with a thin but conspicuous white supercilium behind the eye and separated from the olive upperparts by a prominent orange collar. The plain olive female lacks bright markings except for a broad chestnut collar. In both sexes, large areas of white are exposed in the tail tip, most evident when the tail is fanned in flight. Western Spinebill does not overlap with the similar Eastern Spinebill. It is a common bird from north of Perth around to Esperance within woodlands and heaths that contain stands of banksias.

1
2
♂ 3
imm. 3
♂ 4
♀ 4

1 BROWN HONEYEATER *Lichmera indistincta* 12–16cm/4.75–6.25in

Brown Honeyeater is an unspectacular bird, uniformly olive brown with indistinct yellow-green edging on the wings. A small yellow triangle of bare skin behind the eye is its only prominent feature. This distinctive yellow area, along with the olive flash in the wing, helps differentiate this species from similarly nondescript female or immature myzomelas. This active bird is usually encountered around flowering shrubs in parks or gardens, where it is often vocal, conspicuous, and acrobatic in its pursuit of nectar. It occurs in many habitats, including rainforest edges, mangroves, parklands, gardens, heaths, vegetated watercourses, and eucalypt forests, from WA east to QLD and south to n. NSW. Generally, it is the most numerous honeyeater in its range.

2 DUSKY MYZOMELA *Myzomela obscura* 13–15cm/5–6in

A large myzomela, uniformly chocolate-brown save for some rusty colouration on the underparts. The females of Scarlet and Red-headed Myzomelas are similar to both sexes of Dusky but smaller and paler, especially on the lower underparts. Those two species often also show some red colouration around the base of the bill, which Dusky never shows. Dusky Myzomela is found in the Top End of the NT and along the e. coast from Cape York south to s. QLD, becoming rare south of the tropics. In the north, it is a very common bird of rainforests, monsoon scrub, mangroves, and melaleucas.

3 SCARLET MYZOMELA *Myzomela sanguinolenta* 10–13cm/4–5in

Males are dazzling birds: bright scarlet except for contrasting blackish wings and tail and a white lower belly and vent. The female is inconspicuous, grey-brown except for a subtle reddish wash on the chin, which differentiates it from Dusky Myzomela. Females of Scarlet and Red-headed Myzomelas are very similar. Scarlet Myzomela is a blossom nomad in coastal e. Australia from north of Cairns south to e. VIC. It can be remarkably common at flowering times, when its distinctive descending calls ring out often from the trees, although it can also be completely absent from these areas when blossoms are absent.

3 ♂

4 RED-HEADED MYZOMELA *Myzomela erythrocephala* 11–13cm/4.25–5in

A charcoal-black honeyeater with a prominent crimson head and rump and a black eyepatch, lores, and bill. Males differ from Scarlet Myzomela males in the red being restricted to only the head and rump. Females are a mid-brown and show a red faint wash on the throat and forehead. This is a common bird of mangroves and monsoon forests from tropical WA across to the QLD Gulf Country. It is readily found around Darwin and nearby Buffalo Creek in the NT.

1
2
♂ 3
♀ 3
♂ 4
♀ 4

1 LEWIN'S HONEYEATER *Meliphaga lewinii* 19–22cm/7.5–8.5in

Lewin's is a common and highly vocal eastern honeyeater, all plain olive in colouration except for a bold yellow crescent-shaped ear patch and a strong yellow gape line behind the bill that extends under the eye. Its loud call, a vibrating, machine-gun-like rattle, is a characteristic sound of the e. forests and can be heard at any time of the day. Lewin's Honeyeater is found in wet habitats, mainly rainforests and wet eucalypt forests, but will also wander into densely vegetated gardens. It is common all down the e. coastal strip of Australia from n. QLD south to e. VIC. An inquisitive bird, it is especially common in Lamington NP (QLD).

1

2 YELLOW-SPOTTED HONEYEATER *Meliphaga notata* 17cm/6.75in

An olive-brown honeyeater that is only slightly smaller than the very similar Lewin's but has a proportionately longer bill and a yellow ear patch with a rounded diamond shape. It differs from Graceful Honeyeater in being larger and having a different-shaped ear patch. Yellow-spotted Honeyeater is a common bird of rainforests, mangroves, and wet sclerophyll from Cape York Peninsula to Townsville (QLD), generally occurring at lower elevations than Graceful Honeyeater.

3 GRACEFUL HONEYEATER *Meliphaga gracilis* 15–16.5cm/6–6.5in

This is the smallest of the Australian *Meliphaga* species. It has a similar olive-green plumage to both Lewin's and Yellow-spotted Honeyeaters and differs in being much daintier. The ear patch is a more rounded, semicircular version of Lewin's crescent, differing from Yellow-spotted's diamond. A common bird of rainforests, mangroves, and wet sclerophyll from Cape York Peninsula to Cairns and the Atherton Tableland (QLD), it is generally found at higher elevations than Yellow-spotted.

4 WHITE-LINED HONEYEATER *Meliphaga albilineata* 19cm/7.5in

An uninspiring-looking honeyeater with medium grey-brown upperparts and off-white underparts mottled with mid-brown. It has a black mask and a very distinctive white line extending from the yellowish gape, under the eye, to above the black ear patch. Immatures lack this line. The species is restricted to the w. Arnhem Land escarpment, where it is found in monsoon forests and adjacent eucalypt scrub. Within this very small area it is rather common and easy to locate.

5 KIMBERLEY HONEYEATER *Meliphaga fordiana* 19cm/7.5in

A species previously considered a subspecies of White-lined Honeyeater. It too has medium grey-brown upperparts with a greenish wash to the primary feathers; off-white underparts mottled with mid-brown; a black mask; and a very distinctive white line extending from the yellowish gape, under the eye, to above the black ear patch. Nearly identical in the field to the White-lined, this species is distinguished by range and call rather than distinctive plumage variation. However, in Kimberley Honeyeater the facial line is much straighter, not bending under the eye, and the wing colour is more uniform. Kimberley Honeyeater is found in monsoon vine forests and adjacent eucalypt scrub in the sandstone gorges of the Kimberley region of WA.

1
2
3
3
4
5

1 BRIDLED HONEYEATER *Lichenostomus frenatus* 20–22cm/8–8.5in

Bridled Honeyeater is a very local and striking species of the wet forests of ne. QLD. The head pattern is distinctive: all sooty black with a prominent pinkish-yellow line (or bridle) that runs from the base of the bill and curves up to form a border underneath the eye. The base of the black bill is also yellow. Just above the eye is a small white smudge. Within the bird's range this unique combination of facial features is the best ID feature. Bridled Honeyeater is fairly common from Cooktown to Townsville in wet eucalypt forests and mountain rainforests, where it is most likely to be encountered in pairs or small groups around picnic areas where the birds have become habituated (e.g., Mt Hypipamee NP, QLD).

2 EUNGELLA HONEYEATER *Lichenostomus hindwoodi* 17cm/6.75in

A dark grey honeyeater with faint streaking on the underparts. A white line passes from the black bill under the eye and wraps around the eye. Eungella is superficially similar to the much more common Bridled Honeyeater found farther north, but the Bridled is larger and much brighter on the face. This species is highly restricted to rainforests and wet sclerophyll forests of the Clarke Range just west of Mackay (QLD). It is uncommon in its very confined area of distribution.

3 YELLOW-FACED HONEYEATER *Lichenostomus chrysops* 17cm/6.75in

A distinctive olive-brown honeyeater of se. Australia. It has a black mask split by a yellow line of skin that goes from the bill, thickens to almost a wattle under the eye, and extends back to the ear. This is a common species from e. Cape York Peninsula around the e. coast and rangelands to Adelaide (SA). It occupies a wide variety of habitats in this area, excluding rainforest, but tends to be most common in the s. dry sclerophyll forests.

4 WHITE-GAPED HONEYEATER *Lichenostomus unicolor* 18–20cm/7–8in

A chunky, plain mid-brown honeyeater with an olive wash on the wings. The upper gape has a very distinctive off-white semicircle extending up to the lores. The species is common through the wetter regions of tropical Australia, from the Kimberley (WA) across to Townsville (QLD). It is most common in the Top End of the NT and is a garden bird of Cairns (QLD).

5 YELLOW-THROATED HONEYEATER *Lichenostomus flavicollis* 20cm/8in

A stunning honeyeater in its simplicity. It has moss-green upperparts and a grey head and underparts. The face and crown have a black wash that contrasts dramatically with a glistening yellow throat and a small yellow mark behind the eye. A common endemic honeyeater to TAS, it is found in heathlands, open woodlands, and wet and dry sclerophyll forests.

1
2
3
3
4
5

1 VARIED HONEYEATER *Lichenostomus versicolor* 19–22cm/7.5–8.5in

Varied Honeyeater has all-yellow underparts finely streaked with grey, olive-green upperparts, and a prominent black mask with a yellow streak below. It is much brighter overall than Singing and Mangrove Honeyeaters. Varied is a common, strictly coastal honeyeater from Cape York Peninsula to Townsville (QLD). Less confined than the Mangrove, it is found in adjoining eucalypt, monsoon, and melaleuca forests.

2 MANGROVE HONEYEATER *Lichenostomus fasciogularis* 19–22cm/7.5–8.5in

Mangrove Honeyeater is within a complex of three honeyeaters that separate out geographically and prefer different habitats. They are all grey-brown birds with a black mask, yellow on the face, an indistinct yellow flash in the wing, and brown-streaked underparts. Mangrove Honeyeater can be differentiated from Singing and Varied by its all-yellow throat, which is bordered below by a thick grey horizontal smudge across the chest that separates the throat from the heavily streaked underparts below. It is a specialised species confined to coastal mangroves from Townsville (QLD) south (including Brisbane) to n. NSW.

3 SINGING HONEYEATER *Lichenostomus virescens* 18–22cm/7–8.5in

Singing Honeyeater is a subdued inland sister species of Mangrove and Varied. It has grey-brown upperparts; fawn underparts streaked with brown; and a black mask with a yellow plume under the mask merging to a white small ear patch. It is similar to the coastal Mangrove Honeyeater, although Singing is much more subdued, shows a smaller yellow plume, has no yellow on the throat, and lacks the grey breast band of the Mangrove. Singing is a common honeyeater throughout dry areas of the mainland, though absent from the e. coastal areas. It also does not occur commonly in the Kimberley region, Cape York Peninsula (QLD), or the n. Top End NT, but everywhere else is one of the most common honeyeaters in all wooded habitats, including mallee, mulga, and open woodland, and in sw. Australia also inhabits coastal heaths.

4 YELLOW HONEYEATER *Lichenostomus flavus* 17–18cm/6.75–7in

Yellow Honeyeater is well named: an all-plain yellow honeyeater with an indistinct mask, little more than a charcoal smudge on the yellow. It has no other conspicuous markings, which in fact distinguishes it from all other species in this family. Although moderately common, it has a small range, confined to ne. QLD, where it can be found in eucalypt forests, the edges of tropical rainforests, mangroves, and parks and gardens. It is most likely to be found in pairs or singly in gardens or parks around towns (e.g., Cairns or Townsville). It is a regular and easily found bird along the Cairns Esplanade.

1
2
3
4

1 FUSCOUS HONEYEATER *Lichenostomus fuscus* 15cm/6in

A dull, inconspicuous canopy honeyeater that is easily overlooked. It is olive brown above with faint yellow primary edging on the wing and paler below. It shows smudging on the face and has a small yellow ear plume, faintly bordered black on the front. It has a dark eye that is ringed yellow in non-breeding plumage. Fuscous Honeyeater is found from Cairns (QLD) around to Adelaide (SA), though is most common in the rangelands of NSW. It prefers dry sclerophyll and eucalypt woodlands and is a blossom nomad.

2 WHITE-PLUMED HONEYEATER *Lichenostomus penicillatus* 15–17cm/6–6.75in

A small, plain olive honeyeater with bold white plumes on the neck and a distinctive yellow cast to the head. It is a famously pugnacious species, often observed fighting larger honeyeaters around nectar sources. White-plumed is found across most of c. Australia and absent from the wetter areas of sw. Australia, the tropical north, and wetter parts of e. Australia. It is found in a variety of dry sclerophyll, open woodland, and arid woodland habitats and is most common close to water. In many inland towns it is a common garden bird.

3 YELLOW-TINTED HONEYEATER *Lichenostomus flavescens* 15cm/6in

This honeyeater has an olive back with a faint yellowish wash and yellow edging to the primaries, and paler, duller underparts. A very strong yellow wash on the face contrasts with a moustachial stripe that merges into a thick black ear plume. The duller White-plumed Honeyeater is similar, but Yellow-tinted has an all-black plume and a stronger yellow wash overall. This species is very boisterous and moves in family parties. It is a common bird of the tropical grassland and savannah belt from the Kimberley (WA) across to c. Cape York Peninsula (SA). Small numbers are always present, though large flocks move as blossom nomads.

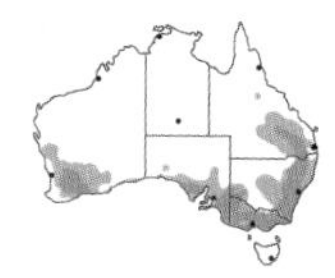

4 WHITE-EARED HONEYEATER *Lichenostomus leucotis* 22cm/8.5in

A stunning honeyeater with olive-green upperparts and yellow-green underparts. The sooty-grey cap and black face, throat, and upper breast give it a hooded appearance, which contrasts with a large white ear patch. In young birds the cap is green, and the black parts are dark green. This bird is widespread in forests, woodlands, and heaths of se. and s. Australia, from the coast to the alpine ranges and in some areas into the drier inland. It is patchily distributed, common in some areas and uncommon in others.

4

1
2
3
4

1 YELLOW-THROATED MINER *Manorina flavigula* 22–28cm/8.5–11in

An inland miner similar in appearance to the common Noisy Miner of the east. However, Yellow-throated shows a bright white rump and has a grey-brown crown (similar in colouration to the rest of the upperparts) and a yellow wash on forehead and neck. It occurs from WA eastward across c. Australia into much of QLD, w. NSW, and VIC. It is largely absent from the e. coastal belt occupied by Noisy Miner. It is also a generalist, found in a variety of drier inland habitats (relative to Noisy Miner), from open woodlands (including mulga and mallee) to eucalypt-lined watercourses to gardens and golf courses. It is usually observed in noisy family parties.

2 BLACK-EARED MINER *Manorina melanotis* 25cm/10in

Very similar to the much more numerous Yellow-throated Miner, with which it interbreeds. Black-eared differs in having darker upperparts that contrast more strongly with the underparts. It also has slightly more extensive black on the ear coverts, lacks yellow on the forehead, and lacks a yellow wash to the sides of the neck. Hybrids with Yellow-throated occur throughout Black-eared Miner populations. This honeyeater requires extensive patches of mallee, which exist in only a few areas of nw. VIC, sw. NSW, and se. SA. It is very rare and becoming rarer through interbreeding with Yellow-throated Miner.

3 NOISY MINER *Manorina melanocephala* 24–28cm/9.5–11in

A very familiar bird in e. Australia. It has a black face and crown, which contrast with its clean white forehead. It is uniformly grey down the back, including the rump, unlike Yellow-throated, which has a white rump. Noisy Miner occurs in a variety of open habitats, (e.g., open woodlands, eucalypt forests, gardens, and parks) but is absent from closed forests and densely covered areas. Generally, it is found in wetter habitats to the east of Yellow-throated Miner. It is common in the east from n. QLD south to VIC, TAS, and around to Adelaide (SA).

4 BELL MINER *Manorina melanophrys* 18–20cm/7–8in

Very distinctive in being the only green member of this subgroup. The bird is uniform green with blackish wing edges, an orange bill and legs, a black frons and partial moustachial stripe, and bright yellow lores. It is gregarious and noisy, living in large colonies, and the tinkling-bells call can be heard from a long distance. Although vocal, Bell Miners can be quite hard to see as they forage high in the canopy. They are common in the mountains and foothills of coastal se. Australia from near Brisbane (QLD) to e. VIC, and prefer wet sclerophyll forests.

1
2
3
4

1 LITTLE WATTLEBIRD *Anthochaera chrysoptera* 27–34cm/10.5–13.5in

Also known as Brush Wattlebird. It is a coastal honeyeater found from se. QLD south through e. NSW into coastal VIC and se. SA. It also occurs on TAS. Smaller than the other eastern wattlebirds, it is dark grey-brown and covered in fine white streaks. It does not overlap in range with the similar Western Wattlebird, which is found in only sw. WA. Little Wattlebird is most abundant in coastal heaths but also occurs in eucalypt forests, scrubby thickets, and parks and gardens. It is quite common in Royal NP south of Sydney (NSW).

2 WESTERN WATTLEBIRD *Anthochaera lunulata* 27–34cm/10.5–13.5in

A small, dark grey-brown wattlebird with fine white streaking below and white tail tips. Concentrated pale streaking on the ear merges with the moustachial streak and gives the impression of a white patch at a distance. This species was thought to be conspecific with the Little Wattlebird of e. Australia; it differs by having a diffuse moustachial streak and a less streaked crown. A common bird in most habitats of the south-west, from north of Perth to Esperance, it is particularly common in coastal heathlands.

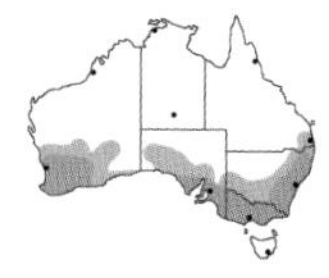

3 RED WATTLEBIRD *Anthochaera carunculata* 33–37cm/13–14.5in

A large, long-tailed honeyeater, dull grey-brown heavily streaked with white. It has a yellowish wash to the vent and belly, a pale patch under the eye, and a small whitish cheek patch with a small pale reddish wattle. The combination of indistinct dull fleshy-pink wattles, red eyes, pale whitish cheeks, and yellow belly distinguishes this honeyeater from all others. This common species forages for nectar in gardens, often in groups that aggressively defend their blossoms from other honeyeaters. It occurs in most eucalypt forests, open woodlands, mallee, and heathlands in s. Australia from s. WA eastward to VIC and north along the e. coast to Brisbane (QLD).

4 YELLOW WATTLEBIRD *Anthochaera paradoxa* 43–50cm/17–19.5in

This massive honeyeater replaces the similar Red Wattlebird on TAS. It is the largest honeyeater species. Another brown, heavily streaked bird, it has unique fleshy yellow wattles that hang down from the ears, a whitish face, and a lemon-yellow wash on the belly. It occurs in a broad range of habitats, including coastal heaths, dry eucalypt forests, open woodlands, parks, and native gardens. Yellow Wattlebirds are common and noisy birds, so are easily seen in their restricted range of TAS, often sighted in towns and suburban areas, where they aggressively defend their territories.

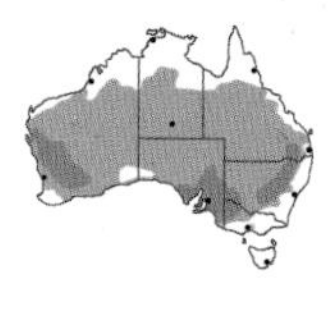

5 SPINY-CHEEKED HONEYEATER *Acanthagenys rufogularis* 22–27cm/8.5–10.5in

A distinctive honeyeater with a rich peachy breast and bold black-and-white stripes down the side of the face. The white stripe contains the spines that give the bird its name, although they are difficult to see in the field. The bill has a salmon-pink base. The head and bill pattern makes it unlike any other honeyeater. Spiny-cheeked Honeyeater is widespread across most of arid Australia, absent from the wetter coastal regions of the east and the tropical north. It is common in mulga, mallee, open woodlands, and farmlands with scattered trees or scrub, and also turns up in Outback gardens.

se. 1
TAS 1
2
3
4
5

1 BAR-BREASTED HONEYEATER *Ramsayornis fasciatus* 14cm/5.5in

A distinctive honeyeater with white underparts heavily marked with black scalloping that forms multiple bands. Upperparts are olive brown with large blotches of black (streaks in immature birds). The face, ears, and throat are white, broken by a very thin, dark moustachial stripe. The species occurs in denser habitats, such as melaleuca woodlands, dry rainforests, and scrubby thickets, often close to water. It occurs from near Broome (WA) across tropical n. Australia to Cairns and coastally south to Mackay (QLD). It is fairly common in the NT, but uncommon to rare in QLD, where Brown-backed Honeyeater occurs.

1

2 BROWN-BACKED HONEYEATER *Ramsayornis modestus* 12cm/4.75in

A small honeyeater with a distinctive buff line under the eye. It is uniform olive brown above with slightly darker smudging around the face and ears. The throat is bright white, and the rest of the underparts are buff with very diffuse scalloping that forms indistinct pale brown bars on the breast and upper belly. Brown-backed Honeyeater occurs in ne. QLD from Cape York Peninsula south to Townsville and is primarily a summer visitor; some of the population apparently migrates to Papua New Guinea. Its preferred habitats are melaleuca woodlands, mangrove edges, and eucalypt scrub, particularly close to water. It is common around Cairns in summer.

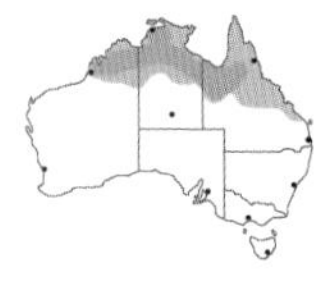

3 RUFOUS-THROATED HONEYEATER *Conopophila rufogularis* 13.5cm/5.25in

At a distance, a dull olive-green honeyeater. Closer inspection shows yellow edges to the tail, yellow edges to the wing feathers forming an indistinct patch on the wings, and a prominent rufous throat that contrasts with the otherwise grey-buff underparts. The immatures lack the rufous throat and can be confused with immature Rufous-banded Honeyeaters. Rufous-throated occurs across a broad swathe of n. Australia but is common only in a band of tropical savannahs and woodlands from Broome (WA) across to the QLD Gulf Country.

4 RUFOUS-BANDED HONEYEATER *Conopophila albogularis* 12–14cm/4.75–5.5in

This distinctive honeyeater of the tropical north has a broad rufous band across the chest, a bright white throat and chin (which contrasts with the grey head above and the darker chest band below), and a clean white belly. Otherwise this honeyeater is quite plain; its only other bold markings are a broad yellow flash in the wing and outer tail. Rufous-banded Honeyeater is confined to the tropical north from n. NT east into extreme n. QLD; it is a very common bird within tropical savannahs, monsoon forests, mangroves, and along vegetated watercourses. It is very conspicuous around Darwin (NT). It is most often encountered in small, active parties that frequently fight with other honeyeaters over nectar sources.

1
2
3
4

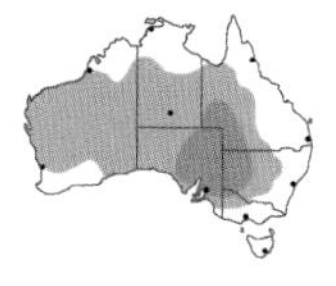

1 CRIMSON CHAT *Epthianura tricolor* 10–13cm/4–5in

Male Crimson Chats are very striking birds: vivid scarlet below with a contrasting white throat, and brown above with a black mask through the eye and a vermilion cap. Females are much more subdued, all greyish brown with a hint of crimson in a wash on the breast and belly. Crimson Chats are wide-ranging, nomadic birds within inland Australia, occurring on plains, scrub, mallee, and open woodlands across the continent, though absent from the tropical north and the coastal zone of the east. They are most often encountered in active flocks feeding low to the ground, or on the ground; they are often recognised by their white-tipped tails or vivid red colours in flight.

1

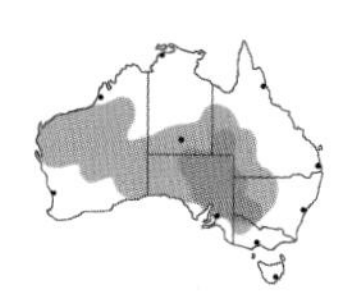

2 ORANGE CHAT *Epthianura aurifrons* 11cm/4.25in

The male, with his black mask and throat contrasting with the orange body, is unmistakeable in Australia and reminiscent of an African weaver (*Ploceus*). The far less distinctive female is mottled brown above with a black tail and orange rump, and has dull pale orange underparts, brown smudging on the ear, and an indistinct fawn eyebrow. Orange Chat ranges from w. NSW across to the c. WA coast on salt lakes and sparse shrublands. The core range of this nomad is the gibber plains of ne. SA north to Birdsville (QLD).

3 WHITE-FRONTED CHAT *Epthianura albifrons* 12cm/4.75in

A stunning bird of s. Australia. The male has a bright white forehead, face, ears, and underparts bordered by a broad jet-black band from the back of the crown to the nape and down the sides to form a broad breast band. The back is a uniform grey-brown, and the wings and tail are black. The female is much duller, with the black replaced by dirty grey (except on the breast) and with grey smudging around the face and ears. These birds are widespread and quite common in s. Australia, where they prefer low vegetation, often around wetlands, in barren open areas, or along the coast.

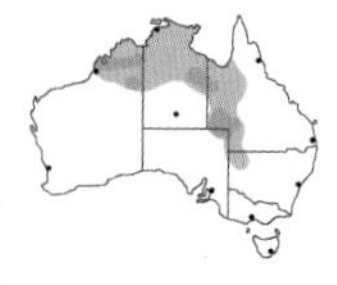

4 YELLOW CHAT *Epthianura crocea* 11cm/4.25in

Although broadly distributed, this bird is hard to locate. The male has a cream eye, yellow-orange head, underparts, and rump, and a very distinct black crescent band on the breast. The back is grey mottled with brown, and the wings and tail are black. Females are much duller and lack the black breast band. The species has a very wide but extremely patchy distribution largely in the nw. third of the continent. It is generally associated with wet areas and can be found in sedges, grasses, and rushes around some tropical wetlands and also farther inland around remote wells. It is probably most regularly found around Broome (WA).

5 GIBBERBIRD *Ashbyia lovensis* 12cm/4.75in

The longest-legged and most pipit-like of the chats. This is a dull bird with very subdued colouring. The male is sandy brown on the back, blacker on the wings and tail, and has a very dull yellow head and underparts. The female is a more washed-out version with a sandy breast. This is an uncommon bird with a distribution centred on the Birdsville and Strzelecki Tracks of ne. SA. It occurs in the same type of gibber-plain environment as Orange Chat and Inland Dotterel.

♂ 1
♀ 1
♂ 2
♀ 2
♂ 3
♀ 3
♂ 4
♀ 4
♀ 5
♀ 5

1 BLACK-HEADED HONEYEATER *Melithreptus affinis* 13–14cm/5–5.5in

A honeyeater with bright olive upperparts, white underparts, and a prominent black head. Black-headed Honeyeater is confined to TAS, where it overlaps with only one other similar species Strong-billed Honeyeater, from which it can be told by its all-black head, solid black throat, and a fine white arc of skin over the eye. It is a common honeyeater, found mainly in the e. half of the island across a range of habitats including dry eucalypt forests, woodlands, temperate rainforests, and coastal heathlands. Like all *Melithreptus* honeyeaters, it is most likely to be encountered in small groups.

2 STRONG-BILLED HONEYEATER *Melithreptus validirostris* 15–17cm/6–6.75in

The Tasmanian version of the Black-chinned Honeyeater. Strong-billed differs from the only other similar sympatric species, Black-headed Honeyeater, in showing a clear white band across the nape and possessing a white throat, small black chin, and pale blue eye skin. Within this group it is also unique in its feeding habits: It searches for insects among strips of bark, much in the manner of a sittella or of a nuthatch (*Sitta*) from Eurasia and North America. It is a Tasmanian endemic, less common than Black-headed Honeyeater and more restricted in habitat, usually found in taller, wetter forests, often with a scrubby understorey.

3 BLACK-CHINNED HONEYEATER *Melithreptus gularis* 16cm/6.25in

This honeyeater is a very attractive species. It has a black hood and green back, similar to many others in this group, and a broad creamy nape band and blue eye skin. The black chin on the white throat is indistinct, as is the grey wash to the breast, but they are the best ID features for differentiating it from other *Melithreptus* honeyeaters. It also has a very different call, a strident *prrrp-prrrp-prrrp*, often given by several birds at once. This honeyeater inhabits dry forests and open woodlands across its range, where it is a blossom nomad and thus locally common at times when eucalypts are flowering. It is found in savannahs, open woodlands, and dry sclerophyll forests throughout e. Australia, from c. QLD around to Adelaide (SA). It is generally uncommon, and its stronghold seems to be the inland slopes and ranges of se. Australia, particularly forests of box and ironbark (*Eucalyptus* species).

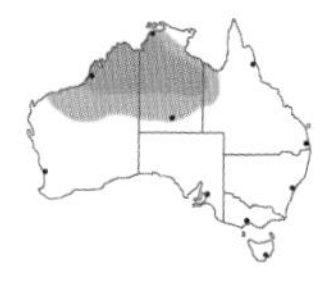

4 GOLDEN-BACKED HONEYEATER *Melithreptus laetior* 16cm/6.25in

This uncommon species is sometimes considered a subspecies of the similar Black-chinned Honeyeater. It differs in having a lime-green eye ring (not blue), a golden-yellow back, and paler underparts, but has a similar call. Their ranges meet in nc. QLD, where there are a number of intermediate-plumaged birds. The only confusion species within its range is the White-throated Honeyeater, which has a pure white throat and underparts. Golden-backed Honeyeater is found in the tropical savannahs across n. Australia and is generally uncommon. With patience and effort it can usually be found in the savannahs south of Katherine or around Timber Creek (NT). Listen for the strident calls, which can be heard from a long distance away.

1
2
3
4

1 WHITE-THROATED HONEYEATER *Melithreptus albogularis* 12–14.5cm/4.75–5.75in

White-throated Honeyeater has a black hood, an all-white throat with no shading or dark areas, a green back, rump and tail, a pale blue crescent of bare skin over the eye, and a clear white collar around the nape. The combination of white throat and blue crescent eliminates all other similar species. This bird is most often found in small active parties around blooming trees. It is a common honeyeater in the north (n. WA, n. NT, and n. QLD) and east (e. QLD south into ne. NSW); it occurs in tropical savannahs, eucalypt forests, open woodlands, vegetated watercourses, and also in parks and gardens.

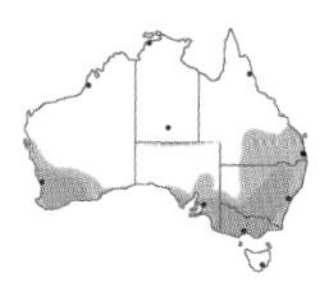

2 BROWN-HEADED HONEYEATER *Melithreptus brevirostris* 12–14cm/4.75–5.5in

Perhaps the most distinctive member of this group, it appears like a sepia-toned version of the other birds. Where the others have black on the plumage, this species has brown; it has a brown hood, a thin cream nape and eye ring, buff underparts, and brownish-olive upperparts. Immature birds have much paler underparts and a greyer hood. This bird is common in most wooded habitats throughout sw. and se. Australia and seems most prevalent in the drier inland areas, where it is frequently found in dry sclerophyll forests, open woodlands, and mallee.

3 SWAN RIVER HONEYEATER *Melithreptus chloropsis* 14–15cm/5.5–6in

Previously considered a subspecies of White-naped Honeyeater, this bird has only recently been split as a full species. It is very similar to White-naped, with which it does not overlap in range, differing primarily in its white or pale green eye skin. The only similar species found in its range is Brown-headed Honeyeater, which is quite different, generally browner and duskier. Swan River Honeyeater occurs in a variety of habitats, including all types of eucalypt woodlands, forests, and even coastal heaths. It is fairly common in its limited range in far sw. Australia from north of Perth to around Esperance. It can usually be found in the Dryandra Woodland south-east of Perth.

4 WHITE-NAPED HONEYEATER *Melithreptus lunatus* 13–14cm/5–5.5in

Another black-hooded, green-backed, and pale-breasted honeyeater. The very thin white nape band is not as prominent as in other species and stops well behind the eye. The eye crescent, when seen well, is distinctive in its red colour, although this bird is often seen high in the canopy so it can be difficult to discern. The most obvious feature separating this species from all but Swan River Honeyeater is the beginning of a black breast band, which stops just past the shoulder and contrasts with the pure white breast. White-naped Honeyeater is found along the e. coast from the Atherton Tableland (QLD) to Adelaide (SA), and is most common from around Brisbane (QLD) to Melbourne (VIC). Though found in a most types of forest habitats, it shows a strong preference for wet sclerophyll forests; it is a blossom nomad.

1
2
3
4

1 BLUE-FACED HONEYEATER *Entomyzon cyanotis* 25–32cm/10–12.5in

A large, distinctive honeyeater with a black head, a bold patch of royal-blue skin on the face, a prominent white stripe down the side of the throat, and a thin white band on the nape. The upperparts are all bright olive green, and below the prominent black throat the underparts are crisp white. This large golden-olive honeyeater is like no other and is easily recognised; it is equally distinctive when it flies, with its deep undulating flight. In n. Australia, flying birds show massive white wing patches. Blue-faced Honeyeater occurs in a variety of habitats, including open woodlands, eucalypt forests, tropical savannahs, farmlands, wooded watercourses, golf courses, and gardens. It is common in the tropical north of Australia and down the e. side of the mainland from n. QLD south to nw. VIC and se. SA.

1

2 MACLEAY'S HONEYEATER *Xanthotis macleayanus* 17–20cm/6.5–8in

A distinctive, boldly marked honeyeater confined to the wet forests of ne. QLD. It is an olive bird, heavily streaked with white over much of the body, with a neat black cap and a large area of buff-coloured bare skin around the eye. These honeyeaters are acrobatic feeders, often observed feeding upside down among the foliage of flowering trees. The species is endemic to the Wet Tropics of ne. QLD and is common within this small range. It is found mainly in forest and woodland habitats, including parks and gardens, and in some areas is quite common, for example on the Atherton Tableland and around Paluma.

3 TAWNY-BREASTED HONEYEATER *Xanthotis flaviventer* 19cm/7.5in

A honeyeater with medium-brown upperparts with the feathers edged in buff. Underparts are a rich tawny with a grey throat and white vent. The face shows a small buff skin patch under and behind the eye. Overall it has a similar but much more subtle pattern than Macleay's Honeyeater. Tawny-breasted is limited to the n. part of Cape York Peninsula (QLD), where it can be found in all existing habitats including rainforests, mangroves, monsoon forests, and savannah woodlands.

4 WHITE-STREAKED HONEYEATER *Trichodere cockerelli* 17cm/6.75in

A curious honeyeater with thick, whitish brush-like plumes on the throat and breast. The head is dark grey, and the back is dark brown with yellow on the wings. On the face, the lores and ear coverts are dark grey with very faint speckling, contrasting with golden plumes behind the ear coverts. Found in the n. half of Cape York Peninsula (QLD), it is a common bird of melaleuca forests, open heathlands, and eucalypt scrub.

1
imm. 1
2
2
3
4

1 REGENT HONEYEATER *Anthochaera phrygia* 20–23cm/8–9in

A striking honeyeater unlike any other bird. Red warty skin around the eye contrasts with the clean black head. The back is black with golden and yellow scales all the way to the rump. The tail is black with bright yellow edges and corners, and the wings are black with large patches of bright yellow. The breast and belly are black, with pale gold arrow marks that blend to solid pale gold on the lower belly and vent. These birds are often found in small groups moving from one tree to the next. Regent is a rare and declining honeyeater of sub-coastal se. Australia from west of Brisbane (QLD) to e. VIC. It is a bird of dry sclerophyll forests, especially the box- and ironbark-dominated eucalypt woodlands of the inland slopes. Most regular sightings are from the Capertee Valley (NSW) and around Chiltern–Mt Pilot NP (VIC).

2 STRIPED HONEYEATER *Plectorhyncha lanceolata* 23cm/9in

The distinguishing feature of this dull-plumaged honeyeater is its striking black-and-white-striped head and nape, which contrasts with a pure white throat. The back and tail are grey-brown, and the lower breast and belly are off-white with very faint, thin, intermittent black stripes not noticeable at a distance. The bluish bill is straight and pointed. This is a highly mobile honeyeater that flies over the canopy, moving large distances between trees. It is a common and nomadic honeyeater in e. Australia from c. QLD to Adelaide (SA). It is common in drier open woodlands of the inland, but around se. QLD and ne. NSW is also found on the coast.

2

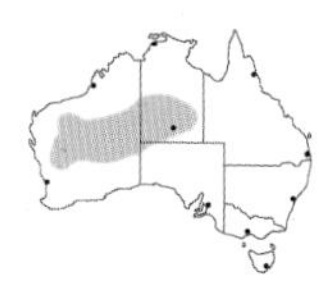

3 GREY HONEYEATER *Conopophila whitei* 11.5cm/4.5in

A very small and nondescript honeyeater, of interest mostly because it is so difficult to locate. The small size, subdued uniform dull grey plumage, pointy straight bill, and almost complete lack of field marks make this bird look more like an Old World warbler than a honeyeater. When the bird is close its fine pale eye ring becomes obvious. A very rare and apparently highly nomadic honeyeater of the inland part of WA, this is one of the most difficult Australian birds to locate. It is usually found in extensive healthy mulga woodlands and is occasionally seen in the vicinity of Alice Springs (NT).

1

2

3

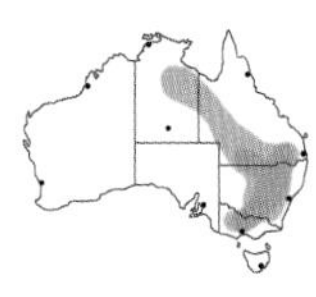

1 PAINTED HONEYEATER *Grantiella picta* 15–17cm/6–6.75in

A striking, clean-cut honeyeater that is black above (brown in females) and bright white from chin to vent. It shows a bold golden-yellow flash in the wings and tail and has a bright pink bill. All other black-and-white honeyeaters with extensive yellow in the wings have bold markings below, lacking Painted's clean white underside, and none has a bright-coloured bill. Painted Honeyeater is an uncommon, highly nomadic honeyeater with a very wide but extremely patchy distribution in e. Australia. Its movements are mysterious; it appears to winter in inland and n. QLD and e. NT, and migrates to se. Australia to breed. It occurs in a variety of open woodlands, including both acacia and eucalypt associations, but there is always a strong association with mistletoe, on which it feeds. It can be found readily in the Leeton–Griffith region (NSW) in spring.

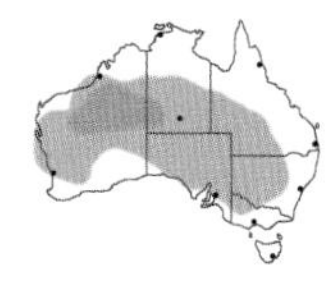

2 PIED HONEYEATER *Certhionyx variegates* 16–19cm/6.25–7.5in

Males are large, neat black-and-white honeyeaters: black above except for a contrasting white rump, clean white below, and with an all-black head. They show bold white flashes in the wing and a broad white base to the tail, the former being conspicuous even when perched. They also possess a unique pale blue semicircle of bare skin below the eye. No other pied honeyeater has an all-dark hood with clean white underparts below it, and no other shows a broad white flash in the wing or blue eye skin. Females are dull grey-brown above and paler off-white below with soft brownish speckling across the chest. The similar though much smaller female Black Honeyeater lacks these markings below. An extremely nomadic bird that may turn up in large numbers after years of absence, Pied Honeyeater is found throughout most of inland Australia, from w. NSW to the WA coast. It may occur in almost any arid open woodland, including mulga and mallee, particularly when *Eremophila* and other shrubs are in flower.

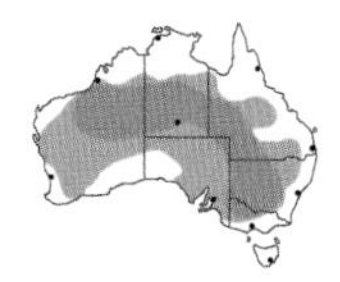

3 BLACK HONEYEATER *Sugomel nigrum* 10–13cm/4–5in

A small pied honeyeater with a short, sharply down-curved beak, all sooty-black upperparts, a solid black hood, and a unique black triangle below its black throat that narrows into a line running vertically down the plain white underparts. The female is a nondescript, grey-brown bird with paler, plainer underparts, which have a darker wash where the male has black but lack the clear speckling of female Pied Honeyeater and lack the clean band of female Banded Honeyeater, with which it might be confused. This is a widespread though highly nomadic inland honeyeater that may be found across much of the continent's interior.

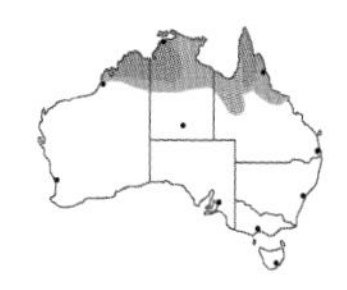

4 BANDED HONEYEATER *Cissomela pectoralis* 11.5–14cm/4.5–5.5in

The male is a clean-cut pied honeyeater with mostly black upperparts and all-white underparts except for its defining feature, a broad black band across its chest, which distinguishes this honeyeater from all others. The female is grey-brown where the male is black but still possesses this distinctive chest band. Banded is a common honeyeater of monsoonal regions of tropical Australia from Broome (WA) across to the Cape York Peninsula (QLD). Common in tropical savannahs and eucalypt woodlands, it is a blossom nomad, often covering long distances. It is far more nomadic in the east of its range, where it can be seasonally absent.

4 ♂ imm.

1
1
♂ 2
♀ 2
♂ 3
♀ 3
4
imm. 4

1 WHITE-CHEEKED HONEYEATER *Phylidonyris niger* 16–20cm/6.25–8in

A striking pied honeyeater with bold yellow markings on the wings and tail that is most similar to New Holland Honeyeater. Unlike that species, White-cheeked has a dark eye, a massive white cheek patch (smaller in sw. WA but still more prominent than in New Holland), and lacks white corners on the tail tip. It is found patchily in e. QLD, e. NSW, and sw. WA in eucalypt forests and heaths. In s. QLD and n. NSW it tends to be restricted to coastal heaths, whereas the Cairns region (QLD) population tends to have wider habitat preferences, including paperbark and *Banksia* stands in tropical savannahs. In WA this common honeyeater can often be found in the same areas as New Holland.

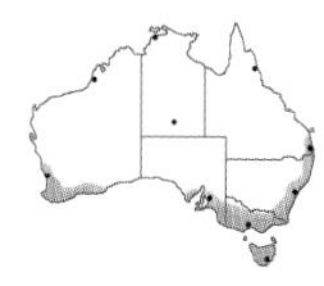

2 NEW HOLLAND HONEYEATER *Phylidonyris novaehollandiae* 16–20cm/6.25–8in

New Holland Honeyeater has a piercing whitish eye, a small patch of white on its face, and white markings on the tail tip, unlike White-cheeked Honeyeater. New Holland is a southern species (with some overlap with White-cheeked in e. NSW and sw. WA) found from se. QLD through s. VIC to se. SA and also TAS. Another population exists in sw. WA. This striking honeyeater is delightfully common on heathlands in the far sw. and se. portions of its range and also occurs widely in eucalypt forests, especially where there is a thick understorey of shrubbery, such as banksias and grevilleas. It is usually found in active groups. On some areas of s. coastal heathlands (e.g., Royal NP near Sydney and in TAS) it can be the most common bird in the area.

2

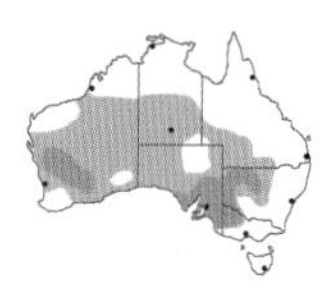

3 WHITE-FRONTED HONEYEATER *Purnella albifrons* 17cm/6.75in

A very distinctive honeyeater of inland Australia. The massive white lores, white frons, and white moustachial stripe on the face contrast with the black throat and breast and with a black head with fine scaling. The back is dark grey scalloped with paler grey, the wings and tail are sooty black edged with orange. The upper belly is white with black streaking fading into a white lower belly and vent. A blossom nomad, this honeyeater is possible throughout most of interior Australia. It shows a strong preference for mallee when trees are blooming and is a regular species in the mallee belt of e. Australia. In the west it is also regular in the open eucalypt woodlands north of the mallee belt.

4 CRESCENT HONEYEATER *Phylidonyris pyrrhopterus* 15cm/6in

A dark grey honeyeater with contrasting yellow in the wings and on the tail. Both males and females have a very fine fawn supercilium above an indistinct dark eye line. The throat is light grey with faint darker grey striping, and the rest of the underparts are a uniform mid-grey. The most conspicuous feature of this species is the pair of prominent broad black crescents contrasting with the grey breast. In the female this crescent is less distinct, and the overall colour is much browner than in the male. An uncommon bird in coastal se. Australia and a common resident of TAS, Crescent Honeyeater occurs in heaths, open woodlands, and sclerophyll forests with thick undergrowth. The usual method of finding this honeyeater is by listening for the loud *ee-JIK* call. It is fairly easy to find around Canberra (ACT) in winter and on Mt Wellington behind Hobart (TAS) in spring and summer.

e. 1
w. 1
2
3
♂ 4
♀ 4

1 HELMETED FRIARBIRD *Philemon buceroides* 32–36cm/12.5–14in

One of the largest Australian honeyeaters, Helmeted Friarbird has a distinctive hammer-shaped head with a tuft of feathers forming a slight crest on the nape. Its stout and long bill has a prominent knob on the top that slopes back towards the eye, and the bird has a mainly dark face of deep blackish-blue facial skin and a uniform grey tail without a pale tip. There are two subspecies in Australia, both occurring in the NT. The Sandstone subspecies, *ammitophila*, lacks the knob, is larger, and is a potential future species split, given its distinct habitat preferences; it is found in woodlands and forests in the vicinity of sandstone escarpments throughout Arnhem Land. Subspecies *gordoni* has a more prominent bill knob and is found in mangroves, woodlands, and forests along the coast and islands of the NT.

2 HORNBILL FRIARBIRD *Philemon yorki* 36cm/14in

A recent split from Helmeted Friarbird of the NT, this species is very similar, although its range doesn't overlap with that species. Hornbill Friarbird differs in having a larger bill knob and a much paler nape. Within the n. part of its range it could be confused with the much smaller Silver-crowned Friarbird, although the two tend to occur in different habitats, with the much rarer Silver-crowned preferring more open, possibly drier woodlands. Hornbill Friarbird is a common, noisy, and aggressive honeyeater found in most wooded habitats, including rainforests, mangroves, parks, and gardens. It is often encountered in pairs or small groups fighting over nectar around flowering eucalypts. It occurs mostly in the lowlands from Weipa in w. Cape York Peninsula south to near Rockhampton (QLD).

3 SILVER-CROWNED FRIARBIRD *Philemon argenticeps* 25–32cm/10–12.5in

A friarbird of the tropical north occurring only in n. WA, NT, and QLD. It is like a smaller, shorter-billed version of Helmeted Friarbird, but in Silver-crowned the knob on the bill is higher and more prominent, the underparts are paler and whiter, and the large area of dark blue facial skin comes to a point behind the eye (the edge is rounded in Helmeted). This bird is quite common in the tropical savannahs and other wooded habitats of the NT and the Gulf Country, including mangroves and monsoon rainforests. It is much rarer in ne. QLD, where it seems to occur in more open habitats than the much larger Hornbill Friarbird.

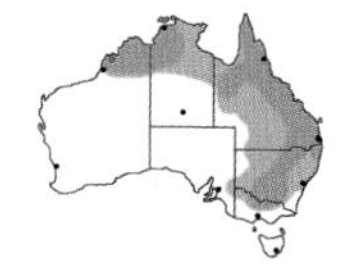

4 LITTLE FRIARBIRD *Philemon citreogularis* 25–29cm/10–11.5in

The smallest and least friarbird-like of the group: It lacks a knob on the bill, has a very limited amount of bare skin on the face below the eye, and lacks the red eye of all the other adult friarbirds (although note that young friarbirds of all species are dark-eyed). It also has a narrow pale tip to the tail. In short, it is a plain friarbird lacking any of the prominent facial features seen in the others and is one of only two that has a pale tip to the tail (along with Noisy). Little Friarbird is found across the tropical north of Australia from WA east to QLD and down the e. side of Australia to the Riverland area of e. SA in a variety of wooded habitats, tropical savannahs, dry sclerophyll forests, open woodlands, parks, and gardens.

5 NOISY FRIARBIRD *Philemon corniculatus* 30–36cm/12–14in

A distinctive friarbird with a dark-skinned head that is completely bald, including the top, which is feathered in all other friarbirds. It possesses a prominent high knob on the top of the bill that sticks straight up (rather than sloping backward as in others). The Noisy is also one of only two friarbirds (along with Little) that have a narrow pale tip to the tail. Found in a variety of wooded habitats in the e. half of QLD, NSW, and VIC, it is incredibly noisy and conspicuous, often seen chasing other friarbirds and honeyeaters away from flowering trees.

gordoni 1
2
3
4
5

1 GREY-CROWNED BABBLER *Pomatostomus temporalis* 25–29cm/10–11.5in

Grey-crowned Babbler can be distinguished from the other species pictured here by its bold yellow eye and pale grey crown. In parts of its range the breast is reddish brown, unlike the pale breasts of other babblers. It occurs mainly in n. and e. Australia from c. WA eastward to e. QLD and south to VIC. In the north of its range, where noisy bands are most likely to be found scampering across the ground or flushing up from roadside verges, it is a common and confiding species (unlike most other babblers). The species is found in open woodlands, tropical savannahs, scrublands, and farmlands with scattered trees. In the south-east of its range, where much of its habitat has been cleared, it is much less common.

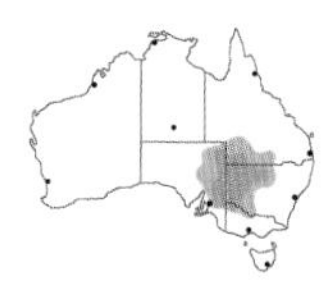

2 CHESTNUT-CROWNED BABBLER *Pomatostomus ruficeps* 23cm/9in

The most distinctively marked of the Australian babblers. It has a dark chestnut crown, which can be difficult to see as the bird rarely sits still for long. It also has a thin buff supercilium, grey-brown back with faint scalloping, dark brown wings with two very thin wing bars, and a chocolate-brown tail with white corners. The distinctive white bib extending all the way to the belly makes this the most striking of the species. This babbler is locally common in open woodlands, mulga, mallee, and shrublands of w. NSW, nw. VIC, e. SA, and sw. QLD. It is fairly common in sw. QLD, where it is regularly seen at Bowra Station.

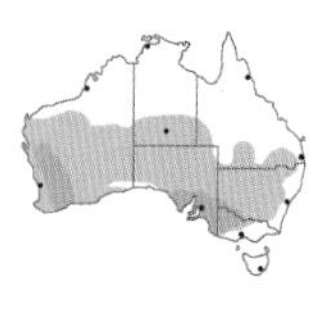

3 WHITE-BROWED BABBLER *Pomatostomus superciliosus* 19–22cm/7.5–8.5in

A small, widespread southern babbler. It is a grey-brown bird with a dark brown tail tipped white, a very thin buff supercilium, and a white throat and breast forming an indistinct bib that merges gradually onto a chocolate belly and vent. It is quite similar to Hall's Babbler, and the two are often confused. Hall's has a much broader, whiter supercilium, and its plumage is generally a darker chocolate brown. Found throughout much of the s. half of the mainland, although absent from e. coastal areas, White-browed is generally uncommon across its range. It occurs in drier woodlands, including dry sclerophyll, open woodlands, mulga, and mallee.

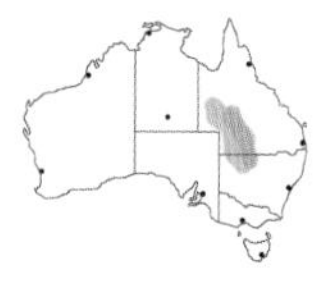

4 HALL'S BABBLER *Pomatostomus halli* 22–24cm/8.5–9.5in

Very similar in appearance to the more widely seen White-browed Babbler, Hall's differs in displaying a broader supercilium and in having a smaller, better-defined white bib. It is also generally darker, more of a chocolate brown. Both of these similar babblers are more secretive than Grey-crowned and more likely to flush and move off rapidly. Hall's Babbler is likely to be encountered in groups foraging on the ground or moving low through the understorey. It is a rare bird, confined to the interior of w. QLD and n. NSW, where it shows a very close association with mulga and is found almost exclusively in this habitat. The best place to see it is Bowra Station in sw. QLD.

1
2
3
4

LOGRUNNERS (ORTHONYCHIDAE)

1 CHOWCHILLA *Orthonyx spaldingii* 25–28cm/10–11in

Chowchilla is the larger, northern version of the better-known Australian Logrunner and is endemic to the mountain rainforests of the Wet Tropics in ne. QLD. It behaves and is shaped much like its southern cousin: a pot-bellied terrestrial bird most often encountered scratching in the shady forest leaf litter for prey. A plainer species, lacking the logrunner's intricately patterned upperparts, it is uniform blackish or dull rufous above and has a variably coloured underside. Female Chowchillas have an orange throat and clean white underparts, while males have a white throat and underparts. Both males and females possess a bold cerulean-blue eye ring. The Chowchilla's call is a very distinctive *Chow-chow-chow-chow-choo-choo* that once learned is never forgotten and is usually the best indication of the presence of this often elusive bird. Unlike the logrunner, Chowchilla is best approached when it is calling, as it is often oblivious to human presence and at its most confiding then. It is best found in the early mornings, when at its most vocal.

2 AUSTRALIAN LOGRUNNER *Orthonyx temminckii* 17–20cm/6.75–8in

This bird is mottled brown, looking almost bark-like, on the back and has several bold pale grey bands on the wing, a paler grey face, and bold underparts that differ in males and females. The female logrunner has a rich burnt-orange throat and white underparts, while the male is all white from the throat down to the vent. As the name suggests, these birds do feed around logs and are often seen running on fallen trees, which they also use as calling posts. Although they are sometimes elusive birds, they rely on their camouflage and can be approached very closely when feeding, if you walk the rainforest trails quietly. At this time they can often be detected by the sound of rustling in the leaf litter. However, when these birds are calling their nature changes markedly and they can become quite furtive. Australian Logrunners are forest-floor dwellers of subtropical and temperate rainforests in se. Australia from se. QLD to s. NSW. They are common in the rainforests west of Brisbane, including Lamington NP (QLD).

2 ♂

♂
♀ 1
♂ 2
♀ 2

QUAIL-THRUSHES AND WHIPBIRDS (PSOPHODIDAE)

1 SPOTTED QUAIL-THRUSH *Cinclosoma punctatum* 25–28cm/10–11in

The male is a gorgeous bird with a bold head pattern showing a brown crown, a prominent white eyebrow, a pale blue-grey face, and a black throat patch that contrasts strongly with a large white teardrop below the eye. The underside pattern changes from blue-grey on the breast to a boldly spotted white underbelly. On top the bird is brownish with bold black streaks from the nape to the uppertail. The female is similar but has a pale peach throat and a plainer head, still possessing a white supercilium but lacking the bold markings of the male. Spotted Quail-Thrush is found locally within eucalypt forests with a grassy understorey, often on rocky ridges. The main population range is from near Rockhampton in c. QLD down the ranges and coastal zone of e. Australia through NSW to VIC and into TAS. The species is widespread but rarely common. A disjunct small population has recently been discovered near Ravenshoe in ne. QLD.

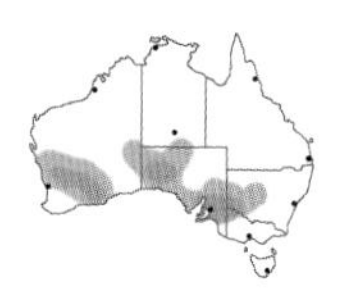

2 CHESTNUT-BACKED QUAIL-THRUSH *Cinclosoma castanotum* 23–25cm/9–10in

This bird has uniform dark olive-brown upperparts and tail, a deep chestnut rump and lower back, and a white eyebrow and moustachial stripe. The male has a glossy black throat and upper breast, which contrasts strongly with the white lower breast and belly. Along the sides of the breast and flanks is a silver-grey area with a line of black scales extending to the vent. The female lacks the black of the male and has an attractive uniform grey throat and breast contrasting with the white belly. The species is widespread, though sporadic, from the mallee areas of inland NSW west through the arid open woodlands of c. Australia and in the mallee and open woodlands of sw. WA. Although shy, with patience it can be seen in many of the mallee reserves of NSW and nw. VIC.

3 NULLARBOR QUAIL-THRUSH *Cinclosoma alisteri* 18cm/7in

Very similar to Cinnamon Quail-Thrush. The upperparts are uniform sandy brown except for a black shoulder patch that is spotted white in the male. Males have a thin buff eyebrow and white moustachial stripe, as in Cinnamon, but the throat and breast are all black. The lower breast and belly is clean white, with black stripes separating it from the flanks. The females have a black shoulder patch with buff edges and a sandy-grey throat and breast but differ from Cinnamon in having a thicker creamy moustachial stripe. It has a very limited range in the gibber-plain, bluebush, and saltbush sedgelands of the Nullarbor Plain centred on the s. WA–SA border. *(See photos of following species)*

4 CINNAMON QUAIL-THRUSH *Cinclosoma cinnamomeum* 20–24cm/8–9.5in

The upperparts are uniform sandy brown except for a black shoulder patch that is spotted white in the male. Males have a thin buff eyebrow and white moustachial stripe. The throat is black, and the chest is off-white with a thick black breast band below it. The lower breast and belly are clean white with black stripes delineating them from the rufous flanks. The males differ from Nullarbor Quail-Thrush males in having a complex breast pattern compared to the simple black throat and breast of Nullarbor. Cinnamon Quail-Thrush females, which have a black shoulder patch and a sandy-grey throat and breast blending gradually into a white belly, blend in perfectly with their terrain. This species is found on the sparsely vegetated gibber plains, rocky ridges, and sand dunes of c. Australia. It can often be found in suitable habitat along the Strzelecki and Birdsville Tracks of ne. SA and sw. QLD.

5 WESTERN QUAIL-THRUSH *Cinclosoma marginatum* 19–23cm/7.5–9in

The upperparts of this quail-thrush are chestnut except for a black shoulder patch edged in white. The male has a broad white eye stripe and moustachial stripe. But it is the underparts that make this bird so attractive: The black throat, the thick chestnut upper breast band fringed with cream on the upper and lower edges, and the thick black lower breast band contrasting with the white belly are very distinctive. Females have a uniform rufous throat and breast and an indistinct buff moustachial stripe and appear brighter than any other female quail-thrush. Habitat preferences vary greatly for this bird, which may be found in mulga scrub, spinifex grasslands, and arid open woodlands, often on stony ground. It inhabits a rarely visited part of Australia, with its distribution centred on arid inland WA and only just reaching into sw. NT and nw. SA.

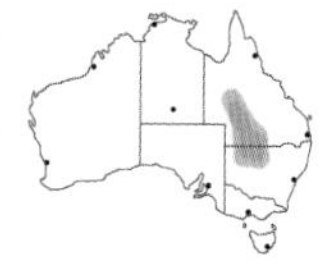

6 CHESTNUT-BREASTED QUAIL-THRUSH *Cinclosoma castaneothorax* 20–24cm/8–9.5in

Very similar to Western Quail-Thrush. Upperparts are chestnut except for a black shoulder patch edged in white. Males have a broad white eye stripe and moustachial stripe. The black throat meeting the thick chestnut breast band, which is bordered by a thin black lower breast band contrasting with the white belly, are very distinctive and separate this species from Western Quail-Thrush. Females have a uniform rufous throat and breast and an indistinct buff moustachial stripe and appear duller than Western Quail-Thrush females. This bird is found in the arid inland of sw. QLD and nw. NSW. It is often associated with rocky ridges. Bowra Station in sw. QLD is a good location to find this species.

♂ 1
♀ 1
♂ 2
♀ 2
♂ 4
♀ 4
♂ 5
♀ 5
♂ 6
♀ 6

QUAIL-THRUSHES AND WHIPBIRDS (PSOPHODIDAE)

1 EASTERN WHIPBIRD *Psophodes olivaceus* 25–30cm/10–12in

Eastern Whipbird has a remarkable and strident call that sounds like the loud cracking of a whip and draws attention when heard, especially for the first time. The male is a large, long-tailed songbird with a black head and breast, a prominent crest on the top of the head, and a broad white moustachial stripe down the side of the face that forms its most striking feature. A brief view may yield nothing more than this flash of white as it scampers away. The back is dull greenish. The female has the same general shape, although with a less-pronounced crest, and is dull greenish all over. This is a species of rainforests and woodlands with dense undergrowth and is common through its range. A skulking bird, usually seen on or near the ground, it is more often heard than seen. It is found in the coastal zone of e. Australia from ne. QLD down to e. VIC.

2 WESTERN WHIPBIRD *Psophodes nigrogularis* 22–25cm/8.5–10in

A serious skulker, far more likely to be heard than seen and one of the few Australian species for which you need to know the call (a warbling yet grating *double-up whit-tu-whit-tu*) to have a chance of seeing it. It is dull olive above and has a long, wedge-shaped tail diffusely tipped with white that is not held cocked. It has a black bib, an obvious white moustachial stripe, and an indistinct crest. The breast is dull olive grey fading into a white belly. Western Whipbird is found in isolated populations along the w. edge of the s. WA coast, extending inland to some mallee area. This bird is both rare and extremely difficult to get a decent look at; expect to spend a great deal of time waiting in known territories in its mallee or heathland habitat before a sighting.

3 MALLEE WHIPBIRD *Psophodes leucogaster* 22–25cm/8.5–10in

Another very secretive species, very closely related to Western Whipbird. It is grey-olive above and has a long, wedge-shaped tail with a diffuse subterminal black band tipped in white. It has an indistinct crest and a black bib bordered by a white moustachial stripe. The breast is dull olive grey blending into a white belly. This species differs from Western Whipbird in having the moustachial streak bordered above by the olive-grey cheek and neck, not a black border. The species is very local in mallee areas from Eyre Peninsula to e. SA. *(See photo of preceding species)*

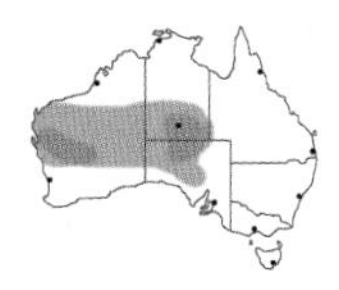

4 CHIRRUPING WEDGEBILL *Psophodes cristatus* 19–22cm/7.5–8.5in

The two wedgebill species are nearly identical, and both are found in the arid inland, usually in sparsely vegetated open areas. Both species have a small upright crest and a longish tail and are plain brown above and paler below. Chirruping Wedgebill also has very faint streaks on the breast. The species are very difficult to separate on sight alone, but luckily there is very little overlap in range, so they can usually be identified based on where they are seen. They also have quite different calls that although varied have a distinct quality; Chirruping Wedgebill's call is a high-pitched and repeated *tip-TSshiep, tip-TSshiep*. This species is widespread and common in e. SA, w. NSW, and sw. QLD in low shrubs such as bluebush and saltbush. It occurs in the same habitat as White-winged Fairywren, so use that species as a guide.

5 CHIMING WEDGEBILL *Psophodes occidentalis* 20cm/8in

Chiming Wedgebill is much shyer than Chirruping Wedgebill and is much more difficult to see well, tending to skulk away. It is very similar in plumage but has a plain rather than faintly streaked breast. Like Chirruping it is also found in arid scrub, often with low shrubs, on which it will sit to sing. The song of Chiming is quite distinctive, a repetitive and rhythmic series of descending notes that sounds like a squeaky wheel continually turning. It replaces Chirruping Wedgebill from c. SA and s. NT west to the coast in similar habitat.

CUCKOOSHRIKES (CAMPEPHAGIDAE)

1 BLACK-FACED CUCKOOSHRIKE *Coracina novaehollandiae* 33cm/13in

A widespread, large powdery-grey bird with a bold black face and throat that forms its most prominent feature. It has a noticeable and distinctive habit of shuffling its wings awkwardly when it lands. Immature birds lack a black throat. The all-black throat and pale grey crown separate adult Black-faced from the smaller White-bellied Cuckooshrike. Black-faced has a sweeping undulating flight, while White-bellied's flight is quite direct. Black-faced Cuckooshrike is one of the most widely distributed and regularly seen Australian birds and can be found in a variety of wooded habitats right across mainland Australia and TAS. It is common in parks and gardens, where it is often seen hunting prey from exposed overhead wires. It is the only cuckooshrike on TAS.

1

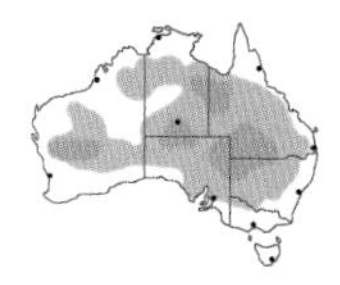

2 GROUND CUCKOOSHRIKE *Coracina maxima* 33–37cm/13–14.5in

Ground Cuckooshrike is a distinctive species of dry habitats within inland Australia. It is the largest cuckooshrike and the only one that possesses a forked tail. This tail is quite long, and gives the bird a distinctive silhouette in flight. It also has a piercing yellow eye and a distinctive white rump that is barred with black and has this same pattern on the breast. It is terrestrial in its habits, unlike other cuckooshrikes, and is usually encountered in small groups foraging on bare, open ground. It is widespread throughout much of the interior of mainland Australia in tropical savannahs, arid open woodlands, and plains. It is thinly distributed and also nomadic, making it a difficult species to reliably locate.

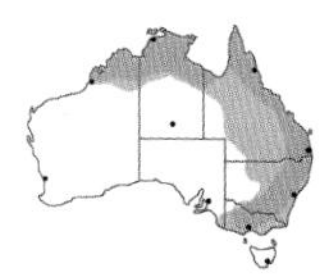

3 WHITE-BELLIED CUCKOOSHRIKE *Coracina papuensis* 25–28cm/10–11in

This small, plain-coloured bird is pale grey with a white underside, a contrasting small black face mask from the bill to the eye, and an inconspicuous white eye ring (lacking in other cuckooshrikes). Immature Black-faced Cuckooshrikes, which have a reduced mask, can be very similar, although White-bellied is markedly smaller, and its black mask does not extend onto the cheek as it does in young Black-faced. A rare se. Australian dark morph of White-bellied has extensive dark markings around the head, but they form more of a hood than a mask and are never as crisp as the black mask of Black-faced. White-bellied is found across tropical n. Australia, where it is a common bird of the tropical savannahs, greatly outnumbering Black-faced. It is also found throughout e. Australia, where it is less common and occurs in a range of wooded habitats including dry sclerophyll and open woodlands. Birds are often seen in singles or pairs perched prominently.

4 BARRED CUCKOOSHRIKE *Coracina lineata* 25cm/10in

A small, dark cuckooshrike. The male is slate grey with a bright yellow eye. The belly and vent are finely barred with white. This barring is very fine and can be difficult to see, and the bird sometimes appears all slate grey unless seen well. Immature birds are similarly patterned but duller. This bird could be confused with Oriental Cuckoo but is much smaller and darker and has a shorter tail and that stunning yellow eye. It could be confused with Common Cicadabird, which also lacks the yellow eye and barring below. A shy bird, Barred Cuckooshrike tends to fly away before an observer can get too close. It is an uncommon canopy bird of rainforests, wet sclerophyll, and other thickly wooded habitats such as gallery forests, and is best seen around fruiting trees, which it visits to feed. It is found along the coastal strip from Cape York Peninsula (QLD) south to n. NSW, becoming progressively rarer south of the tropics.

1
imm. 1
2
2
3
4

1 COMMON CICADABIRD *Coracina tenuirostris* 25cm/10in

A small cuckooshrike that is quite shy and difficult to see well. The male is slate grey with black smudging around the face, black markings on the wing coverts and edges, and black outer tail feathers. The female has a dirty-brown back, fawn underparts finely barred with chocolate brown, and a narrow eyebrow and a parallel line under the eye that give the appearance of a broken eye ring. Common Cicadabird is a resident across the n. Kimberley (WA) and Top End (NT) found in monsoon forests and more densely wooded habitats. It is also found down the e. coast from Cape York Peninsula (QLD) to e. VIC; there are resident populations in the north, but it is a summer migrant in se. Australia.

2 VARIED TRILLER *Lalage leucomela* 19cm/7.5in

A striking black-and-white cuckooshrike with a buff eyebrow, buff vent, and white wing bars. The male differs from breeding male White-winged Triller in having duller and faintly barred underparts, a rich buff vent, and a pale supercilium (breeding male White-winged lacks an eyebrow). Furthermore, the white on the wing of Varied Triller forms two distinct wing bars (unlike the large patch on White-winged). The female Varied is rather like a subdued, greyer-brown version of the male. This species has very different habitat preferences from White-winged Triller, occurring in dense, wet coastal habitats, including rainforests, monsoon forests, wet sclerophyll, and dense woodlands. It is often encountered in small parties, usually moving through the canopy foraging for fruit and invertebrates. Varied is a bird of the coastal zone in n. WA, n. NT, and the e. coast of QLD and NSW.

2 ♀

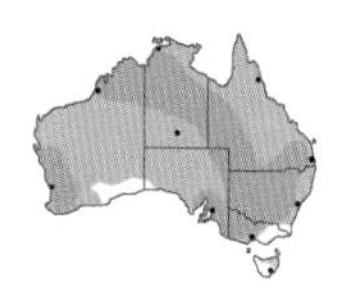

3 WHITE-WINGED TRILLER *Lalage tricolor* 18cm/7in

Breeding male White-winged Trillers are striking black-and-white birds with black upperparts, white underparts, and a large white patch in the wing. White-winged Trillers can be distinguished from Varied Trillers by their lack of a supercilium and by their crisp all-white underparts and vent. The females and non-breeding males are duller, brownish in comparison, and can show a supercilium and therefore appear closer to Varied, although both these White-winged plumages lack barring on the underparts and have a pale vent that is concolourous with the rest of the underparts. White-winged Triller inhabits a vast array of dry habitats where it may be encountered both foraging in the trees and on the ground. It is a widespread species found commonly throughout the mainland and occasionally in n. TAS. It is a seasonal migrant, with birds moving to s. Australia in the summer to breed and wintering in the savannahs of n. Australia.

♂ 1
♀ 1
♂ 2
♀ 2
♂ 3
♀ 3

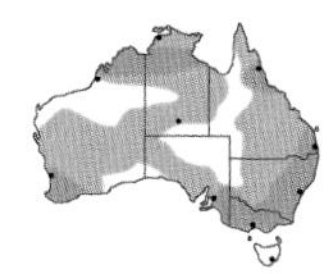

1 VARIED SITTELLA *Daphoenositta chrysoptera* 10–12cm/4–4.75in

As the name suggests, this is a variable species encompassing five distinct subspecies found throughout Australia. It is one of three species that make up the family Neosittidae. All Varied Sittellas are generally grey above with streaked upperparts and have paler underparts that can be unstreaked or streaked and either a dark cap, a hood, or a completely white head, depending on subspecies. These birds are very active, constantly on the move and regularly flitting from branch to branch or trunk to trunk. In these short flights they display a broad pale bar in the wing, ranging from rufous in some populations to grey in others, a white rump, and extensive white in the tail. Varied Sittella is widespread, occurring across the entire mainland although absent from TAS. It can be found in almost any wooded habitat except rainforests, including eucalypt forests, open woodlands, mallee, and also parks and gardens. It usually prefers rougher-barked trees and areas where there is plentiful standing dead wood.

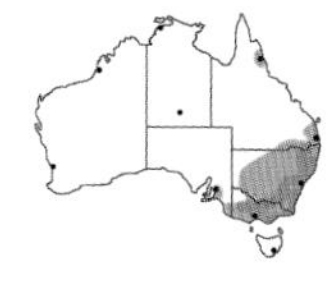

2 EASTERN SHRIKETIT *Falcunculus frontatus* 16–19cm/6.25–7.5in

This species of the whistler family (Pachycephalidae) forms a complex with Northern and Western Shriketits; all are remarkably similar but easily separated by range. All three species feed by tearing bark to search for hidden insects. This bird is boldly patterned with a striking black hood, white frons, thick white eyebrow behind the eye, and large white moustachial patch. It has mid-green upperparts, black primaries, and yellow underparts. The black chin and throat of the male is replaced by green in the female. Young birds have the same head pattern but a dirty-white throat and a very washed-out browner body pattern. Eastern Shriketits are rarely abundant but are widespread in most wooded habitats throughout se. Australia; an isolated population exists in ne. QLD. The best means of tracking them down is to learn the distinctive mournful whistle they often give as they move around, often in pairs.

3 NORTHERN SHRIKETIT *Falcunculus whitei* 15cm/6in

Extremely similar to Eastern Shriketit and regarded by some as conspecific. It differs in being smaller, having brighter yellow underparts and yellower upperparts with yellow wing edges and a yellow wash to the primaries. A rare bird, Northern Shriketit is patchily distributed between the e. Kimberley (WA) and the e. Top End (NT), occurring in tropical eucalypt woodlands. This habitat is very widespread, and the limiting factor to this species' distribution is not understood. Most recent sightings have come from the area south of Katherine, including the Central Arnhem Highway and Warloch Ponds near Mataranka (NT). *(No photo)*

4 WESTERN SHRIKETIT *Falcunculus leucogaster* 17–19cm/6.75–7.5in

This species is similar to Eastern and Northern Shriketits; the primary difference is the broad white band across its belly. It is found only in sw. WA, where like its relatives it occurs in a range of wooded habitats. Like Northern Shriketit it is sparsely distributed and can be quite difficult to find. Good places to search for it include the Dryandra Woodland, Stirling Range, and other areas of extensive forest.

5 CRESTED BELLBIRD *Oreoica gutturalis* 20–22cm/8–8.5in

This shy bird of the whistler family (Pachycephalidae) can be quite difficult to see and is usually spotted perching on an exposed branch giving its haunting, ventriloquial call. It has an olive-brown back and paler brown underparts. The male has a grey head with a black cap, a white frons and white chin patch, and a contrasting black bib below the throat. The female is very dull with olive-brown upperparts and paler olive-brown underparts and buff lores and throat. Very vocal, this species calls insistently through the day in an easily imitated four-note whistle; it is highly ventriloquial, and the call is very difficult to locate. The crest is raised when the bird is calling or excited. In females and young birds all black areas except crest are replaced by brown, giving the whole bird a much more subdued look. This common species occurs in a wide variety of arid shrublands, including mallee and mulga, and arid open woodlands throughout much of coastal WA to the rangelands of e. Australia.

c. 1
se. 1
ne. 1
nw. 1
2
4
♂ 5
♀ 5

1 WHITE-BREASTED WHISTLER *Pachycephala lanioides* 20cm/8in

A striking large whistler of the northern mangroves. The male has a black hood that contrasts starkly with the pure white throat and black breast band. It has a grey back, black tail, and creamy breast and belly. A chestnut nape separates the hood from the back and extends onto the edge of the breast. The female is much like the female Rufous Whistler, with greyish upperparts, buff underparts, a white throat, and faint streaking over most of the underparts, and is best distinguished by its larger size. An uncommon bird, White-breasted Whistler is patchily distributed on seaward sides of mangroves from tropical coasts of WA to the nw. QLD gulf coast. It is best seen around Broome (WA).

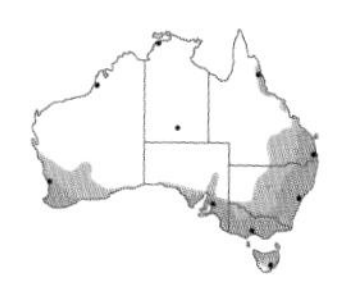

2 AUSTRALIAN GOLDEN WHISTLER *Pachycephala pectoralis* 15–17cm/6–6.75in

The male Australian Golden Whistler is a gorgeous bird displaying a black hood bordered with a bright canary-yellow collar, a bold white throat, and vivid golden-yellow underparts. The female is subdued—olive-toned with an indistinct wash of yellow on the belly and few bold markings—and best identified as a whistler by structure and from other whistlers by an accompanying male or by the yellow wash below. Calls of Australian Golden Whistler are loud and often draw considerable attention. The species is found in a wide variety of forested habitats in e. and s. Australia and is often quite common.

2 ♂

3 MANGROVE GOLDEN WHISTLER *Pachycephala melanura* 15–17cm/6–6.75in

A very similar bird to the much more common Australian Golden Whistler. Structurally the two species can be separated by Mangrove's smaller body size and relatively longer bill. The male Mangrove is brighter yellow below and has a broader yellow collar and a paler back than the male Australian. The female Mangrove varies, showing bright yellow underparts in some populations and off-white to buff underparts in others, but in all forms has brighter green upperparts than the female Australian Golden Whistler. Mangrove Golden Whistler is patchily distributed in mangrove forests and monsoon forests from the tropical coasts of WA to ne. QLD, although it is quite rare in ne. QLD.

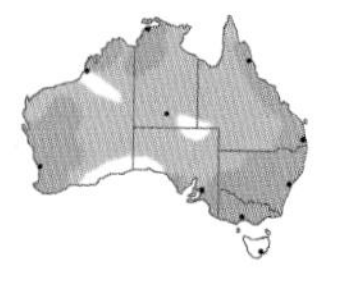

4 RUFOUS WHISTLER *Pachycephala rufiventris* 16–18cm/6.25–7in

Rufous is the common dry-country whistler. Boldly marked vocal birds, the males are greyish above and have a striking white throat patch surrounded by a thick black border and rufous underparts. The inconspicuous females, which look very different from the males, are also greyish but lack the conspicuous head pattern; their most useful ID features are faint streaks on the upper breast and a pale rufous-washed belly. Rufous Whistler is commonly found throughout Australia in most habitats, except the driest parts of the interior and the s. coast, and is absent from TAS. Its rich, varied song and often-repeated *ee-chong* call are quintessential sounds of the Australian spring.

♂ 1
♀ 1
♂ 2
♀ 2
♂ 3
♀ 3
♂ 4
♀ 4

1 GREY WHISTLER *Pachycephala simplex* 15cm/6in

This whistler is quite small, nondescript, and often puzzling to identify. Many new birders or those who have not seen it before will rack their brains and page through their field guides trying to figure out what it is before realising it is a Grey Whistler. It is plain grey-brown above, with paler underparts and a greyish head. The subspecies from ne. QLD (*peninsulae*) has a pale yellow belly; it is widespread and found in rainforests and monsoon and gallery forests, and the birds can sometimes be difficult to see as they forage actively high in the canopy. The NT subspecies (*simplex*) is fairly easy to spot in and around Darwin at East Point and Lee Point and Buffalo Creek. The vocalisations of the two subspecies are different, and the species may be split in the future.

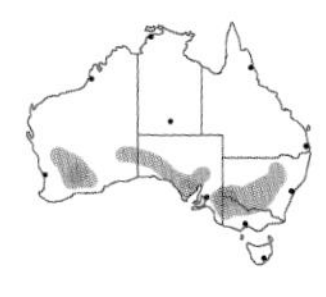

2 GILBERT'S WHISTLER *Pachycephala inornata* 19–20cm/7.5–8in

This species is fairly common across a variety of dry woodland habitats, particularly mallee, throughout dry s. Australia. Males are slate grey above and have a pale belly and an orange-rufous throat. Females and immatures are similar but lack the orange throat, and the immature is slightly browner. This is a shy species but is quite vocal and fairly common in suitable habitat. The loud, repeating *chong-chong-chong* call is often heard in many of the mallee reserves in se. Australia. These whistlers are quite inquisitive and will often come to investigate squeaking.

2 ♂

3 RED-LORED WHISTLER *Pachycephala rufogularis* 19–22cm/7.5–8.5in

This whistler is very rare within its limited distribution. It is restricted mostly to mallee and is very local, absent from seemingly suitable habitat in some areas. A shy bird, it is usually located when singing; unfortunately, it seems to call for only a brief period of the year prior to breeding. The male is large and plain grey with a buff belly; it has an orange-rufous throat that extends up onto the face, which distinguishes it from the similar and much more common Gilbert's Whistler. The female is similar to the male but paler, and the immature is browner and has a rufous eyebrow. The stronghold for this species seems to be the mallee of nw. VIC and se. SA, where it can be seen in some parts of Hattah-Kulkyne NP and Gluepot Reserve. There is another much rarer population centred around Round Hill and Nombinnie NR (NSW).

4 OLIVE WHISTLER *Pachycephala olivacea* 20–22cm/8–8.5in

This large whistler has an interesting distribution: In the north of its se. Australian range it is restricted to high-elevation rainforests, but farther south it is less confined and in TAS is found in a wide range of habitats from the coast to the mountains. It has olive-brown upperparts, buff underparts, a grey head, and a white throat. A fairly shy bird, it tends to stay down low in thick undergrowth but will investigate pishing or squeaking. Olive Whistler is nowhere common but can be found more easily in TAS than elsewhere.

NT 1
QLD 1
♂ 2
♀ 2
♂ imm. 3
4

1 GREY SHRIKETHRUSH *Colluricincla harmonica* 24cm/9.5in

The most widespread shrikethrush in Australia, this species (unlike the others on this page) is largely plain grey with pale off-white underparts. There are several different subspecies, some with a plain grey mantle matching the rest of the upperparts and head, and others with a rufous cast to the mantle. Grey Shrikethrush is common and widespread, found throughout the mainland and TAS and absent from only a few small sections of the continent. It occurs in most wooded habitats and therefore is the most likely shrikethrush to be encountered, often seen in suburban parks and gardens. Grey Shrikethrush is also famed for its huge repertoire of songs.

2 LITTLE SHRIKETHRUSH *Colluricincla megarhyncha* 19cm/7.5in

Also known as Rufous Shrikethrush. The smallest shrikethrush, Little is a brown bird with rich rufous underparts and light streaking sometimes evident on the breast. In the north-east Little Shrikethrushes have pale horn-coloured bills, while in the rest of the species' range the bills are dark. The very similar Sandstone Shrikethrush is significantly larger and longer-tailed, and is a local species confined to sandstone escarpments where Little does not occur. Little Shrikethrush is also much more arboreal and regularly found within trees. It inhabits rainforests, coastal forests, thickets (e.g., lantana), paperbark stands, and mangroves. It occurs in the coastal zone of n. and ne. Australia from the north of WA through n. NT and n. QLD and south along the coastal region down into n. NSW.

3 BOWER'S SHRIKETHRUSH *Colluricincla boweri* 22cm/8.5in

A small, chunky shrikethrush with a bullish head. It has a slate-brown back and rufous underparts with fine grey streaks on the throat and breast. The chunky appearance and streaking on the breast best tell it from Little Shrikethrush. This species is endemic to the Wet Tropics of ne. QLD where it is most common in the higher-elevation rainforests of the Atherton Tableland and Paluma region, although it occasionally moves to the lowlands, especially in winter. It can usually be seen readily in any of the large areas of rainforest in ne. QLD, including Curtain Fig near Yungaburra, Mt Hypipamee, and Mt Lewis.

4 SANDSTONE SHRIKETHRUSH *Colluricincla woodwardi* 23–25cm/9–10in

The largest and most specialised shrikethrush, confined to sandstone escarpments in n. Australia. It is like a large, long-tailed version of Little Shrikethrush: brown on the upperparts, with rich rufous underparts and light mottling or streaks on the breast. Little Shrikethrush is distinctly more arboreal and regularly observed foraging within trees. Sandstone Shrikethrush is very restricted in its range, always associated with sandstone escarpments from n. WA across the north of the NT to the QLD border. It is fairly common within its restricted habitat, usually encountered singly or in pairs.

se. 1
c. & w. 1
NT 2
QLD 2
3
4

FIGBIRD AND ORIOLES (ORIOLIDAE)

1 AUSTRALASIAN FIGBIRD *Sphecotheres vieilloti* 27–29cm/10.5–11.5in

Australians Figbirds are colourful flocking orioles with three subspecies. Males in the south (Green Figbird) have a green underside, while the two northern subspecies (both called Yellow Figbird) have lemon-yellow underparts. Males of all subspecies have olive upperparts, a black head with bright red facial skin, and a stout, black bill. The bright yellow or green unstreaked underparts and vivid facial skin help to separate figbirds from other orioles. Females have dull olive-brown upperparts and pale cream underparts with very heavy olive-brown streaking. They have the same area of facial skin as males, though it is dull brown. Females can be told from immature Olive-backed Orioles by the figbird's stubbier bill and the facial skin. The two Yellow Figbirds are found coastally in the tropical north from n. WA around to n. QLD; Green Figbird occurs from the c. QLD coast south to e. VIC. Figbirds are common in lowland rainforests, parks, and gardens, and are easily seen in suburban areas throughout their range.

1

2 GREEN ORIOLE *Oriolus flavocinctus* 25–29cm/10–11.5in

A very familiar bird of tropical Australia, Green Oriole has a warm olive-toned body, a rich yellow wash to the belly, and fine dark streaks all over the body that give the bird an unkempt look. It has a prominent reddish eye and a conspicuous reddish bill. It is a canopy bird usually spotted singly after one tracks it down by its loud songs. Green Oriole is a tropical species found in rainforests, mangroves, woodlands, parks, and gardens, from n. WA eastward into n. QLD. Although it is common, it generally perches high in trees, making it hard to see. The loud, far-carrying, rollicking calls are a distinctive backdrop to the tropical dawn chorus, and the species is more often heard than seen.

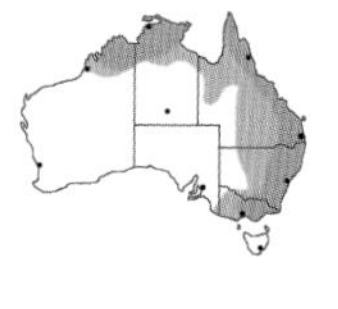

3 OLIVE-BACKED ORIOLE *Oriolus sagittatus* 28cm/11in

A solitary canopy oriole, Olive-backed has a bright olive head and back and grey wings and tail. Underparts are a clean white finely streaked with black and showing a strong green wash on the upper breast and throat. The bright red bill and eye are always obvious. Young birds have the same pattern but in grey tones; they can be confused with female figbirds but have a longer black bill and lack the facial skin figbirds show. Although generally uncommon, Olive-backed Oriole can be locally abundant across tropical n. Australia from the Kimberley east to QLD and south through e. Australia to SA. It is found in most wooded habitats including rainforests, sclerophyll forests, open woodlands, mulga, and mallee. In drier areas it is the only oriole species.

NT 1
n. QLD 1
c. QLD 1
♀ 1
2
imm. 2
3
♀ imm. 3

1 WHITE-BREASTED WOODSWALLOW *Artamus leucorynchus* 17cm/6.75in

White-breasted Woodswallow is the most familiar and frequently encountered species of woodswallow in n. Australia. It is easily recognised by its grey-brown upperparts, uniform except for a bold white rump patch, its all-grey hood, and pure ghostly white underparts. Male and female are alike. When the bird is in flight the dark-hooded appearance is obvious and a good ID feature, along with the white rump patch and, from the underside, a plain black tail. As it is a generalist, this bird can be found in many habitats throughout its range. Across tropical n. Australia and in coastal areas along the e. coast it is resident, while in the se. part of its range it is more seasonal and nomadic, arriving in spring. It is common around towns, where flocks are often found perched on overhead wires.

1

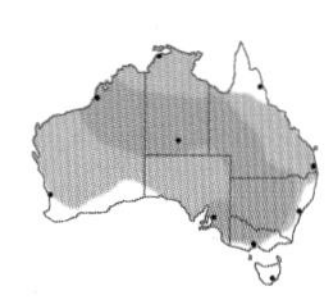

2 MASKED WOODSWALLOW *Artamus personatus* 19cm/7.5in

A distinctive woodswallow with a bold black mask and throat that contrast strongly with the pale grey crown and nape. It is superficially similar to White-breasted, although that species differs in having a dark grey hood that is uniform in colour on the throat, crown, and nape. In flight Masked shows no pale rump and has a white undertail, different from the White-breasted's white rump and black undertail. This species regularly mixes with flocks of White-browed Woodswallows too. Masked Woodswallow is found in a variety of dry, open habitats. A nomadic bird of inland Australia, it only very rarely appears around the coasts. Although it can be found over much of the mainland, it is a scarce species that wanders widely and is generally not easy to find.

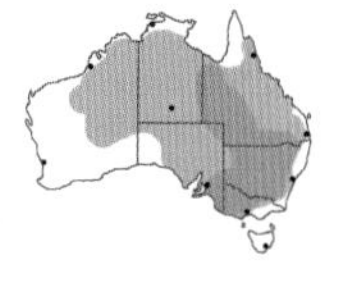

3 WHITE-BROWED WOODSWALLOW *Artamus superciliosus* 18–19cm/7–7.5in

A gorgeous blue-grey bird with deep chestnut underparts and a bold white eyebrow, unlike any other woodswallow species. In flight this striking bird shows a clean ivory-white underwing that contrasts with the chestnut underside of the body, quite different from any other species. This woodswallow may feed on nectar when blossoms occur, although it also regularly hawks insects on the wing, as do all woodswallows. Like Masked Woodswallow, it also inhabits a range of dry open habitats, including open forests, woodlands, farmlands, mallee, and mulga. White-browed Woodswallow is highly nomadic over its range in most of c. and e. Australia.

1
♂ 2
♀ 2
♂ 3
♀ 3

1 BLACK-FACED WOODSWALLOW *Artamus cinereus* 18cm/7in

A pale ash-grey woodswallow with a small, inconspicuous dusky-grey mask around the eye and the base of the bill. This face mask is not as extensive or as bold as the larger jet-black mask of Masked Woodswallow. The pale grey-bodied Black-faced Woodswallow is also duller-coloured below but has a striking tail pattern—black on the underside with two bold white marks at the tip—whereas Masked has an all-whitish tail. This open-country species is found throughout large areas of the mainland from the w. side of WA east to NSW and QLD, although absent from extreme se. Australia and TAS. It is abundant across the arid interior and one of the most commonly encountered birds in inland Australia.

2 DUSKY WOODSWALLOW *Artamus cyanopterus* 18cm/7in

A dark chocolate-brown southern woodswallow, similar in appearance to Little Woodswallow. Dusky Woodswallow shows pure white underwings in flight that contrast strongly with the chocolate-brown body, while Little's underwings are dusky brown and do not contrast so strongly with the body colour. When perched Dusky Woodswallow also displays a thin white leading edge to the closed wing, absent in Little. Both show a black tail in flight with bold white tail spots. Dusky Woodswallow is often seen sallying regularly to and from a chosen perch, usually a dead snag. A southern species, it is a breeding migrant to TAS and occurs year-round over most of NSW, s. WA, and se. QLD. It can be seen in both open woodlands and closed forests.

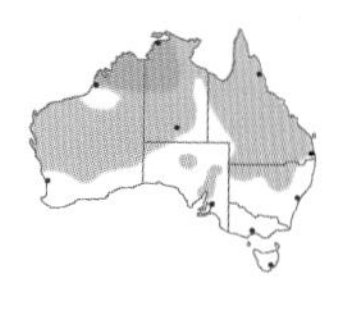

3 LITTLE WOODSWALLOW *Artamus minor* 12cm/4.75in

The smaller, northern and inland cousin of Dusky Woodswallow. Also chocolate-coloured, it shows entirely dark brown closed wings when it is perched, lacking the distinct white edge on the closed wing that Dusky exhibits. In flight Little's underwings contrast with the body less, as they are washed with brown, unlike the clean white underwings of Dusky. Little Woodswallow occurs across much of inland Australia, though it avoids the far south of the country. The species is found in a variety of habitats, including tropical savannahs, arid open woodlands, and arid grasslands. It shows a definite preference for rocky gorges and can usually be found in such habitat throughout its range, particularly in the tropical north.

1
1
2
2
3
3

BUTCHERBIRDS AND CURRAWONGS (CRACTICIDAE)

1 PIED BUTCHERBIRD *Cracticus nigrogularis* 32–34cm/12.5–13.5in

The most widespread of the butcherbirds, Pied shows a complete black hood, solid black back, a white collar around the nape, and extensive white wing markings. The black throat differentiates it from Black-backed, Grey, and Silver-backed Butcherbirds. Its black back separates it from Grey and Silver-backed. Pied Butcherbird is a common species throughout the mainland but absent from deserts and rainforests and does not occur in the far south-east and TAS. The bird inhabits open country, including farmlands, open woodlands, scrubby areas, and grasslands, and is most easily seen perched prominently on roadside trees or wires while hunting.

2 BLACK-BACKED BUTCHERBIRD *Cracticus mentalis* 28cm/11in

A pied bird, which at first glance looks like a Pied Butcherbird but has the same general pattern of Grey Butcherbird. It can be told from Pied by its smaller size, white throat, more white in the wing, and black nape. It is told from Grey by its black lores, black back, pure white underparts, and pied wings. Black-backed is a common bird over the drier parts of Cape York Peninsula (QLD) in tropical savannahs and also farmlands and townships. It is usually easily located around Weipa.

3 SILVER-BACKED BUTCHERBIRD *Cracticus argenteus* 24–28cm/9.5–11in

A restricted-range species of the Top End. It is very similar to Grey Butcherbird and is sometimes considered a subspecies of it. Silver-backed is smaller and has a much paler silver-grey (not mid-grey) back and black (not white) lores. It is patchily distributed throughout the Kimberley (WA) and Top End (NT), found in savannah woodlands, parks, and gardens, but is nowhere common.

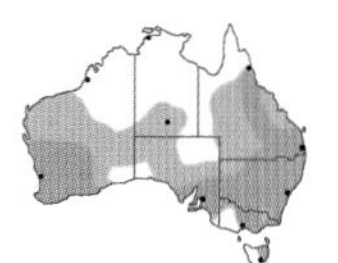

4 GREY BUTCHERBIRD *Cracticus torquatus* 28–32cm/11–12.5in

This butcherbird has a black hood, grey back, black wings and tail, white throat, and light grey wash on the breast. It is very similar to Silver-backed Butcherbird, which does not overlap in range, but this species can be distinguished by its darker grey back and white lores. Within its range Grey Butcherbird can be separated from Pied by its white (not black) throat and grey (not black) back. Immature birds are shades of brown and look much like Pied immatures, though the Grey's basic pattern remains and the hood is clearly visible. This species occurs through most of s. Australia, northward in e. QLD to the Cairns region. It is found in a range of habitats including sclerophyll forests, monsoon forests, open woodlands, farmlands, parks, and gardens. It is a common bird of many s. Australian towns and cities.

5 BLACK BUTCHERBIRD *Cracticus quoyi* 39–43cm/15.5–17in

Black Butcherbird is the only butcherbird that has an all-black body. This may make it more likely to be confused with crows or currawongs than other butcherbirds. The silver bill with a black, hooked tip is distinctive; all currawongs and crows show unhooked, all-blackish bills. There are several very similar forms of Black Butcherbird; the immatures of one form are all rufous, but the bill again provides the best indication of the species. This is a tropical bird, found in mangroves, monsoon forests, and rainforests in the Wet Tropics of ne. QLD and in the tropical north of the NT. It is not uncommon and can be seen most easily in tropical town parks such as those within Cairns (QLD) and Darwin (NT).

1
imm. 1
2
3
4
imm. 4
5
5

BUTCHERBIRDS AND CURRAWONGS (CRACTICIDAE)

1 AUSTRALIAN MAGPIE *Gymnorhina tibicen* 36–44cm/14–17.5in

A large, long-legged pied bird, extremely familiar bird to most Australians. While highly variable in its large range, the magpie always has a solid black underside and a large pale to white nape patch, which differentiate magpies from all the butcherbirds. It is a widespread and common species found widely through the mainland and TAS. However, it is generally much rarer in humid tropical areas and is absent from both the driest deserts and sw. TAS. It is adaptable, occupying most habitats where there are trees, including many urban areas and cities; it can be found commonly in most cities south of Cairns (QLD) and often frequents parks and gardens.

2 PIED CURRAWONG *Strepera graculina* 41–51cm/16–20in

A predominantly black bird with yellow eyes, a white rump, a white patch in the wing, white tail tips, and a white vent. Pied is the only currawong with a broad white base to the tail, which distinguishes it from Grey Currawong. These are predatory birds that eat nestlings, lizards, and insects, and also berries. Pied is the most frequently encountered currawong, common through e. mainland Australia, although absent from TAS. It can be found in rainforests, eucalypt forests and woodlands, farms, parks, and gardens, and is a familiar bird in many urban areas. Its call is a very familiar sound of se. Australia.

3 GREY CURRAWONG *Strepera versicolor* 44–53cm/17.5–21in

There are several subspecies of this large bird ranging in shade from glossy black to mid-grey. The general pattern remains the same: It is a mostly blackish bird with a white vent and a white patch of varying thickness on the wing in flight. All forms have a dark rump, which separates this species from Pied Currawong, and on TAS the white vent and white line in the wing separate it from Black Currawong. Grey Currawong occurs in a range of habitats, including open woodlands, farms, scrub, and heaths, and is very rarely seen in urban areas. A southern species found across the mainland from s. WA to se. NSW and TAS, it overlaps with Pied in s. NSW and VIC, and with Black on TAS.

4 BLACK CURRAWONG *Strepera fuliginosa* 46–48cm/18–19in

A mainly glossy black bird endemic to TAS. This species is unique in this group in lacking a white vent and being all black below. Grey Currawong on TAS shows large white flashes on the primaries of the upperwing, which are obvious in flight, unlike the inconspicuous crescent on the base of the upperwing primaries of Black Currawong. Forest Raven, another large black bird on TAS, may also cause confusion, although currawongs have longer, pointed bills and display a white tail tip, whereas the raven is all black. Black Currawong is found in a variety of habitats over most of TAS and is easily seen.

n. & e. 1
s. & w. 1
2
2
se. 3
s. 3
sw. 3
4

FANTAILS (RHIPIDURIDAE)

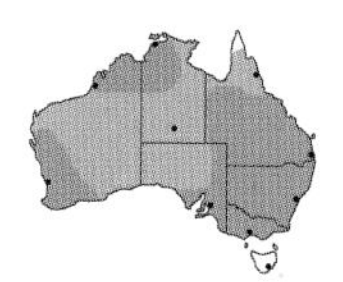

1 WILLIE WAGTAIL *Rhipidura leucophrys* 19–22cm/7.5–8.5in

This bird is familiar to most Australians, thanks to its bold and conspicuous nature. It is the only black-and-white fantail and is all black except for a clean white belly and a narrow white line above the eye. It is very widespread and adaptable, found all over Australia (vagrant to TAS) in almost every habitat except for rainforests. These birds are often encountered in gardens, parks, and lawns, where they often hop around on the ground, chattering frequently and wagging their tails conspicuously.

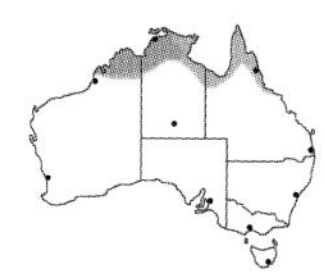

2 NORTHERN FANTAIL *Rhipidura rufiventris* 16–18.5cm/6.25–7in

Northern Fantail has grey upperparts with a white-sided tail, pale off-white underparts, a broad horizontal band across the chest, and a thin white fleck above the eye. It is similar to Grey Fantail but shows thin white streaks running vertically down the dark grey breast band, and structurally it is larger-billed and shorter-tailed than Grey. Furthermore, Northern behaves quite differently: It is relatively inactive as it sallies after insects from exposed perches. It usually sits quietly and upright in the manner of a monarch flycatcher (family Monarchidae), fanning its tail much less often than the hyperactive Grey. Northern Fantail is confined to the tropical north (n. WA, NT, and QLD only), where it occurs in open eucalypt forests and woodlands, mangroves, monsoon forests, and on the edges of tropical rainforests. Although it is common, it is inconspicuous and less often encountered than Grey Fantail.

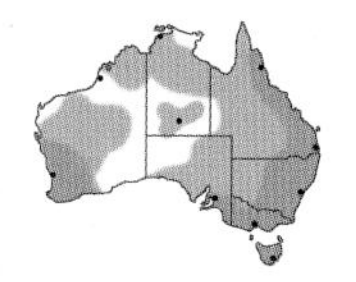

3 GREY FANTAIL *Rhipidura albiscapa* 15–17cm/6–6.75in

A widespread species similar to Northern Fantail but displaying markedly different behaviour: Grey is a hyperactive bird that regularly flares its tail and chases insects around actively, rarely sitting still for long periods. It also lacks the light streaking within the chest band of Northern and shows bolder white wing bars. Furthermore, Grey's tail is thinly tipped with white and not merely edged white on the sides as in Northern. There are several subspecies of Grey Fantail that vary in their colouration; generally darker birds occur in the forests of e. Australia, and paler birds are found in warmer, drier areas. Grey Fantails often join mixed feeding flocks, and this is often when they are at their most visible. The species occurs throughout the continent (including TAS), absent mainly from e. interior WA. It is a generalist found in most habitat types, both on the coast and inland.

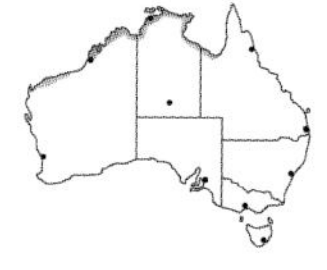

4 MANGROVE FANTAIL *Rhipidura phasiana* 15–16cm/6–6.25in

A medium-grey fantail confined to mangroves across nw. Australia. It is extremely similar to the very variable Grey Fantail but generally paler. It is mid-grey and has a fawn belly, as does the subspecies of Grey Fantail that also occurs in n. Australia, but Mangrove Fantail differs in having the white in the tail limited to the edges. Mangrove Fantail is not a common species and requires extensive mangrove forests. It is found from n. WA across to the Gulf Country of QLD.

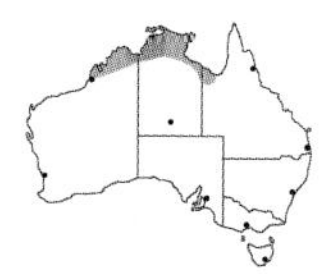

5 ARAFURA FANTAIL *Rhipidura dryas* 15–16cm/6–6.25in

Arafura Fantail is very similar to the much more widespread Rufous Fantail but has the strong rufous colour restricted to the lower back and rump. Arafura's tail is grey-brown with large bright white tips. This species replaces Rufous Fantail from nw. QLD across to the Kimberley region (WA). It is found in a variety of dense habitats, including mangroves, dry rainforests, and gallery forests. It is occurs regularly around Darwin in the Top End (NT), where it can be spotted in many patches of monsoon forest such as those at Buffalo Creek and Fogg Dam.

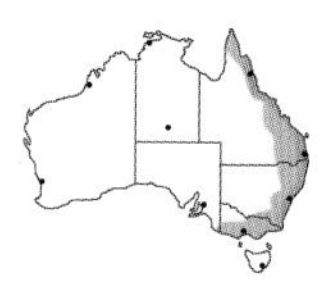

6 RUFOUS FANTAIL *Rhipidura rufifrons* 15–16cm/6–6.25in

A distinctively coloured fantail, this is a brownish bird with a rich rufous rump and uppertail, a small rufous smudge above the eyes, and a bold black band across the breast. Like Grey Fantail it is a hyperactive bird, always fanning its distinctive tail. It is fairly common in wet forests of e. Australia from n. QLD south to s. VIC, where it is usually encountered low in the shady understorey. Birds in the s. part of the range are migratory, arriving in spring to breed and spending the winter in n. QLD and Papua New Guinea.

1
2
3
4
5
6

CROWS (CORVIDAE)

1 TORRESIAN CROW *Corvus orru* 48–53cm/19–21in

This is the only crow species found in far n. Australia and is the most common down the e. coast of QLD. It is the only species likely to be seen in both Brisbane (QLD) and Darwin (NT). There is little to separate it from other corvids by sight; in flight the wings are slightly more rounded, and it doesn't have pronounced throat hackles. A common call is a rapid *uk-uk-uk*, higher and more clipped than the call of Australian Raven and higher and less hoarse than that of Little Crow. It often gives more drawn-out calls, which are difficult to separate from those of other species, so listen carefully to recognise the 'typical' call.

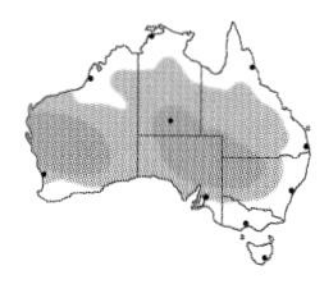

2 LITTLE CROW *Corvus bennetti* 46–48cm/18–19in

Although this is Australia's smallest crow, it is only marginally smaller than the other species. It is a bird of the arid inland and is the species most likely to be seen in the dry, stony deserts of the Red Centre (NT) and inland WA. It is quite sociable and often forms large flocks. The call is short and clipped, like that of Torresian Crow, but is a hoarser and deeper *ark-ark-ark*.

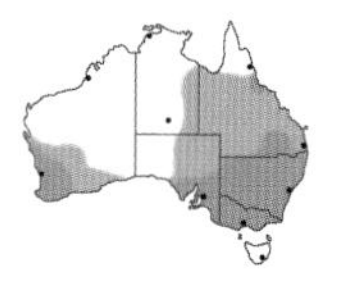

3 AUSTRALIAN RAVEN *Corvus coronoides* 51–53cm/20–21in

This corvid is common in s. Australia, often seen in cities, including Sydney (NSW), Perth (WA), and Melbourne (VIC). The species has pronounced throat hackles, which can often be seen hanging from the throat as the bird calls or when it is perched or foraging on the ground, if views are clear. It is a large bird and has a drawn-out call, sometimes compared to a baby's crying: *aah-ahh-aaaahhhh*. It is probably the most commonly encountered corvid in s. Australia and is found in a wide range of habitats, particularly farmlands.

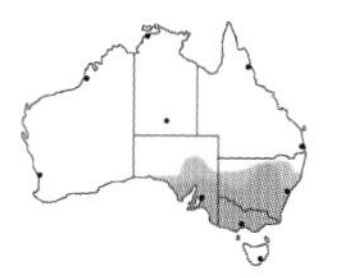

4 LITTLE RAVEN *Corvus mellori* 48–51cm/19–20in

This corvid is only slightly smaller than Australian Raven, with which it is most likely to be confused. Its call is a harsh and clipped *ark-ark-ark-ark*, more rapid and lower than that of Australian Raven. When it calls, Little Raven flicks its wings rapidly in time with its call, a good feature to aid in identification. It is fairly common in farmlands of se. Australia and is the most common species in Melbourne (VIC) and Adelaide (SA). This raven is quite sociable, particularly in winter, often forming large flocks. In flight it has obviously quicker wing beats than Australian Raven, but this feature takes some experience to notice.

5 FOREST RAVEN *Corvus tasmanicus* 52–55cm/20.5–21.5in

This is the only corvid found in TAS, so it is easy to identify there, where the only confusion species are Grey and Black Currawongs. Forest Raven is also found along the coast to se. SA and has is an isolated population on the nc. coast of NSW. It is Australia's largest corvid and has a relatively short tail. It has throat hackles, but they are not as pronounced as in Australian Raven. The call is similar to that of Australian Raven but is deeper and slower. The name is not a useful guide to identification as Forest Raven is found in a wide range of habitats.

1
2
3
4
5

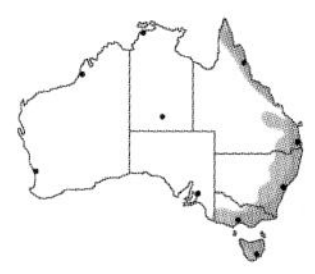

1 SATIN FLYCATCHER *Myiagra cyanoleuca* 16cm/6.25in

The male is glossy blue-black above from hood to back to tail. Its lower breast, belly, and vent are pure white, giving this bird a very glossy, clean pied look. Females are duller above and have a rufous throat and breast; their plumage shows more contrast than that of both Leaden and Broad-billed Flycatchers, though this is hard to discern in poor light. Satin Flycatcher sits high in the canopy and often calls after completing a successful sally. It is found from the Gulf Country of n. QLD around the e. coast to SA in rainforests, wet sclerophyll, and mangroves. It breeds in the se. mainland and TAS, wintering in n. QLD. It is much more common than Leaden Flycatcher in VIC and TAS and much less common in the rest of their shared range.

2 BROAD-BILLED FLYCATCHER *Myiagra ruficollis* 15–17cm/6–6.75in

A furtive flycatcher. Male and female adults are similar; they have a uniform glossy slate head and upperparts, a bright rufous throat and breast, and a sharply contrasting white lower breast to vent. The bird has a very flat-headed appearance with a sharp angle from the peak of the head to the neck. It is very similar to the female of the more common and widespread Leaden Flycatcher, but Broad-billed adult has a slightly broader bill that is bowed outward along the edges, darker blue upperparts, and deeper orange colouration on the throat and chest. Broad-billed also does not habitually quiver its tail, which Leaden does with regularity. Broad-billed Flycatchers inhabit monsoon forests, mangroves, and gallery forests into tropical woodlands and savannahs. They are found from near Broome in WA around the n. coast to near Cairns in QLD.

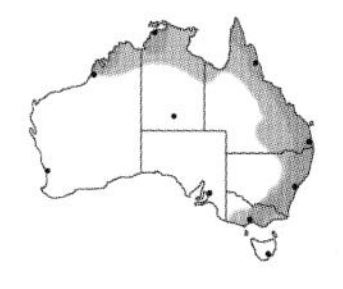

3 LEADEN FLYCATCHER *Myiagra rubecula* 15–17cm/6–6.75in

Leaden Flycatcher regularly quivers its tail and is strikingly dimorphic: Males are bi-coloured birds with slate-blue upperparts, including head and chin, which gives them a hooded appearance. Females are slate grey above and have a peach-coloured chin and throat. The males are similar to Satin Flycatcher, which is glossier and more pied-looking. Females are remarkably similar to female Satin Flycatcher, which is glossier above than Leaden, and to adult Broad-billed, which generally has darker, deeper blue upperparts relative to female Leaden, and a deeper and brighter orange throat and chest. Leaden Flycatcher occurs in eucalypt forests, open woodlands, mangroves, along the banks of vegetated rivers, and in tropical savannahs but generally avoids dense forests, where it is replaced by Broad-billed and Satin Flycatchers. Leaden is a common species found across the tropical north of Australia from WA east to QLD, and in e. Australia from n. QLD south into s. and e. VIC. It is a vagrant to TAS.

3 ♂

♂ 1
♀ 1
2
2
♂ 3
♀ 3

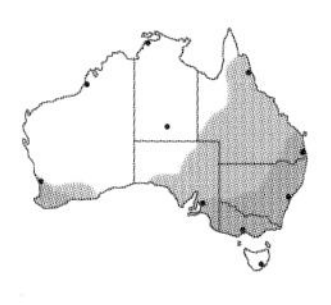

1 RESTLESS FLYCATCHER *Myiagra inquieta* 19–20cm/7.5–8in

Restless Flycatcher has all-black upperparts and head and is solidly pale below, mostly white except for a subtle buff tinge to the throat, which differentiates it from the extremely similar Paperbark Flycatcher. It prefers open wooded habitats, often occurring along eucalypt-lined watercourses (especially those with Red Gums) but also found in wooded parks and even golf courses. It is a fairly common species that often draws attention with its scissor-grinder calls. The species is found in much of e. Australia from e. Cape York Peninsula southward throughout NSW, VIC, and e. SA, as well as in sw. WA.

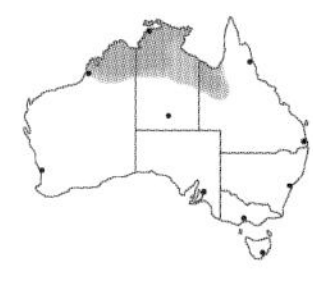

2 PAPERBARK FLYCATCHER *Myiagra nana* 17cm/6.75in

Paperbark Flycatcher is solid glossy black above and pure white below. Remarkably similar to Restless Flycatcher, of which it was until recently regarded a subspecies, it differs in its clean white underparts lacking any hint of buff tones. Young Paperbark birds may have a dirty smudge to the underparts. Found from the Kimberley (WA) across the Top End (NT) to the QLD Gulf Country, Paperback Flycatcher is a common bird at most billabongs and river edges in the Top End.

3 SHINING FLYCATCHER *Myiagra alecto* 15–18cm/6–7in

The male is a small, all glossy blue-black bird. Its most striking feature is a bright orange gape, evident only when its bill is open. The female is strikingly different: rich rufous above, with a glossy blue hood and a white underside. The male is a much smaller bird than any alternative blackish bird and forages much lower down than others. Shining Flycatchers are coastal species of the tropical north of Australia found from n. WA across the Top End (NT) and from the n. Cape York Peninsula down the QLD coast. They are most likely to be encountered on boat cruises along river channels in their range (e.g., Daintree River cruises in QLD and Kakadu NP boat trips in the NT).

3 ♂

4 YELLOW-BREASTED BOATBILL *Machaerirhynchus flaviventer* 11.5cm/4.5in

A gorgeous and very distinctive little flycatcher; unlike the other species on this page, which are in the family Monarchidae, this species is in the single-genus family Machaerirhynchidae. A chunky, large-headed bird with a very large boat-shaped bill, it has black upperparts with white wing and tail edges and a bright yellow eyebrow. The throat is white, grading into the bright yellow underparts. Females are only slightly duller than males, with the true black replaced by a sootier black. The species is an uncommon rainforest, monsoon thicket, and mangrove bird from Cape York Peninsula to south of Cairns (QLD). It responds very well to whistles and playback of its call.

1
2
♂ 3
♀ 3
♂ 4
♀ 4

MONARCHS (MONARCHIDAE)

1 SPECTACLED MONARCH *Symposiachrus trivirgatus* 14–16cm/5.5–6.25in

Spectacled Monarch is greyish above with a black face mask, a black throat, and a rich, deep orange breast. It has a boldly marked white-tipped black tail, which it fans when it is sallying for insects. It is a fairly common bird of coastal e. Australia, occurring from Cape York Peninsula in extreme n. QLD south to the c. NSW coastal zone, where it inhabits the understorey and gullies within rainforests and along densely-vegetated watercourses.

2 BLACK-FACED MONARCH *Monarcha melanopsis* 17–19cm/6.75–7.5in

A very attractive flycatcher with mid-grey upperparts, head, and breast, and rich rufous underparts from lower breast to vent. The black forehead, lores, and bib give it a masked appearance, and the dark eyes have a large black ring that makes them appear outsized. The wings and tail are slightly darker grey than the back, but this is noticeable only in good light. Black-faced Monarch shares range and habitat with Spectacled Monarch and differs in having grey on the upper breast and no white on the underparts, and lacking white tail tips. It is an uncommon bird in rainforests, wet sclerophyll forests, and thick gullies in the coastal regions from Cape York Peninsula (QLD) to e. VIC.

3 BLACK-WINGED MONARCH *Monarcha frater* 18–19cm/7–7.5in

A powder-grey flycatcher with a black forehead, lores, and throat. The black wings and all-black tail contrast with the powder-grey rump. The breast is also powder grey, while the lower breast to vent is rufous. Black-winged shares its restricted distribution with the more wide-ranging Black-faced Monarch, with lacks the jet-black wings and tail. A summer breeding migrant from Papua New Guinea to the rainforests of far ne. Cape York Peninsula (QLD), Black-winged Monarch is best located at Iron Range NP.

4 PIED MONARCH *Arses kaupi* 14–15cm/5.5–6in

Pied is a dazzling monarch, black above except for a prominent white collar and a large white crescent across the back. It is white below with a broad black band across the chest. It also displays a beautiful cerulean-blue eye ring. When excited the bird puffs up the frills on the nape, reducing the appearance of the black of the crown and making the bird look much more like Frill-necked Monarch (which lacks the black breast band). This eye-catching rainforest bird has the strange habit of feeding in treecreeper-like fashion, searching for insect prey by hopping up and down trunks, flicking its tail, and opening its wings conspicuously when it does so. It is found in only a small area of the Wet Tropics of ne. QLD and best located at Curtain Fig.

5 FRILL-NECKED MONARCH *Arses lorealis* 15cm/6in

A beautiful pied flycatcher, Frill-necked is crisp white below and has a black hood and chin and fluffy white feathers on the nape. The blue skin around the eyes is very obvious. The mantle, back, wings, and tail are jet black except for a broad arc of pure white across the upper back. It looks similar to Pied Monarch but lacks the black breast band and has more-erect white feathers on the nape, which the Pied displays only when excited. This Cape York Peninsula (QLD) species is a common bird of the rainforests in Iron Range NP and an uncommon bird of monsoon forests west to Weipa.

6 WHITE-EARED MONARCH *Carterornis leucotis* 14cm/5.5in

A very distinctive bird among the flycatchers, White-eyed Monarch is the only one with grey underparts and also differs from others in its facial pattern. It has a black hood marked with a white eyebrow starting above the eye, a white spot below the eye, and a white ear patch. The back and wings are black with two white wing bars joined along the wing coverts by a white line, which gives the impression of one large wing patch at a distance. The base of the white-tipped black tail is mid-grey, the same colour as the underparts. The throat is white speckled with black. Less obvious than other monarchs, and still an active bird, White-eared Monarch tends to do more gleaning and make fewer sallies than other monarchs. It is a very uncommon bird in lowland rainforests, wet sclerophyll forests, mangroves, and thick gullies in the coastal regions from Cape York Peninsula (QLD) to n. NSW.

1
2
3
♂ 4
5
6

1 MAGPIE-LARK *Grallina cyanoleuca* 27cm/10.5in

An odd, oversized long-legged monarch usually seen on the ground. It has black upperparts, a horn-coloured bill, and a piercing yellow eye, and its underparts are largely white apart from a black-bordered white throat in females and a solid black throat patch in males. Magpie-Lark is a very vocal bird, nicknamed 'Peewit' after its loud calls. Although now considered an unusual member of the Monarchidae, it was formerly considered more akin to the Australian mudnesters family, as it constructs a large bowl-like nest from dried mud, which leads to its other nickname, 'Mudlark'. Magpie-Larks are extremely common wherever there are trees but like to forage in many open habitats, including lawns around airports and towns on mainland Australia.

1 ♀

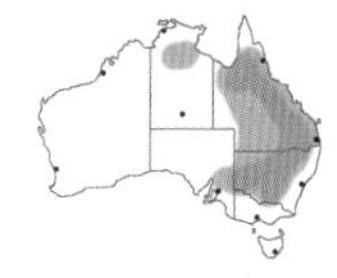

2 APOSTLEBIRD *Struthidea cinerea* 29–32cm/11.5–12.5in.

Apostlebird is all greyish brown with few markings, save for lighter flecks all over the body and the darker brown wings. A member of the Australian mudnesters family, Corcoracidae, it is a stout bird, reminiscent of Old World babblers (Timaliidae), with a heavy, short, thick bill. It is highly social and almost always seen in large boisterous groups that forage mainly on the ground and make lots of harsh scratchy calls to one another. It is a bird of open dry habitats, including open woodlands (e.g., mulga), scrubby country, and farmlands. There is hardly a golf course within its range that does not have a resident group. Apostlebird is common in the c. NT, the s. three-quarters of QLD, and much of NSW except the coastal strip. It also occurs in se. SA and nw. VIC. It is a conspicuous bird wherever it occurs; groups are most likely to be encountered feeding along the verges of inland roads fringed with open woodlands and scrublands.

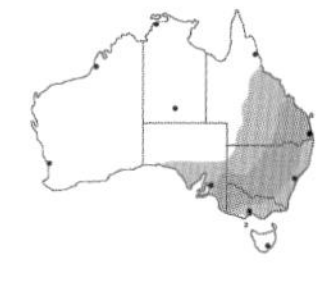

3 WHITE-WINGED CHOUGH *Corcorax melanoramphos* 43cm/17.5in

A predominantly black bird with white panels in its wing that appear as massive white wing flashes in flight. It has bright red eyes and a slim, down-curved bill that along with the white wing markings separate it from the currawongs. Members of the Corcoracidae family, Choughs are highly social cooperative nesters that make conspicuous, large mud nests. They are terrestrial, almost always seen in groups feeding on the ground. Choughs are dry-country birds of mallee, scrublands, and open woods such as brigalow found in the se. corner of SA north-eastward through VIC, most of NSW, and se. QLD.

♂ 1
♀ 1
2
3
3

1 VICTORIA'S RIFLEBIRD *Ptiloris victoriae* 23–25cm/9–10in

Male is very similar in appearance to Paradise Riflebird: a stout-bodied, short-tailed bird with a long, decurved bill and striking iridescence. It has a glistening green gorget and brow, and an olive-glossed belly apparent only in strong sunlight. Females are brownish above and have a conspicuous off-white eyebrow and lightly scalloped warm buff underparts. Victoria's Riflebird has an extraordinary display in which the male uses a large snag to dance for females. He fans his wings over the top of his head, opens his bill to reveal a vivid yellow gape, and then jerks his head from side to side, fluffs out his iridescent throat, making it catch the sunlight, and puffs up his breast feathers, accentuating the gloss on his belly also. He often does this while circling a female or he performs alone on a stump, emitting a loud, far-carrying rasping call to draw the attention of local females. The species does not overlap with other riflebirds in its range, which is confined to rainforests, mangroves, and wet sclerophyll forests in tropical ne. QLD. It is frugivorous and most likely to be seen in a fruiting tree or probing the bark of a forest tree in the manner of a treecreeper, or observers may visit one of the well-known display sites on the Atherton Tableland.

2 MAGNIFICENT RIFLEBIRD *Ptiloris magnificus* 28–34cm/11–13.5in

The largest of the riflebirds. The male is similar to the others, a largely blackish bird with stunning iridescence in its plumage and a powerful, down-curved bill. It differs from the other two species (with which it does not overlap) in its deep purple belly and a reddish band running horizontally below the gorget, visible only in strong sunlight, such as when it is displaying. It is also unique in possessing flank plumes, which are erected in a display similar in general style to that of the other riflebirds: It holds its fanned wings up, raises its beak to expose the gorget, and calls, revealing a striking, pale yellow gape. Common in rainforests, monsoon forests, and mangroves of n. Cape York Peninsula (QLD), it is a noisy and easy-to-locate bird within its range and is readily seen near Weipa.

3 PARADISE RIFLEBIRD *Ptiloris paradiseus* 28–30cm/11–12in

The male is a plump-bodied, blackish bird-of-paradise with a powerful and long down-curved bill, a short tail, and shimmering, iridescent plumage. Its blackish appearance changes suddenly when the light catches it (especially when the bird is displaying) and transforms the black throat into a large, glistening green bib and reveals a shimmering green forehead and a soft gold iridescent cast across the belly. Its displays involve upward fanning of the wings, flicking the head from side to side with the neck pointed up, and revealing its bright green gorget and bright yellow inner throat while calling. The loud, far-carrying, rasping calls draw attention and aid detection. Females are brown birds with rufous wings, pale buff underparts heavily scaled with brown, and a prominent buff brow. The species is very similar in appearance to Victoria's Riflebird, which occurs farther north. Paradise Riflebird is confined to rainforests and wet sclerophyll forests of se. QLD and ne. NSW. It is moderately common, but difficult to locate away from display trees, around O'Reilly's, within Lamington NP.

4 TRUMPET MANUCODE *Phonygammus keraudrenii* 28–32cm/11–12.5in

A long-necked, long-tailed, blackish bird-of-paradise that resembles a drongo more closely than a 'BOP'. Its blackish feathers are glossed green and it has beady red eyes. It is far longer-tailed than the other Australian BOPs—the noticeably short-tailed riflebirds—and its squared tail differentiates this BOP from the drongos. Trumpet Manucode possesses loose neck shackles that it fluffs up, along with its back feathers, when it is displaying, while it stretches its long neck out and throws its head back, calling loudly and flapping its wings. In flight it appears clumsy, with slow deliberate flapping, broad wings, and an obviously long-necked and long-tailed appearance. Males and females are alike. Confined to the rainforests and monsoon forests of n. Cape York Peninsula (QLD), it is common within Iron Range NP and around Weipa.

♂ 1
♀ 1
♂ 2
♀ 2
♂ 3
4

1 LEMON-BELLIED FLYROBIN *Microeca flavigaster* 14cm/5.5in

A relatively indistinct flycatcher of the family Petroicidae, this bird is usually seen on a prominent perch high in the canopy. It is unobtrusive and quiet, so it can be difficult to track down, and the bright, melodic song is often the first cue to its presence. It is dull olive brown above and has a faint eyebrow, a pale throat, and a yellowish breast and belly. It is fairly common across far n. Australia and easiest to see in the NT, where it can be found around Darwin and in Kakadu NP. It seems less common in ne. QLD but can still be seen fairly easily in open woodlands on the Atherton Tableland.

2 YELLOW-LEGGED FLYROBIN *Microeca griseoceps* 12cm/4.75in

This small flycatcher is found only on far n. Cape York Peninsula (QLD), where it occurs in the outer canopy on the edges of rainforests and monsoon forests. It is quite similar to Lemon-bellied Flyrobin but is smaller and has a greyish head and bright yellow legs. It also has a pale lower mandible, which may sound difficult to see but, as the bird perches high in the canopy and is often viewed from below, can be a good feature to know. The species is unlikely to be found without searching.

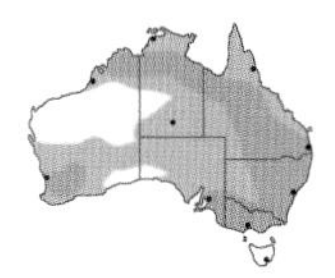

3 JACKY WINTER *Microeca fascinans* 14cm/5.5in

This flycatcher is plain dull brown above with a paler breast. It has a dark tail with white outer edges, which are the only distinctive feature you can see on the bird. It is usually seen sitting quietly on a prominent low perch, such as a stump or fence post, watching for insects. It often wags its tail from side to side, especially after landing. This plain little flycatcher is very widespread, found in dry open woodlands all across Australia but particularly in some of the arid inland areas. It is quite common in s. inland NSW and is regularly seen in the Capertee Valley and similar places.

4 SPANGLED DRONGO *Dicrurus bracteatus* 28–32cm/11–12.5in

Drongos are a tropical family (Dicruridae) of large black insectivorous birds that sit bolt upright on open perches in search of airborne insect prey. They exhibit long, distinctively shaped tails; Spangled Drongo has a deeply forked tail with slightly upturned tips. These are acrobatic birds, agile in their pursuit of insects on the wing, when they often draw attention to themselves. They are also highly territorial and often detected when fighting with rival drongos or other birds. Spangled Drongo is usually found alone or in pairs, normally sitting high on an open branch. It is partially migratory and found in n. WA, n. NT, n. and e. QLD, and e. NSW, in rainforests, woodlands, gardens, mangroves, and densely wooded watercourses.

4

1 RED-CAPPED ROBIN *Petroica goodenovii* 12cm/4.75in

A dazzling dry-country robin. The male has black upperparts, white panelling on the wings and tail, and a bright red cap. The throat is black, the upper and lower breast are red. Though superficially similar to male Scarlet Robin, Red-capped possesses a bright red cap and is the only robin in Australia with this feature. The duller female still has a hint of red on the forehead, if only as a subtle wash, lacking in all other similar robins. The species is widespread and common over the s. half of mainland Australia in more arid regions than the other red robins, occupying dry habitats such as open woodlands, mulga, brigalow, and mallee. It is most likely to be encountered sitting on a low open branch from which it launches regular forays to the ground to capture insects.

1 ♂

2 SCARLET ROBIN *Petroica boodang* 13cm/5in

The male Scarlet Robin is a very striking bird with a black mantle and wings, a bold white wing flash, and a shocking scarlet breast. It is similar to Red-capped Robin, but the two don't normally overlap in range, Scarlet preferring wetter habitats. Scarlet displays a black crown with a bright white forehead, while the forehead and crown are both scarlet in male Red-capped. The mostly brown female Scarlet is more colourful and contrasting than other similar female robins, showing a distinct red wash on the breast and a prominent white forehead spot, lacking in all other similar species. Scarlet Robin occurs in forests and woodlands in s. WA and the se. mainland, as well as TAS.

3 FLAME ROBIN *Petroica phoenicea* 14cm/5.5in

Another beautiful south-eastern robin. The male Flame Robin, although still an undeniably striking bird, is a little more subdued than Scarlet: It has sooty-grey (not black) upperparts, a reduced white forehead spot, and a crimson (not scarlet) breast. Females differ from other female robins in having a pronounced L-shaped white panel on black wings, a very reduced white patch on the lores, and a brown wash to the breast. Flame Robin is found within woodlands, forests, and scrub from sea level into the mountains (up to nearly 1,800m/5,900ft) in the se. mainland and TAS.

4 ROSE ROBIN *Petroica rosea* 11–13cm/4.25–5in

The most subdued of the small red robins, this is predominantly a canopy bird. Males are slate grey above, and have a slate-grey throat, a rose breast, and a white belly. The long tail is bordered in white, and the wings lack white panelling. Females are a dull steel grey with a white belly, a rose wash to the breast, and indistinct white panelling on wings and tail edges. Uncommon in mountain rainforests and wet sclerophyll forests from c. QLD around e. Australia to Adelaide (SA), this species rarely comes below mid-canopy and is seldom seen without searching. Is rather common on the Border Trail of Lamington NP (QLD).

5 PINK ROBIN *Petroica rodinogaster* 13cm/5in

The male Pink Robin has black upperparts, throat, and upper breast. The rest of the breast and upper belly are bright pink, the vent is white, and most birds show a small white frons just above the bill. Females are a uniform olive to almost chocolate brown with a buff belly and wing panels. Pink Robin nests in wet sclerophyll forests and rainforests of VIC, NSW, and TAS, and migrates to more open lowland habitats when not breeding. Males are assertive, but even in known territory this can be a difficult species to get a good look at and is easy to miss.

♂ 1
♀ 1
♂ 2
♀ 2
♂ 3
♀ 3
♂ 4
♀ 4
♂ 5
♀ 5

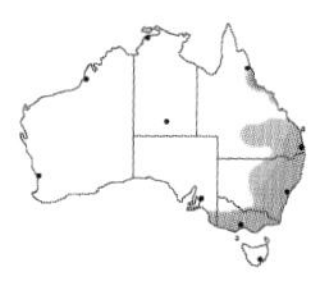

1 EASTERN YELLOW ROBIN *Eopsaltria australis* 15cm/6in

Eastern Yellow is a bright yellow robin of the east with a dull ashy-grey head and upperparts and bright canary-yellow underparts from the throat down. This robin has two different subspecies: in the north-east, Eastern Yellow Robins show a vivid canary-yellow rump, while the rump is olive in the south-eastern form. The species is most similar to Western Yellow Robin but has an extensive yellow throat and upper breast unlike its grey-throated western cousin. In Eastern Yellow's range, Pale-yellow Robin is the main confusion species; it possesses a pale face between the bill and the eye, while Eastern Yellow has a uniformly grey head and face. Pale-yellow also lacks a vivid yellow rump. Eastern Yellow Robin behaves in a similar fashion to Pale-yellow Robin and is most often observed perched low down in trees or on the sides of trunks, from which it frequently hops down to the ground to snatch insects. It is a common bird in e. rainforests, woodlands, and thickets from ne. QLD south to se. SA.

1

2 PALE-YELLOW ROBIN *Tregellasia capito* 13cm/5in

Pale-yellow Robin is the most subdued of the yellow robins. It has dull yellow underparts and a pale buff or white face with no conspicuous pale markings around the eye. Pale-yellow Robin is also uniform olive on the head and upperparts, with none of the smoky-grey or black colouration the others possess on the head. It is a rainforest robin found only locally in the fragmented rainforests of the east, where it is fairly common in its small range in the tropical and temperate rainforests of ne. QLD, s. QLD, and n. NSW. It is frequently found perched below eye level on vertical trunks, from which it often flits down to the ground to capture prey.

3 WESTERN YELLOW ROBIN *Eopsaltria griseogularis* 15cm/6in

The western version of Eastern Yellow Robin. There are two subspecies of Western Yellow Robin: The eastern form displays an olive-green rump, while the western shows a bright yellow rump. The species differs from Eastern Yellow in having a pale grey throat and chest and is otherwise similar, but the two species do not overlap in range. Western Yellow Robin is found only in the south (extreme s. SA) and west of Australia (sw. WA). It occurs in open forests and woodlands and mallee and other scrublands, and it is fairly common. Western Yellow often forages higher in trees than the other yellow robins.

4 WHITE-FACED ROBIN *Tregellasia leucops* 13cm/5in

The most distinctive of the yellow robins, this species has olive-green upperparts, yellow underparts, and a diffusely bordered black head that contrasts with an extremely conspicuous white face. The black eyes accentuate the white, giving the face an owl-like appearance. Like all juvenile yellow robins, the young are uniform olive brown with some subtle mottling. This robin is frequently found perched below eye level on vertical trunks. The species has a very small range in Australia concentrated on the rainforests of Iron Range NP, where it replaces Pale-yellow as the common ground-feeding robin.

se. 1
ne. 1
2
2
w. 3
4

AUSTRALASIAN ROBINS (PETROICIDAE)

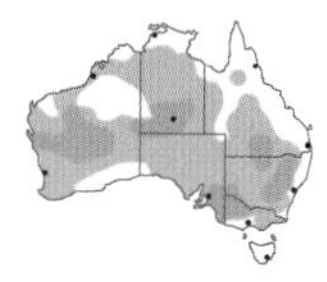

1 HOODED ROBIN *Melanodryas cucullata* 16cm/6.25in

The male Hooded Robin is a striking pied bird with a jet-black hood and crisp white underparts. It is all black above save for a white flash in its wing and a white line on the shoulder. The female is grey all over with darker wings that also show white bars; displaying a distinct bar on the upperwing in flight, as well as a white flash on the sides of the uppertail, the female is the only plain grey bird with this bold wing pattern and can therefore be identified by that alone. Hooded Robins are often found sitting up on exposed perches (such as dead snags) watching for prey. They occur in dry-country habitats such as mulga, mallee, eucalypt forests, tropical savannahs, and open woodlands but avoid wetter areas such as rainforests.

2 DUSKY ROBIN *Melanodryas vittata* 16.5cm/6.5in

This large robin is a Tasmanian endemic found in a range of habitats. It is quite dull: plain brown above and pale grey below. Young birds can be confusing, as they are heavily mottled dark brown and white, but are often found in the company of adults. Dusky Robins are often found in pairs or small family groups in open areas close to dense vegetation, where they sit on prominent perches such as fence posts or stumps watching the ground for insects. A moderately common bird of TAS, the species prefers open dry sclerophyll forests, heaths, and fringe areas of wet sclerophyll.

3 MANGROVE ROBIN *Peneoenanthe pulverulenta* 15–16cm/6–6.25in

This bird is fairly easy to identify in its habitat: It is restricted to dense mangroves around the coast and inlets and creeks across n. Australia. It is boldly marked, dark slate grey above and white below. The tail is black with small white outer panels at the base. This robin can be easy to see in some areas but is very shy in others. The distinctive vocalisations include a mournful whistle that is easily imitated and a musical chatter. It occurs in extensive areas of mangroves across the tropical Australian coast. Though generally uncommon, it is regular around Darwin (NT) and Cairns (QLD), where it may leave the mangroves to feed on local open ground.

4 WHITE-BREASTED ROBIN *Eopsaltria georgiana* 15–16cm/6–6.25in

A plain grey robin confined to the coastal strip of sw. WA, it is all grey above with indistinct white tail corners and whitish below, with the only bold marking a horizontal white bar on the wing visible only in flight. It is most similar to female Hooded Robin, but White-breasted does not show the broad wing bar when perched that Hooded does. In flight they are remarkably alike, but White-breasted lacks the white uppertail sides of Hooded. White-breasted Robin is an uncommon species found in wet eucalypt forests, thickets, and along vegetated creeks. It is most likely to be encountered perched low in the understorey or clinging to the sides of a vertical trunk. It can be found in Porongurup NP and also in forests on the edges of Perth (e.g., Wungong Dam Reserve).

4

♂
♀
1
2
2
3
4

1 GREY-HEADED ROBIN *Heteromyias cinereifrons* 17–18cm/6.75–7in

A grey robin confined to highland rainforests in ne. QLD. Although largely grey, it is a striking species with a bold white bar in the wing (visible when perched), jet-black lores, and a pale grey head with a warm wash to the ear coverts. The back is rufous, becoming bright rusty-coloured on the rump and tail. Underneath it is contrastingly very pale grey. It is a confiding, terrestrial robin in its tiny range (within which it is fairly common), where it is usually encountered hopping around on the ground, on tables, or perched on small posts within picnic areas (e.g., Mt Hypipamee NP and Curtain Fig) and often allows very close approach. Like many other the Australasian robins, Grey-headed may be observed perching low on the side of a vertical tree trunk.

2 WHITE-BROWED ROBIN *Poecilodryas superciliosa* 15–17cm/6–6.75in

An arboreal robin with grey-brown upperparts, a black mask, and a very strong white eyebrow. The underparts are lightly washed in grey, and the black wings have a prominent white patch on the primaries and white edging to the secondaries. The black tail, which is often held in a cocked position, is tipped white. The white throat extends to under the dark grey ear patch. It looks very similar to Buff-sided Robin but is more subdued overall and lacks the rufous wash to the sides of the breast. White-browed feeds higher in trees than other robins and comes to the ground much less frequently. A scarce species found from Cape York Peninsula down the tropical coast of QLD, it occurs in a wide variety of habitats but prefers lush, thick overgrown creeks, regardless of surrounding habitat. It is found regularly along shrubby creeks near Lake Mitchell.

3 BUFF-SIDED ROBIN *Poecilodryas cerviniventris* 15–17cm/6–6.75in

A gorgeous arboreal robin worth searching for, Buff-sided has brown upperparts, a black mask, and a very strong white eyebrow. The underparts are lightly washed in grey, but the sides of the breast, flanks to vent, are washed in a rich rufous. The black wings have a prominent white patch on the primaries and white edging to the secondaries. The black tail, which is often held in a cocked position, is tipped white. The white throat extends to under the black ear patch. Buff-sided looks like a bolder and more saturated version of White-browed Robin, which shows less contrast on the face and lacks a rufous wash below. A scarce and secretive robin, Buff-sided is found from the Kimberley (WA) across the Top End (NT) to nw. QLD. It prefers monsoon thickets and thick gallery forests in otherwise dry landscapes. It is easily found around Timber Creek township and at Plum Creek in Kakadu NP, both in the NT.

4 SOUTHERN SCRUB ROBIN *Drymodes brunneopygia* 22–23cm/8.5–9in

This shy bird is both vocal and inquisitive, so with patience it can be tracked down and will respond cautiously to pishing. The call is variable, but a common rendition is *chip-chp weeeet*, with an upward inflection on the *weeeet*. It is fairly nondescript: dull grey-brown above with an indistinct vertical black mark through the eye and paler below. It has a long rufous tail that is often held cocked and dark wings with pale edging on many of the feathers. Southern Scrub Robin is a moderately common species throughout the mallee regions of s. Australia, though not totally restricted to mallee, occurring in other dry scrublands in very low numbers. In areas of healthy mallee, this bird responds well to whistling, and will likely call from the top of a near tree.

5 NORTHERN SCRUB ROBIN *Drymodes superciliaris* 22cm/8.5in

Quite rare in Australia, this bird is found only very locally in a small area on far n. Cape York Peninsula (QLD). It forages on the ground in the dense understorey of rainforests and vine thickets, and can be quite difficult to see. It is inquisitive, though, and with patience may be observed closely. Northern Scrub Robin is distinctively plumaged: rusty brown above with a vertical black stripe through the eye and pale below. The wings are black with white tips to some of the feathers forming a double white wing bar.

3

1
1
2
3
4
5

GRASSBIRDS; LARKS; PIPIT AND WAGTAIL

1 BROWN SONGLARK *Megalurus cruralis* 18–25cm/7–10in

Songlarks have fluttering spring display flights that make them more visible in this season. Adult breeding male Brown Songlark is plain dark brown all over, unlike the other Australian species. However, females and non-breeding males are brown streaky birds quite similar to Rufous Songlark, although lacking the rich rufous rump patch of that species and having more extensive dark markings on the upperparts. Both songlarks, members of the Locustellidae (grassbirds and allies), can be told from pipits by their plain unstreaked underparts. Brown Songlark is widely distributed in open-country habitats such as farmlands and scrubby country on the s. mainland but is absent from far n. Australia. Males are most likely to be seen in spring singing from the tops of fences or posts with conspicuously cocked tails.

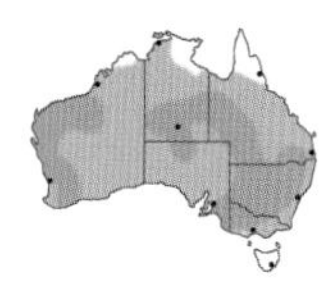

2 RUFOUS SONGLARK *Megalurus mathewsi* 16–19cm/6.25–7.5in

A pale songlark with a bold rusty rump on both males and females. Otherwise, the upperparts are fairly plain, lacking the extensive dark centres to the feathers that female Brown exhibits. Rufous Songlark possesses unstreaked underparts, distinguishing it from Australian Pipit and Horsfield's Bush Lark. It is widespread on mainland Australia, absent only from the Cape York Peninsula (QLD), the Top End (NT), and the Kimberley region (WA), though is much less common in the desert areas of interior e. WA. It tends to prefer habitats with more trees than Brown, occurring in open woodlands and croplands with scattered trees. It is most often encountered in spring, when the males perch high on open branches and call frequently while making conspicuous, sallying display flights.

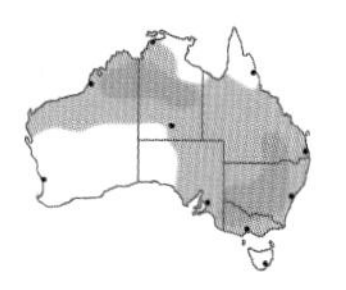

3 HORSFIELD'S BUSH LARK *Mirafra javanica* 13–15cm/5–6in

A stocky, short-tailed member of the lark family, Alaudidae. Upperparts are sandy brown with brown streaks on the back and rufous edges to the flight feathers, visible perched and in flight. Underparts are buff with thick brown streaking on the throat and breast. A buff eye line and darker ear patch give the face a masked appearance. This species resembles the introduced Eurasian Skylark but is shorter-legged, shorter-tailed, and lacks the skylark's crest. Though absent from the ne. Top End (NT) and Cape York Peninsula (QLD), Horsfield's Bush Lark is a very common bird across the savannahs of n. Australia, occurring in decreasing numbers south-eastward through all of NSW and into Eyre Peninsula (SA), and apparently being replaced by Eurasian Skylark in the south-east. It can be found in a variety of natural and man-made short-grass habitats and is fond of airstrips in n. Australia.

4 EURASIAN SKYLARK *Alauda arvensis* 18cm/7in

A pale-billed, long-legged lark (family Alaudidae) with a prominent crest. It has sandy-brown upperparts heavily streaked with brown and a buff breast with heavy brown streaking that is strongly demarked from the cream belly and vent. It appears similar to Horsfield's Bush Lark but looks longer-tailed and longer-legged and lacks the rufous edging to the flight feathers. A moderately common and increasing introduced species, Eurasian Skylark is found from the c. NSW coast across to SA in many cultivated fields and roadside verges, especially in s. NSW and VIC croplands.

5 AUSTRALIAN PIPIT *Anthus novaeseelandiae* 17–18cm/6.75–7in

Pipits are slim-billed, long-tailed, streaked brown songbirds that walk on the ground with a distinctive bobbing gait, regularly wagging their tails up and down in between spurts of running. Australian Pipit (family Motacillidae) can be told from songlarks by virtue of its heavily streaked chest and its white outer tail feathers, which are clearly visible in flight. It is a very widespread species found over all of the mainland and TAS in open habitats. It is unlikely to be found in closed forests.

6 EASTERN YELLOW WAGTAIL *Motacilla tschutschensis* 17–18cm/6.75–7in

A widespread and varied species from Siberia with multiple subspecies that migrate to n. Australia in small numbers. A ground-feeding bird similar to Australian Pipit and also in the family Motacillidae, Eastern Yellow Wagtail is long-legged and long-tailed and performs a bobbing feeding action. In non-breeding plumage, which is most common in Australia, wagtails have an olive-green or brown back, dark wings fringed in buff, a dark brown uppertail with buff outer tail feathers, a prominent buff eyebrow, and a chocolate-brown ear patch. The throat and breast are off-white, and the belly and flanks are washed yellow. In breeding plumage, the white-throated nominate subspecies is easily distinguished from the more common yellow-throated subspecies (*simillima*). This wagtail is a very uncommon but regular summer migrant to n. Australia, with most sightings around Darwin (NT) and Broome (WA), but is reported around the Australian coast.

♂ 1
♀ 1
2
2
3
4
5
6

CISTICOLAS; REED WARBLER; GRASSBIRDS

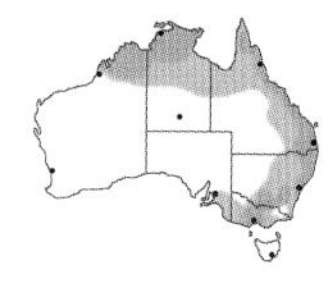

1 GOLDEN-HEADED CISTICOLA *Cisticola exilis* 9.5–11cm/3.75–4.25cm

A small bird of the family Cisticolidae, much warmer-toned than Zitting Cisticola, with mouse-brown upperparts heavily streaked with chocolate brown and fawn underparts with a rich golden wash to the upper sides of the breast and flanks. Females and non-breeding males have a streaked crown, as does Zitting, but have a dull golden nape and a diffuse white throat patch. The breeding male develops a golden crown and nape, quite unlike Zitting. A very common bird across the north of Australia and around the east to SA, Golden-headed Cisticola is found in most wet grasslands, marshes, and irrigation channels. Where the two overlap, it is much more common than Zitting.

1 nbr.

2 ZITTING CISTICOLA *Cisticola juncidus* 10cm/4cm

A small bird with a dull mouse-brown crown, nape, and back, all heavily streaked with black, and unstreaked cream underparts. It has a much colder-toned look than Golden-headed Cisticola. Zitting is patchily distributed across the n. coast from the Kimberley (WA) eastward to the Tropic of Capricorn in QLD and is nearly always rarer than Golden-headed Cisticola. It prefers tall grass and rank herbage. A regular site is the entrance road to Howard Springs near Darwin (NT).

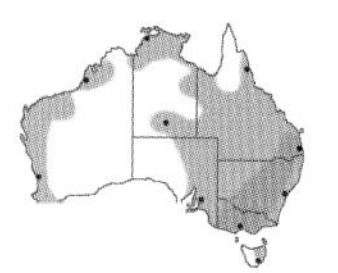

3 AUSTRALIAN REED WARBLER *Acrocephalus australis* 16cm/6.25in

This is a very vocal but otherwise subdued bird of the family Acrocephalidae. It has dull olive-brown upperparts and plain fawn underparts; its only notable mark is a buff eyebrow. The species is common in sw., nw., and e. Australia, including TAS; it is found in reed beds, especially cumbungi (bulrush), on river and lake edges.

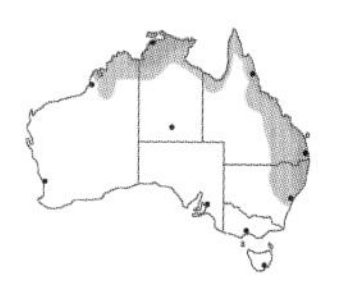

4 TAWNY GRASSBIRD *Megalurus timoriensis* 18cm/7in

This skulker of the Locustellidae family has an olive-brown back heavily streaked with black, and an off-white throat grading into a mouse-brown belly, both unstreaked. The streaked back separates it from Australian Reed Warbler, and the unstreaked underparts distinguish it from Little Grassbird. Tawny Grassbird is found around coastal and sub-coastal Australia from the Kimberley (WA) to s. NSW, though it is most common in the Top End (NT). It is not restricted to wet grasslands and cumbungi (bulrush) but is also found in wet heaths. It is easy to attract with pishing and responds very well to recordings of its calls.

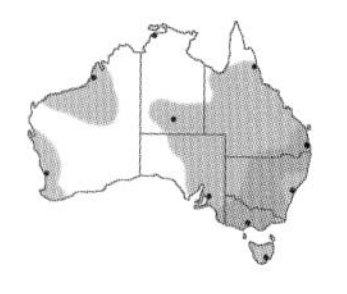

5 LITTLE GRASSBIRD *Megalurus gramineus* 14cm/5.5in

This skulker has an olive-brown back and fawn underparts, heavy black streaking on the back, and faint striations on the breast, upper belly, and flanks. Found through most of e. Australia and w. WA, Little Grassbird is less common and more closely tied to reed beds and cumbungi (bulrush) than Tawny. Though harder to lure by pishing than Tawny, it will respond to recordings of its call, making short sallies to inspect the intruder.

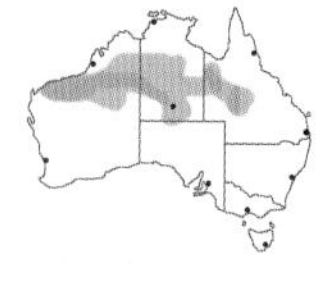

6 SPINIFEXBIRD *Megalurus carteri* 15cm/6in

A desert grassbird, its upperparts are uniform brown, underparts are fawn, and it has a prominent off-white eyebrow and a very long tail. An indistinct but useful field characteristic is the broad barring at the base of the undertail. Found in spinifex grasslands along the Tropic of Capricorn from w. QLD to the w. WA coastline, this bird is both sparsely distributed and a serious skulker, and thus requires a fair effort to locate and see well. It responds to recordings of its call in a seemingly 'half-hearted' manner.

♂ br. 1
♂ nbr. 1
2
3
4
4
5
6

1 ASHY-BELLIED WHITE-EYE *Zosterops citrinella* 11cm/4.25in

An offshore white-eye with pale olive upperparts, off-white underparts, and a lime-yellow chin, forehead, and vent. Like all Australian white-eyes (family Zosteropidae), it has a solid white eye ring. It occurs in Torres Strait islands and small islands off the ne. coast of Cape York Peninsula (QLD). This is a difficult bird to get to but easily found within its territory.

2 CANARY WHITE-EYE *Zosterops luteus* 10–13cm/4–5in

A small, group-feeding bird, this species is uniform bright lemon yellow below and bright yellow-green above and has a solid white eye ring. Canary White-eye is a coastal species in the north, found from the c. WA coast around to the sw. side of Cape York Peninsula (QLD), with an isolated population around Ayr, on the c. QLD coast.

3 SILVEREYE *Zosterops lateralis* 10–12cm/4–4.75in

Silvereye is a subdued white-eye, grey-olive above and with the yellow on the underside confined to the throat; the head may be washed yellow. The distinctive ring around the eye easily identifies this bird. Some Silvereye subspecies, of which there are many, also show a rufous or grey wash along the flanks. This is a widespread species through e. and s. Australia, including TAS, found in many habitats and is a very familiar bird to Australian birders.

4 MISTLETOEBIRD *Dicaeum hirundinaceum* 10–11cm/4–4.25in

The male Mistletoebird (family Dicaeidae) is unmistakable, with a striking red throat, a broad black line running down the breast, a soft rose-coloured vent, and deep blue upperparts. Females are less colourful, grey birds, though they still show the distinctive rose vent, which aids identification. These birds feed on mistletoe, wherever it occurs, across a broad range of habitats. Very widespread, it is absent only from TAS and the driest desert regions.

5 OLIVE-BACKED SUNBIRD *Cinnyris jugularis* 11–12cm/4.25–4.75in

Also known as Yellow-bellied Sunbird; a member of the family Nectariniidae. Both sexes are bright lemon yellow below and olive above, and the male has a striking, metallic-blue throat. The bill shape alone makes these birds identifiable. They are familiar to many people in their range as they often feed on garden flowers, which they defend vigorously, and frequently nest under the eaves of houses. The species is restricted to the coast of QLD from Rockhampton north to Cape York, where it occurs in rainforests, mangroves, and gardens.

1
2
s. 3
ne. 3
♂ 4
♀ 4
♂ 5
♀ 5

SWALLOWS AND MARTINS (HIRUNDINIDAE)

1 WELCOME SWALLOW *Hirundo neoxena* 14–15cm/5.5–6in

Welcome is the common Australian swallow. It is similar to the familiar Barn Swallow of the Northern Hemisphere, glossy blue above with a rusty face, but lacks the dark chest band Barn has. Welcome Swallow has a deeply forked tail and a scattering of pale spots on the uppertail, usually visible only at close range. Frequently encountered over urban areas and in many different habitats, this swallow is numerous and very likely to be seen by any visitor. It is one of the most common birds over all of s. and e. Australia (including TAS), absent only from portions of WA, the NT, and w. Cape York Peninsula (QLD).

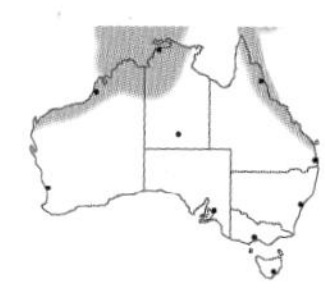

2 BARN SWALLOW *Hirundo rustica* 16cm/6.25in

Very similar to the far more numerous Welcome Swallow, this species differs in having a pronounced black chest band, creamier underparts, and a more deeply forked tail. It is a very uncommon Holarctic summer visitor to n. Australia, where it fills the niche of the much more common Welcome Swallow. In the east this bird may be overlooked among Welcome Swallow flocks.

3 FAIRY MARTIN *Petrochelidon ariel* 12cm/4.75in

Fairy Martin is a swallow with a squared-off, slightly notched tail. It can be identified by the combination of a clean white rump and a rusty top to the head. Also known as 'Bottle Swallows', these birds construct bottle-shaped nests from mud. Fairy Martins are often found around nesting or roosting colonies around culverts, though they are social even away from colonies. The species is common over all of mainland Australia, although absent from TAS.

4 TREE MARTIN *Petrochelidon nigricans* 14cm/5.5in

This species is very similar overall to Fairy Martin, having an off-white rump (with faint streaking) and a squarish tail, except it has a dark head concolourous with the upperparts. Found over most of mainland Australia and TAS, it breeds in the south of the continent and occurs in a wide variety of wooded habitats. In c. Australia it is limited to areas where it can have breeding colonies, such as stands of Red Gums near rivers and lakes.

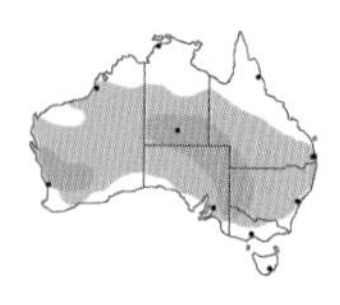

5 WHITE-BACKED SWALLOW *Cheramoeca leucosterna* 15cm/6in

By far the most stunning of the Australian swallows; it is a pied bird with a white head, back, chest, and forewings. It has otherwise black wings, a black belly, and a deeply forked black tail. With its erratic, almost bat-like flight, it is spectacular to watch. A bird of inland s. Australia, the species is patchily distributed in mallee, mulga, spinifex, and sedgelands, but is usually found near areas where the birds can dig tunnels, such as dune ridges and sandy creek sides.

1
1
2
2
3
4
5
5

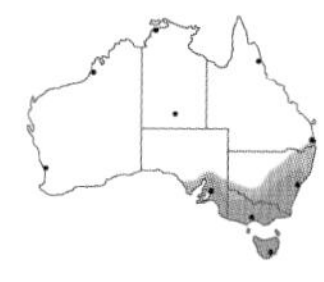

1 COMMON BLACKBIRD *Turdus merula* 25cm/10in

An introduced thrush species from Europe, Common Blackbird has a beautiful melodic song. The male is distinctive, all glossy black with a prominent yellow bill and eye ring. The female is all brown with a duller yellow bill. Within the bird's Australian range it is strongly associated with urban areas and is fairly common in the suburbs of VIC, TAS, and se. NSW, spreading to Sydney and even Brisbane (QLD) and to SA.

2 BASSIAN THRUSH *Zoothera lunulata* 28cm/11in

A thrush with olive-brown upperparts and cream underparts, both heavily scalloped with chocolate brown. This species is nearly identical to Russet-tailed Thrush, from which it differs in not showing white outer tail feathers in flight and in having a proportionately longer tail. It is the common *Zoothera* thrush of e. Australia and has two populations, one from the rainforests of the Atherton Tableland and environs (ne. QLD), and the other from the subtropical rainforests of se. QLD to the wet sclerophyll forests of SA and TAS.

3 RUSSET-TAILED THRUSH *Zoothera heinei* 27cm/10.5in

A thrush with olive-brown upperparts and cream underparts, both heavily scalloped with chocolate brown. This species is nearly identical to Bassian Thrush, from which it differs by showing white outer tail feathers in flight and having less scalloping on the uppertail. This species occurs in the rainforests of the Atherton Tableland and environs (ne. QLD) and also in the subtropical and wet sclerophyll forests of se. QLD and ne. NSW. It occurs at lower elevations than Bassian Thrush, with the zone of overlap between 500m and 725m (about 1,600–2,500ft). An excellent location to study these two birds is the area around O'Reilly's in Lamington NP (QLD).

4 SONG THRUSH *Turdus philomelos* 23cm/9in

This ground-feeding species has mid-brown upperparts, a cream breast, and a white belly arrow-marked with chocolate brown to black. It is an uncommon introduced species, limited to the Melbourne area (VIC) and apparently not spreading.

♂ 1
♀ 1
se. 2
ne. 2
3
4

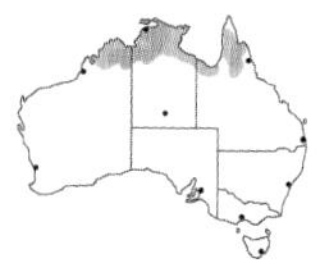

1 MASKED FINCH *Poephila personata* 12–14cm/4.75–5.5in

Very similar in appearance to Long-tailed Finch. Both are largely brown in colour and display a long, pointed black tail and a white rump. However, the smaller Masked lacks the powder-blue hood of Long-tailed and shows a continuous black mask around the face, not only a clean black bib. Masked Finch is also uniformly brown over the body, lacking the warm peach flush to the underside shown by Long-tailed. In the e. part of its range Masked further differs in bill colour: Its relatively heavier bill is yellow, while in this area Long-tailed Finch shows a coral-red bill. Masked is a common finch within grassy woodlands in n. Australia from near Broome (WA) to n. QLD just west of the Atherton Tableland.

2 LONG-TAILED FINCH *Poephila acuticauda* 15–16cm/6–6.25in

This species has a powder-blue head, a long black tail, and peach-washed underparts and displays a large black throat patch. The bill is relatively smaller than that of similar Masked Finch and varies in colour from yellow in the west to coral-red at the e. end of its range. Eastern birds, therefore, can be differentiated from Masked by the colour of the bill. Long-tailed Finches are social birds, usually encountered in pairs or groups and most easily found while visiting waterholes to drink, when their tails are often conspicuous. Long-tailed is common across n. Australia from Broome eastward to nw. QLD; it inhabits grassy woodlands and *Pandanus* groves, usually near water.

3 BLACK-THROATED FINCH *Poephila cincta* 10cm/4in

This looks like a short-tailed version of the preceding two species; it is mouse brown with a grey head and a black bill, throat, tail, and thighs. It is normally found from Cape York Peninsula through semi-arid QLD to n. NSW, but the southern population is now rare, and the bird is regular only from about Emerald (west of Rockhampton, QLD) north. It occurs in both tall and unburnt tropical savannahs and often comes to cattle dams to drink in the mid-morning. The most reliable sites for seeing it are around the se. base of the peninsula at Mt Carbine, Mt Molloy, and south of Charters Towers.

4 CRIMSON FINCH *Neochmia phaeton* 12–14cm/4.75–5.5in

The male Crimson Finch is a gorgeous long-tailed bird with a red bill, a rich crimson face (surrounded by a powdery-grey hood) and breast, and white spotting on the flanks. The tail is similarly red, although the mantle is brown. No other finch has this much red in its plumage. The female is more subdued, with mostly brown colouration and a bold red face and bill; it is similar to Star Finch but lacks the white breast spots of that species. However, these birds are usually seen in flocks, and the presence of males aids identification. Crimson Finch is a tropical species found in n. WA, n. NT, and n. and ne. QLD only, where it inhabits tropical grasslands, areas of sparse paperbark trees, croplands, *Pandanus* groves, and even gardens and grassy roadside verges.

4 ♂ n.–nw.

5 BLUE-FACED PARROTFINCH *Erythrura trichroa* 13cm/5in

Simple yet stunning, a lime-green finch with a bright blue face and forehead and a red tail. Females have less blue, and juveniles lack it completely. A widespread species across Papua New Guinea and w. Pacific islands, in Australia it is highly restricted to the rainforest edges of the n. Atherton Tableland (QLD), concentrated on Mt Lewis and the fragmented forest patches between Julatten and Mt Molloy. A possible second viable population occurs in Iron Range NP. This is the hardest finch to see well, even in a known territory, as it is much shyer than any other of the northern finches.

nw. 1
ne. 1
2
3
♂ n.–nw. 4
♂ far n. QLD 4
imm. 4
5

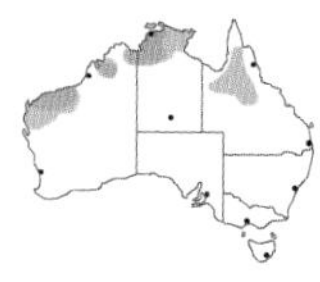

1 STAR FINCH *Neochmia ruficauda* 11cm/4.25in

Star Finch varies in colouration from olive green with white spots on the breast, a white belly, and a bright red face in the nominate eastern subspecies to an over-saturated lime-green back and yellow belly in the north-western subspecies, with the north-eastern birds in between the two. Juveniles lack the spots and the red face. This is a fairly common finch of the tall grasslands in the Kimberley region (WA), the Top End (NT), and the Gulf Country to w. Cape York Peninsula (QLD). It is regularly found in the unburnt edges at the Timber Creek airstrip (NT). The much rarer (probably extinct), more easterly nominate population once occurred on Cape York Peninsula and through semi-arid QLD to the NSW border.

2 PLUM-HEADED FINCH *Neochmia modesta* 11cm/4.25in

The back is dull brown with white spots, and the underparts are white finely barred with brown. The males have a plum crown and throat. Females have a small white brow and white throat. Juveniles have brown back spots, are uniform pale brown below, and lack any purple. Plum-headed is an uncommon and sporadically distributed finch of the grasslands and croplands of semi-arid e. Australia, mainly west of the coastal ranges. It is usually found in grassy creeks or around small dams used for cattle feeding.

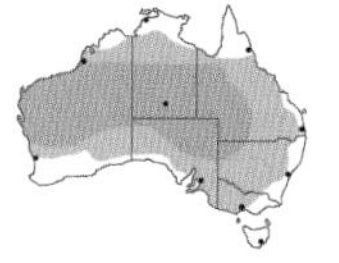

3 ZEBRA FINCH *Taeniopygia guttata* 10cm/4in

Also known as Chestnut-eared Finch. The male is strikingly patterned with a bright red bill, bold chestnut ear patches, finely barred black-and-white breast, and rufous flanks dotted with white spots. The tail is black boldly barred with white and is very obvious when the birds are flying away from the observer. Females are dull greyish birds lacking most of the male's bold markings but still displaying the distinctive tail pattern. Zebra Finch occurs over many of the drier parts of the country in a variety of habitats including grasslands, croplands, saltbush scrub, and other open country.

3

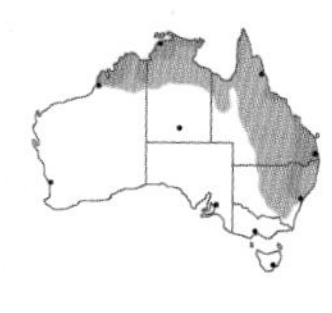

4 DOUBLE-BARRED FINCH *Taeniopygia bichenovii* 10–11cm/4–4.25in

A very strikingly patterned and handsome finch. It is a brownish-backed bird with bright white underparts and two bold black horizontal bars across the chest; the upper bar circles up and forms a border around the white face mask. The black eye is conspicuous, standing out boldly from the white face, which has an owl-like expression. The wings are chocolate brown to black and strongly dotted with white. This is a common species of the tropical north and east. It is the most confiding of the northern finches and can be found in flocks in open woodlands, tropical grasslands, grassy roadside verges, and croplands.

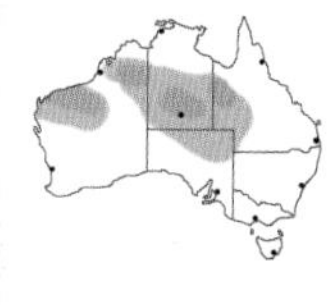

5 PAINTED FINCH *Emblema pictum* 11cm/4.25cm

A stunning bird in its open habitat, with uniform brown upperparts except for the scarlet rump and black tail. Underparts are chocolate brown to black marked with large white scallops. The males have a scarlet face and a red central line down the breast. Painted Finch prefers spinifex-covered rocky ridges with a permanent water supply nearby. It feeds in small groups dispersed across the ridgelines, but large groups congregate at water sources. It is an uncommon bird of the w. interior, found in tropical areas of the w. two-thirds of the continent but in neither the n. Kimberley region nor the Top End.

ne. 1
imm. 1
2
3
4
4
♂ 5
♀ 5

1 RED-EARED FIRETAIL *Stagonopleura oculata* 12cm/4.75in

This largely brown finch has a finely vermiculated back, a thickly scalloped black-and-white belly and undertail, the characteristic scarlet rump and uppertail of all firetails, a small black face mask, and a red patch on the ears. It is a restricted-range species found only in the extreme south of WA in wet areas with substantial undergrowth, such as heaths, well-vegetated watercourses, and eucalypt forests with a thick understorey. It is uncommon, furtive, and reclusive, often feeding low in thick cover, and therefore can be difficult to find.

2 BEAUTIFUL FIRETAIL *Stagonopleura bella* 12cm/4.75in

This finch has olive-brown upperparts and paler underparts, both vermiculated with black, and a bright red rump. It is similar to Red-eared Firetail but lacks the red patch behind the eye and the black-and-white scalloping on the belly. As there is no overlap, the two species can be identified by range alone. This is an uncommon and somewhat secretive firetail of coastal regions of extreme se. Australia and TAS. It occurs in both coastal and alpine heaths and in open woodlands with a thick understorey. Though it rarely comes out from cover, it can sometimes be found feeding at the edges of trails through heathlands.

3 RED-BROWED FINCH *Neochmia temporalis* 11–12cm/4.25–4.75in

The well-named Red-browed Finch has a fiery red eyebrow and bright scarlet uppertail and rump, which are its most prominent features. The eyebrow alone serves to identify this species. These firetails are found in a range of habitats including gardens, golf courses, woodland and forest clearings, grasslands, and croplands, where the bright red rumps of birds in flight often draw attention to the attractive flocks. This species is often around human habitations in its range and sometimes visits feeders. It is a common eastern finch from the Cape York Peninsula in n. QLD south along the coastal strip to VIC and se. SA.

4 DIAMOND FIRETAIL *Stagonopleura guttata* 12cm/4.75in

A stunningly beautiful and scarce finch. Like the other firetails it too has a bright vermilion-red rump and uppertail, but unlike the others it has a pale dove-grey head and white underparts with a contrasting broad black breast band and black flanks, which are dotted with bold white spots. Found in open woodlands, riverside Red Gums, eucalypt forests, and mulga in e. Australia from se. QLD through e. NSW and VIC to far se. SA. This breathtaking finch is most likely to be found in pairs perched low in the understorey or on a barbed wire fence on the edge of a forest.

1
2
3
3
4
imm. 4

1 GOULDIAN FINCH *Erythrura gouldiae* 13–14cm/5–5.5in

A scarce, dazzling multi-coloured finch of the tropical north. Adults have a patchwork of vivid bright colours: bright emerald green on the back, electric blue on the rump, a purple breast, and a lemon-yellow underside. The black tail has thin wispy central feathers that extend to a point. Adults always possess a pale ivory-coloured bill, but the head pattern, while always striking, is variable. There are red-faced, black-faced, and yellow-orange-faced colour morphs, and several morphs can be found in the same flock. Immatures are indistinctive, dull brown birds, confusable with the immatures of the other finches. Gouldian Finch is patchily distributed in n. WA, NT, and QLD in tropical savannahs, grasslands, and other wooded areas near water. Like all finches, it is most likely to be encountered in flocks gathering to drink at waterholes in late afternoon in the drier parts of its range.

2 CHESTNUT-BREASTED MANNIKIN *Lonchura castaneothorax* 11–12cm/4.25–4.75in

A strikingly marked finch with a black face and a bright orange breast bordered below with a thick black line separating it from the clean white belly. Immatures are dull brown birds with dark bills and legs, very similar to immature Yellow-rumped Mannikins in appearance but differing from young Pictorella Mannikins and Gouldian Finches in displaying dark legs, not pink. The species inhabits grasslands, croplands, and rank grasslands bordering wetlands all across n. Australia and in e. Australia to ne. NSW. It is most often seen when a large flock is inadvertently disturbed while foraging in tall grasses, and the birds emerge to perch conspicuously on the tops of the thin stems, which sway prominently under their weight.

2

3 PICTORELLA MANNIKIN *Heteromunia pectoralis* 11cm/4.25in

A stunning finch with a black mask (brown in female) bordered at the rear with a diffuse orange border and below with a scalloped, pied breast. It has a fawn belly and grey-brown upperparts; pale dots on the wings are visible at close range. Immatures are dull brown with pink legs, like immature Gouldian Finches, which are more yellow-toned. Pictorella Mannikin is widespread but uncommon in the Kimberley (WA), n. NT, and nw. QLD. It prefers tall grasslands and tropical savannahs and is best seen in late afternoon when flocks come to the edges of dirt roads to feed. A good area to look is the northernmost 50km (30mi) of the Buchanan Highway (NT).

4 YELLOW-RUMPED MANNIKIN *Lonchura flaviprymna* 11–12cm/4.25–4.75in

The most somber of the Australian finches; it has a dark brown back, a buff head, a dark cream rump and tail, cream underparts, and a black vent. Immatures are mid-brown with a creamy belly and dark bill and legs; the dark legs help to differentiate it from immature Gouldian Finch and Pictorella Mannikin. Generally uncommon though locally abundant from the nw. Top End (NT) to the e. edge of the Kimberley (WA), it prefers tall stands of grass and often moves in flocks with Chestnut-breasted Mannikin. Timber Creek airstrip is a regular site for it when it is not burnt.

5 SCALY-BREASTED MUNIA *Lonchura punctulata* 11–11.5cm/4.25–4.5in

This is a chocolate-brown finch with boldly brown-scalloped white underparts. Upperparts are uniform chocolate brown. Immature birds are mid-brown with paler underparts, darker than the similar immature Chestnut-breasted Mannikin. An expanding introduced species to the tropical and subtropical coastal areas of e. Australia, Scaly-breasted Munia can be especially common in country towns and around cane farms. It often associates with Chestnut-breasted Mannikin.

black-faced ad. & imm. 1
red-faced ad. 1
ad.
imm. 2
♂ 3
imm. & ♀ (lower rt) 3
4
5

1 EURASIAN TREE SPARROW *Passer montanus* 14cm/5.5in

A rather similar species to the male House Sparrow (both are Old World sparrows, family Passeridae). Sexes are alike: streaked brown birds with a rufous cap, rufous nape, black throat, and black bill. The species differs from the male House Sparrow in having black ear coverts separated from the black throat and mask by a grey border. An uncommon introduced species with its population centred on Melbourne and e. VIC, it appears to be slowly spreading.

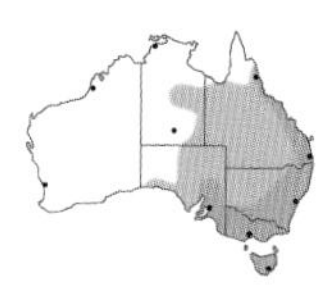

2 HOUSE SPARROW *Passer domesticus* 15cm/6in

A familiar Old World species now found widely across e. Australia. The male is a streaked brown bird with a grey cap, rufous nape, black throat, and chunky black bill. The female lacks the male's bold head pattern, is brown and streaky on the upperparts, and has a brownish head with an obvious pale supercilium. A generalist species, House Sparrow is most likely to be encountered in noisy boisterous flocks within urban areas, although it also occurs widely in farmlands and scrublands outside of these. It is found around almost every town and most farmhouses throughout e. Australia.

2 ♂

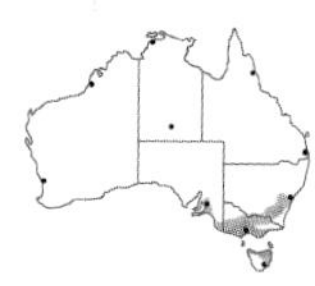

3 EUROPEAN GREENFINCH *Chloris chloris* 14–16cm/5.5–6in

A dull lime-green finch (family Fringillidae), it has black wings and tail with yellow wing and tail panels that are best seen in flight. The immature birds are a very subdued grey-brown with indistinct darker brown streaking; they have the diagnostic black wings and tail of the adult, with indistinct yellow panels. A common introduced species, European Greenfinch is found in farmlands and urban habitats of far se. mainland Australia and TAS.

4 EUROPEAN GOLDFINCH *Carduelis carduelis* 12–13.5cm/4.75–5.25in

A truly stunning bird of the family Fringillidae. The base colour is a dull grey-brown, but it has a scarlet face and forecrown, a black hind crown and neck sides, and a pure white crescent extending from the crown through the ear coverts to the edge of the throat separating the red and the black. The wings and tail are black with very large shining-yellow wing panels, which are distinct from a great distance when the bird is in flight. Immature birds are a very subdued grey-brown with indistinct darker brown streaking; they have the diagnostic black wings and tail of the adult with glowing yellow panels that are far more pronounced than the wing panels of immature European Greenfinches. European Goldfinch is a common introduced species found in farmlands and urban habitats of the se. mainland from Adelaide (SA) to near Brisbane (QLD) and in TAS.

1
1
♂ 2
♀ 2
3
4

STARLINGS; BULBUL

1 COMMON STARLING *Sturnus vulgaris* 20–22cm/8–8.5in

A familiar starling species (family Sturnidae) introduced from Europe to Sydney and Melbourne in the 1800s. In breeding plumage it is a glossy black bird with a bright yellow bill and reddish legs. In winter dress the body is boldly spotted with buff and the bill is dark. Young birds are all brown. Originally introduced to control crop pests, this aggressive species has now come to be a pest itself, consuming valuable farmed fruits and crops and also adversely affecting cavity-nesting native birds, which it outcompetes for nesting sites. An aggressive, adaptive, and familiar bird, it is very common over most of se. Australia and found in a range of habitats. It is most often seen in urban areas or croplands, where it is usually encountered in groups feeding on the ground or perched prominently on overhead wires.

2 COMMON MYNA *Acridotheres tristis* 23–25cm/9–10in

An introduced Asian starling (family Sturnidae) found extensively in e. Australia, Common Myna is a very distinctive brownish bird with bold yellow facial skin behind the eye and a bright yellow bill and legs. In flight it displays broad white flashes in the wing. Common Myna is familiar to many people in their range, as it is an urban species common around towns and cities and along highways; it is most likely to be found foraging in groups on the ground. The species is found patchily in ne. QLD, se. QLD, e. NSW, and VIC and is common around Cairns, Brisbane, and Sydney.

3 RED-WHISKERED BULBUL *Pycnonotus jocosus* 23cm/9in

This bird is very distinctive, displaying a black hood, a red mark above the white ear patch, and very long, pointy crest. The back is a uniform olive brown, and the tail is black above and tipped in white. The throat and breast are white, framed by a black half-collar, and the underparts are grey-brown with a red vent. Red-whiskered Bulbul (of family Pycnonotidae) is a very common introduced species in Sydney (NSW) and surrounding urban areas south to Nowra and north to Coffs Harbour; another population is established in Mackay (QLD).

4 METALLIC STARLING *Aplonis metallica* 23cm/9in

A medium-sized, long-tailed black bird (of family Sturnidae) with a greenish-purple gloss and an obvious red eye. In many ways the female is the more interesting looking, showing a pure white throat and white underparts heavily streaked in black. This is a highly sociable bird usually seen in large flocks swirling over the canopy in search of fruit. It is a common summertime breeding migrant in the lowland rainforests and mangroves of ne. QLD, less common in the upland rainforests. It is often seen around the Cairns Esplanade and Botanic Gardens.

4 ♂

br. 1
imm. 1
2
3
♂ 4
♀ 4

ABBREVIATIONS

♀	female
♂	male
ACT	Australian Capital Territory
ad.	adult
br.	breeding plumage
c.	central
cm	centimetre
e.	eastern
ec.	east-central
ft	foot/feet
ha.	hectare
ID	identification
imm.	immature
in	inch
kg	kilogram
lb	pound
m	metre
mi	mile
mm	millimetre
n.	northern
nbr.	non-breeding plumage
nc.	north-central
ne.	north-eastern
NP	National Park
NR	Nature Reserve
NSW	New South Wales
NT	Northern Territory
nw.	north-western
partial br.	partial breeding plumage
QLD	Queensland
s.	southern
SA	South Australia
sc.	south-central
se.	south-eastern
SF	State Forest
subsp.	subspecies
sw.	south-western
TAS	Tasmania
VIC	Victoria
w.	western
WA	Western Australia
wc.	west-central

GLOSSARY

arboreal Tree-dwelling.

austral Of the Southern Hemisphere, in this book usually referring to the seasons.

Australasia A geographic region comprising Australia, New Zealand, New Guinea, and associated islands.

Banksia A genus of native Australian plants with characteristic brightly coloured cone-like inflorescences; the dominant genus in heathlands. Many *Banksia* species are important food sources for nectar-eating birds such as White-cheeked Honeyeater, and seeds are important for some parrot species.

basal Near or referring to the base of a structure (such as a wing); compare *distal*.

blossom nomad A species that follows flower blooms over vast areas. Such birds appear common in different areas for short periods during flowering and disappear for the rest of the year.

bower A framework made of leaves, sticks, and other plant parts (and often bits of colourful plastic and other items) constructed by male birds of certain species within the bowerbird family (Ptilonorhynchidae) in order to attract females.

brigalow An Australian open woodland habitat with a mix of acacia, cypress-pine (*Callitris*), and eucalypt trees. It typically has a number of 'mallee-form' trees, which are multi-stemmed with each stem growing out from an underground base.

brood parasite A bird that lays its eggs in nests of other species, allowing the host bird to incubate its eggs and raise its young, usually to the detriment of the host's brood.

cere A fleshy covering located at the base of the upper mandible over the nostrils that is present in only some groups of birds, most notably parrots, hawks, falcons, owls, and pigeons.

confusion species A species that is similar to and likely to be confused with the species under discussion.

conspecific Considered part of the same species.

coverts An area of feathers found on various parts of a bird's body. The *wing coverts* are several series of feathers that make up the 'shoulder' area on a closed wing and the leading edge of the wing in flight. *Uppertail coverts* are the feathers on the bird's upperside between the rump and the base of the tail; *undertail coverts* are the feathers on the underside of the tail between its base and the vent. See also *ear coverts*.

crepuscular Active at dawn and/or dusk.

cryptic Hidden or serving to hide, as in cryptic colouring or behaviour that camouflages a bird and allows it to blend in with its surroundings.

dewlap A flap of loose skin under the throat, such as that displayed by Brolga (a crane species).

dimorphic Having two forms; e.g., some birds are sexually dimorphic, with males and females different of colours or sizes.

distal Near or referring to the tip of a structure (such as a wing); compare *basal*.

diurnal Active by day.

ear coverts A set of feathers that cover the ear opening, located on the cheek area, which may sometimes be distinctively or contrastingly coloured, forming an *ear patch*, as in Laughing Kookaburra.

endemic With a native range confined to a certain area; an 'Australian endemic' is native and confined to Australia.

Eremophila A genus of native Australian trees, many of which are pollinated by birds, most notably honeyeaters and woodswallows; includes the fuchsia bushes.

eucalypt A tree (or shrub) of the myrtle family (Myrtaceae) that may belong to one of several related Australasian genera including *Eucalyptus* and *Corymbia*.

flight feathers The rows of quill-like feathers visible along the trailing portion and at the tips of the wings in flight, made up of primaries (in the outer wing, distal to the bend of the wing) and secondaries (in the inner wing, inside the bend of the wing), as well as the tail feathers.

forehead The area at the front of the head immediately above the bill.

frons The area of feathers between the bill and the eye; also known as *lores*.

frontal shield A hardened or fleshy area of a bird's forehead that extends down to the bill, as in some gallinules including coots and moorhens.

frugivorous Fruit-eating.

gallery forest Forest that forms corridors along rivers and other waterways and is usually thicker than the surrounding habitat.

gape The opening of the mouth or the inside of the mouth.

genus In taxonomic classification, a category that ranks below family and that contains one or more species. The genus (plural *genera*) forms the first part of a bird's two-part scientific name.

gorget An area of shining feathers on the chin or throat of a bird, such as that displayed by Victoria's Riflebird.

gregarious Living in groups.

Grevillea A genus of native Australasian flowers that are important for nectar-eating birds. Most are small bushes, although some occur as large trees, such as the Silky-oak.

gular patch An area of bare, featherless skin present on the throat of some birds that extends from the base of the lower mandible to the bird's neck (e.g., the bright red pouch inflated by male frigatebirds during courtship).

Holarctic migrant A bird that migrates from both the Palaearctic (Asia and Europe) and Nearctic (North America) zones of the Northern Hemisphere.

immature In this book may refer to any young or pre-adult bird of a species, including the *juvenile*.

irrupt To undergo a sudden increase in numbers, usually in response to ecological or climatic factors.

Jarrah A w. Australian species of eucalyptus tree (*Eucalyptus marginata*) or woodland dominated by this species.

juvenile A young bird that is wearing its first set of feathers; this and any bird in pre-adult stages may be referred to as *immature*.

Karri A massive w. Australian species of eucalyptus tree (*Eucalyptus diversicolor*) or a form of wet sclerophyll forest dominated by this species.

Kimberley A geologically distinct region located in the far north of WA characterized by sandstone, including ridges, gorges, and mountain ranges, isolated by its location, bordered by the Indian Ocean to the west, Timor Sea to the north, deserts to the south, and the NT to the east.

local Confined to certain usually small and spotty areas.

lores The area of feathers between the bill and eye; adj. *loral*. Also known as *frons*.

mallee A type of scrub-like woodland dominated by several unique species of multi-stemmed eucalypts that form a distinctive structure as each stem grows out from the plant's underground base.

mandible A section of a bird's bill (beak); can refer to the upper or lower section.

mantle The area on the upper surface of a bird that is between the nape and the back.

morph A variation in a species, usually a distinctive plumage colouration (e.g., the pale and dark morphs of some shearwaters and petrels).

moustache An area of differently coloured feathers below the ear coverts that forms a distinct line, rather like a moustache in humans, on a bird's face; also referred to as a *moustachial stripe*.

mulga A type of acacia tree (*Acacia aneura*) or, more commonly, a habitat of inland Australia dominated by this species as well as other acacias.

muzzle The area of feathers surrounding the base of the bill formed from the forehead, frons (lores), and chin. Some birds show distinctive patterns in this area (e.g., Providence, Herald, and Great-winged Petrels), and so it can be an important ID feature.

nankeen A pale rufous (reddish, orange, or pinkish) colour.

nape The back of the neck.

nocturnal Active by night.

nominate subspecies Within a species complex, the subspecies that was first described and therefore is considered the 'type' of the species; it has the same scientific name as the species (e.g., Port Lincoln Parrot, *Barnardius zonarius* subsp. *zonarius*, is the nominate subspecies of Australian Ringneck). It does not equate to abundance.

Old World Geographical region comprising the Eastern Hemisphere (Africa, Asia, and Europe).

Outback Rural country within inland Australia.

Palaearctic migrant A bird that migrates from the Palaearctic zone, which covers the eastern portion of the Northern Hemisphere including n. Africa, Europe, and much of Asia.

Pandanus A genus of palm-like plants that is unrelated to the true palms.

paperbark A group of untidy-looking native Australian trees within the genus *Melaleuca* displaying characteristic flaky bark that looks conspicuously tattered and appears as if it has been partially peeled off.

pectoral Relating to the chest of a bird (e.g., Pectoral Sandpiper is named for its clear, well-defined chest band).

pelagic Ocean-dwelling (e.g., albatrosses are pelagic birds that spend most of their lives at sea).

pied Strongly blotchy or bi-coloured, in birding usually black and white.

pishing Making a 'pish'-like sound similar to the scolding calls of some songbirds, which are attracted to the sound in order to investigate the source of the perceived threat.

plumage The entire covering of feathers over a bird's body.

plume A conspicuous decorative feather, such as the long, fine display feathers that project from the shoulder area of Great Egret in breeding plumage.

primaries The flight feathers that form the outer, distal section of a bird's wing including the tips. On the closed wing these are barely visible, largely hidden by the secondary feathers, although the outer edges of the primaries are usually visible at the front edge of the folded wing, and the tips are often visible extending below the tips of the secondaries.

prion Any of several small Southern Hemisphere petrels of the genus *Pachyptila*.

rump Area on an animal between the tail and the back.

sally In birds, a quick, sudden flight made in order to catch prey. Thus a flycatcher may sally from a perching branch in order to snatch an insect from the air, making a brief, fast foray and returning to its perch with or without the targeted prey.

savannah A sparsely wooded area with an extensive grassy understorey.

scapulars An area of feathers located around the shoulders of a bird.

sclerophyll A term applied to plants that develop hard, often small or thin leaves as an adaptation to harsh conditions with little moisture.

secondaries The section of a bird's flight feathers inside of the bend of the wing, between the bend and the body.

shorebird A species of long-legged waterbird from the oystercatcher, plover, sandpiper, or stilt and avocet family. In Europe these are alternatively called 'waders'.

species A taxonomic division below genus that comprises individuals that successfully breed with one another (and not with other species). Each species is designated by a two-part scientific name consisting of a genus name and a species name.

spinifex A term applied to a range of grasses, mainly from the genus *Triodia*, that have very long, spiky leaves tipped with silica.

subspecies A division of a species geographically separated and often differing in appearance from other populations of that species; subspecies are technically still able to breed successfully with one another.

subterminal Just short of the terminus or tip; a 'subterminal band' is near but not at the very tip of the tail.

supercilium A marking on a bird's face similar to an eyebrow formed by feathers that run from the base of the bill over and beyond the eye.

sympatric Occurring in the same place, as when two species occupy the same range and habitat.

talons The sharp claws possessed by raptors (hawks, eagles, falcons, and kin) and owls.

taxonomy In biology, the system of classifying organisms in accordance with their relationships to one another (adj. *taxonomic*).

terrestrial Ground-dwelling.

Top End A geographical region consisting of the northernmost portion of the NT.

tropical zone In Australia the area north of the Tropic of Capricorn, a line of latitude at 23.5° south that marks the southernmost extent of the sun's overhead position.

understorey The layer of the forest structure below the main canopy down to the ground vegetation.

underwing coverts The feathering on the underside of a bird's wing extending from the 'armpit' to the bend of the wing.

vagrant A species that turns up rarely and is not a regular or predictable migrant (the bird has effectively lost its way).

vent The area of a bird's body below the belly towards the underside of the tail base that surrounds the anus.

Wandoo A type of eucalyptus tree (usually *Eucalyptus wandoo*, but the name may be applied to other species) of w. Australia or a woodland dominated by the species.

wattle A fleshy protrusion of skin that hangs from an area on a bird's head (e.g. Red Wattlebird).

wing linings The underwing coverts of a bird.

wrist The bend in a bird's wing, visible in flight; also referred to as the 'elbow'.

PHOTO CREDITS

Unless indicated otherwise below, all photographs in this guide are the work of and owned by Geoff Jones. Photographs are referenced A–J, reading left to right and top to bottom on a page.

John Barkla 265 F.
A Boyle 97 D, 109 H, 121 E, 123 J, 143 I, 152 D, 161 D, 169 B, 173 B, 175 B, 175 D, 193 C, 199 B, 265 D, 285 D.
Simon Bucknell 158 A, 161 B.
Iain Campbell 7, 10, 13, 14, 16, 17, 18, 19, 20, 21 A, 21 B, 23 A, 23 B, 24, 25, 27 A, 27 B, 28, 29 A, 29 B, 30, 32, 33, 34 A, 34 B, 36, 38, 39, 40, 41 A, 41 B, 43 C, 43 D, 49 E, 48, 53 A, 54, 60, 63 A, 63 B, 63 C, 63 D, 65 B, 65 C, 67 B, 67 G, 67 H, 69 G, 70, 73 D, 73 F, 73 H, 80, 82, 84, 100, 101 C, 102, 104, 105 A, 105 F, 108, 109 D, 111 E, 115 C, 115 E, 117 D, 119 H, 120, 123 A, 126, 136, 139 E, 147 E, 148, 150, 151 B, 159 E, 159 G, 170, 178, 181 A, 181 B, 181 C, 181 D, 182, 183 C, 183 D, 184, 185 A, 187 A, 187 B, 187 D, 189 F, 189 H, 191 A, 191 C, 195 E, 195 F, 196, 197 A, 197 B, 201 F, 202, 203 B, 207 B, 209 B, 213 A, 213 B, 213 F, 216, 221 A, 223 D, 225 D, 225 E, 225 G, 225 H, 227 A, 227 B, 227 C, 229 A, 229 C, 230, 232, 244, 245 E, 247 A, 249 D, 251 C, 252, 253 A, 255 C, 261 A, 261 E, 263 A, 265 E, 267 A, 268, 269 A, 271 D, 279 D, 281 B, 281 C, 287 C, 288, 289 C, 289 D, 293 A, 294, 295 A, 295 B, 295 C, 297 A, 297 D, 297 E, 299 A, 299 D, 300, 303 B, 307 E, 308, 313 A, 313 B, 313 C, 313 E, 317 A, 317 C, 319 A, 319 E, 319 H, 347 D, 349 E, 351 D.
R Clarke 69 E, 69 F, 71 E, 87 C, 113 H, 127 E, 129 D, 143 C, 143 L, 145 D, 153 B, 163 C, 163 D, 163 E, 163 F, 167 F, 169 C, 169 D, 175 G, 175 H, 177 F, 177 G, 183 B, 211 A, 211 F, 217 B, 217 D, 233 D, 235 C, 243 C, 243 D, 245 C, 255 D, 257 A, 285 C, 289 B, 303 H, 305 C, 309 B, 313 F, 315 A, 331 D, 345 H.
Ian Davies 159 H.
Steve Howell 97 A.
Robert Hutchinson 237 A.
Peter Jacobs 303 D.
Nick Leseberg 58, 73G, 146, 176.
Ian Montgomery 271 B, 341 C, 341 D.
Rob Morris 303 G.
Gary Oliver 291 C.
Laurie Ross 131 A.
Brent Stephenson 101 D.
David Stowe 243 A, 261 D.
Christopher Watson 129 I, 129 J, 259 F.
K Willis 143 A.
Sam Woods 44, 112, 162, 166, 173 A, 203 E, 225 C.
Kim Wormald 263 F.

INDEX OF SPECIES